How to Rebuild Your
VOLKSWAGEN
Air-Cooled Engine

By Tom Wilson

HPBooks

Notice: The information contained in this book is true and complete to the best of our knowledge. All recommendations on parts and procedures are made without any guarantees on the part of the author or the publisher. Because the quality of parts, procedures and methods are beyond our control, author and publisher disclaim all liability incurred in connection with the use of this information.

All rights reserved. No part of this work may be reproduced or transmitted, in any form by any means, electronic or mechanical, including photocopying and recording, or by any information-storage or retrieval system, without written permission from the publisher, except in the case of brief quotations embodied in critical articles or reviews.

HPBooks
A division of Penguin Group (USA) Inc.
375 Hudson Street
New York, New York 10014
© 1987 Price Stern Sloan, Inc.

Printed in the U.S.A.

49 48 47 46 45 44

Visit our website at
www.penguin.com

Cover photo courtesy Southwest Import Rebuilders
Photos: Tom Wilson, others noted.

ISBN: 978-0-89586-225-9

Table of Contents

Introduction	4

1. Time To Rebuild? 5
Accumulated Mileage 5
Oil Consumption 5
Poor Performance 7
Diagnostic Tests 13

2. Engine Removal 18
Preparation 18
Beetles & Karmann Ghias (Type 1) 19
Bus & Transporters (Types 2 & 4) 23
Fastback, Squareback, Notchback (Type 3) 26
411/412 26
Porsche/VW 914 27

3. Parts Identification & Interchange 31
Indentification 32
Engine Descriptions 34
Cases 35
Crankshafts 38
Flywheels 43
Connecting Rods 44
Pistons & Cylinders 45
Cylinder Heads 46
Oil Pumps & Camshafts 53
Oil Coolers & Sheet Metal 55

4. Teardown 57
Accessory Removal—Uprights 58
Accessory Removal—Flat 62
Basic Engine 66
Valve Train 66
Cylinder Heads 68
Oil Pump 68
Splitting Cases 70
Crankshaft Teardown 72

5. Crankcase & Cylinder Reconditioning 75
Clean & Inspect Crankcase Parts 75
Crankshaft 82
Pistons & Connecting Rods 88
Oil Pump 94
Camshaft 96

6. Cylinder Head Reconditioning 100
Disassembly 101
Valve Guides & Stems 105
Inspecting & Reconditioning Valves 107
Valve-Seat Reconditioning 110
Rocker-Arm Service 111
Valve-Spring Inspection & Installation 112
Cylinder Head Assembly 114
Intake & Exhaust Manifolds 115

7. Engine Assembly 116
Crankcase Assembly 119
Install Crankshaft 121
Install Camshaft 122
Prep Cylinders 132
Valve Train 134
External Accessories 138
    Type 1 & Pre-'72 Bus 138
    Type 3 143
    411/412, Post-'72 Bus, 914 148

8. Engine Installation, Break-in, Tuneup 156
Transaxle Prep 156
Engine Installation 157
Type 1 160
Type 2 161
Type 3 162
Type 4 163
914 166
First Start! 169
Break-in & Tuneup 170

Index 173

Introduction

The Volkswagen Beetle hardly needs an introduction. In any society with private transportation they're ubiquitous in the extreme; it's difficult to imagine roads without them.

But in the mid-'30s there were no Volkswagens, not even in Germany. In fact, there weren't many cars of any type on German roads, a fact Adolf Hitler said he was going to change. His requirements for an inexpensive, mass-produced, high-cruising-speed car were met (if not without difficulty) by a bright engineer named Ferdinand Porsche. The Volkswagen was presented at the 1939 Berlin Motor Show. A factory was built in Wolfsburg for Beetle production and Germany was about to get its car.

Of course, what Germany got was a long ways from the people's dream of motoring down the autobahn. War brought Volkswagen production only in the transmuted Type 82 military form, now known as the Thing. Although 70,000 Type 82s were built for the Wehrmacht, such a basic design was hardly suitable for popular transportation when hostilities ended, or for the chaos in what was left of Germany. Without a government, currency or economy, it seemed the Volkswagen had been stillborn. But from the rubble of 1945 a few cars were built from spare parts. The British officer in charge of the Wolfsburg factory assisted the German workers in building more cars, and the Beetle was on its way.

Eventually the Volkswagen came to the United States. As in other worldwide markets, the Beetle sold on its economy and superior workmanship. Americans came to respect and ultimately adore the round-backed car, buying it in numbers other import builders could only envy. Bus and Squareback versions followed with equal success.

Now, long after the introduction of faster, quieter and roomier economy cars, air-cooled Volkswagens continue to be popular. Other cars may be more modern, but none offer the old-world craftsmanship or personality of a Volkswagen.

And so we reach the point of this book, rebuilding air-cooled Volkswagen engines. This is an engine that needs step-by-step instructions for rebuilding. It's not that it's so difficult to rebuild. No, with minimal patience, tools and cash, the air-cooled VW is easily overhauled. It's just that the engine is so completely different. In fact, you'd have difficulty coming up with a design more out of the ordinary if you tried. Traditional rebuilding techniques and books based on them don't have much to offer the VW rebuilder.

But just as unfamiliar roads are easily traveled if you have a good map, this book helps make VW engine rebuilding easy. Pitfalls can be avoided if you know about them ahead of time, and like a detailed map, this book points out the hazards.

With the wrong turns clearly marked, rebuilding an air-cooled VW is fun. Just as a Beetle or Bus is fun to own and drive, rebuilding these engines is probably more satisfying than going through other, more common engine styles. Unlike many engines, the VW offers opportunities to measure and adjust basic engine parameters; not just disassemble and assemble engine components.

This book eliminates a lot of legwork for you. I've traveled to machine shops, parts suppliers, manufacturers, racers and other VW specialists to gather the information presented here. The knowledge in these words and pictures represents the combined experience of many people, all experts in their field.

For the camera and personal experience, I rebuilt several air-cooled engines, and I think you'll agree the intricacies of VW engine rebuilding are more thoroughly documented here than anywhere else. Additionally, the "Parts Identification & Interchange" chapter offers considerable money-saving information, not available anywhere else.

A few specific understandings and cautions are appropriate here. First, until you've been around VWs for some time, it's difficult to remember to keep the positions front and back properly oriented. Like left and right, front and back in this book are based on the engine while it is in the chassis. Thus, the flywheel is at the engine's *front*, and the crankshaft pulley is at the *rear*. Cylinders 1 and 2 are on the *right*, and 3 and 4 are on the *left*. This goes against common, ingrained automotive knowledge and takes some time to get used to.

Therefore, you have to keep reminding yourself that the flywheel end is the engine's front. After awhile it becomes natural. Of course, there has to be an exception, and the 914 is it. A mid-engine car, the 914 engine is turned around so its flywheel is at the car's rear. Unless this is important, as in the removal and installation sequences, 914 engines are treated like any other. So when I speak of the front oil seal I mean the one at the flywheel end, whether you have a Bus or 914.

Also, always take the time to double-check your work. Just remember the saying, "There's never time to do it right the first time, but there's always time to do it over." It's a lot faster to double-check than it is to rebuild it twice.

Finally, while VWs are a common sight and don't command high prices, that doesn't mean they are a cheap, throw-away car. A lot of care and thought went into every VW built, and treating the carefully constructed air-cooled engine like an appliance won't pay off. VW engines are full of precision tolerances that respond to cleanliness and careful assembly.

So read ahead, and keep this book on the bench where it will be handy. If this is your first engine rebuild, it may seem like there are too many steps or points to remember. To that I say, get started and handle each point one at a time. Read ahead of your progress in the shop to keep the job in perspective and alert yourself to needed tools or supplies. Soon you'll be listening to your Volks puttering smoothly in the driveway—a sound of wonderful personal satisfaction.

CHAPTER 1
Time To Rebuild?

Solidly built with old-world craftsmanship, air-cooled VWs have earned a loyal enthusiast following. With regular and frequent care, they give years of driving service.

Simple as it may sound, before beginning an engine rebuild, first decide if a rebuild is necessary. Sometimes this is easy, for instance, when two connecting rods are poking through the top of the case. Other times, some diagnosis is needed.

In this chapter, we'll examine some diagnostic steps to determine engine condition. By following them, you'll get an exact "State of the Engine," which will help you decide if a rebuild is required.

Before beginning your efforts, leaf through this book and any other VW literature available, i.e., the owner's manual. Study the pictures and skim the high points of the text. Get a basic comprehension of how air-cooled VWs are bolted together, so you can understand what's going on when they fall apart.

ACCUMULATED MILEAGE

Total mileage isn't a very good yardstick of engine condition. Pay more attention to how the car was operated and maintained during those miles. These are more important than the accumulated miles.

While the air-cooled VW engine, especially Types 1—3, isn't well-known for extreme longevity, it will go more than 100,000 miles between rebuilds. When driven inside its performance limitations, and given proper service, a sedan engine should last this long.

Bus engines and those driven off-road typically last less. Bus engines wear faster because of extra weight, wind resistance and their low gearing. A Bus engine revs higher than a sedan, so the engine "travels farther" than a sedan powerplant for each mile covered. Dirt is the enemy of off-road engines.

Those engines with the shortest lifespan are the poor engines driven hard and given little service. VW engines are susceptible to heat. Driven hard, they can be overheated, especially if the oil isn't changed often. The air-cooled VW just doesn't hold that much oil. It also has no paper oil filter, (except in the late Bus and Type 4), so frequent oil changes are *absolutely mandatory* for long engine life.

OIL CONSUMPTION

Oil consumption is determined by an engine's internal clearances and is an excellent gage of engine condition. When an engine is new, its internal clearances are easily bridged by an oil film. There are tight seals at the piston rings and valves, and high oil pressure at the main, rod and camshaft bearings.

As miles accumulate, these parts wear and dimensions increase. Then, oil cannot bridge the gaps between rings and cylinder walls and is sucked into the combustion chamber and burned. You'll see a wispy trail of blue smoke from the exhaust pipe.

At the crankshaft, excess clearance has less resistance to oil flow, and oil works its way to the ends of the journals where it's flung off. This causes more oil to try and fill the void and

5

Filthy engine and missing air filter will shorten this Bus engine's utility. Lack of service and poor operating practices will destroy any engine, but air-cooled VWs in particular require regular oil changes and spot-on ignition timing.

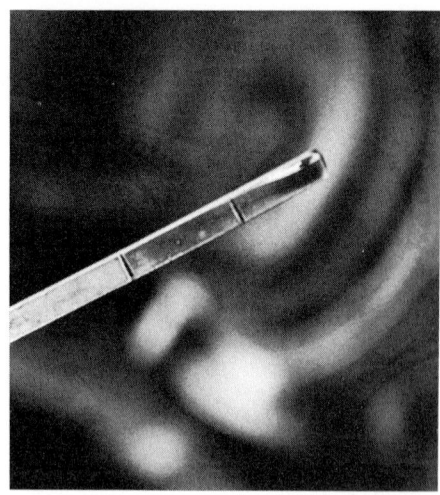

Oil consumption is an excellent indicator of an engine's internal condition. To check, make sure oil is level with top line. Record odometer reading, and read dipstick regularly. When oil level drops to the first line note mileage and subtract first reading from second. Difference is oil consumption rate.

oil pressure drops. It also causes increased oil consumption because more oil is splashed on the cylinder bores. When oil consumption is high and oil pressure low, a rebuild is required to correct excess clearances.

How Much Is Too Much?—Certainly, there's no harm in burning a quart of oil every 1000 miles or more. Between 500—1000 miles per quart indicates a slightly wide clearance somewhere in the engine, but not enough to justify tearing it down. Such oil consumption can result from parts on the loose end of the acceptable-tolerance range, or from partial rebuilds where the cylinders and pistons were not replaced. If a new quart is needed every 500 miles or less, the engine needs attention.

There are two parts that contribute to an engine burning oil: worn rings and valve guides. Both let oil enter the combustion chamber where it is partially burned and sent out the exhaust. A single puff of smoke immediately upon start-up after sitting overnight usually means worn guides and piston rings.

Another good test is to find a long hill to coast down while in top gear. When you reach the bottom, glance in the mirror as you open the throttle. A puff of smoke indicates worn guides *or* rings. But, if the engine lays down a smoke screen, you can bet the rings are at fault.

Blowby—Just as worn rings and cylinders allow oil to enter the combustion chamber, they also let combustion gases pass in the other direction, into the crankcase. These *blowby* gases pressurize the crankcase, contaminating the oil. Telltale signs are blowby vapor blowing out the oil-filler and dipstick holes.

Blowby used to be vented to the atmosphere at the oil-filler hole, but it has been routed to the air filter by a hose and metered orifice (valve) since the mid-'60s. From the air filter, the blowby is burned in the combustion chambers, along with the normal air/fuel mixture. This *positive crankcase ventilation* (PCV) plumbing draws more blowby out of the crankcase than merely venting it to the atmosphere. Oil stays cleaner, and less pollutants are spewed into the atmosphere.

There is no PCV valve on Type 1 and upright Type 2 engines. Type 3 and 4 engines use a *PCV valve* in the hose leading from the crankcase to the intake-air distributor. Fresh air from the air filter enters the engine at the rocker covers on these engines. A flame arrestor is placed in both hoses running from the air filter to the rocker covers to stop backfires from reaching the crankcase.

Valve Guides—Some oil passage past the valve stems and piston rings is normal. After all, if the rings and guides were sealed oil-tight, they would wear out in less than 10 miles from metal-to-metal contact.

Excessive oil loss through the guides occurs when guide-to-valve clearance is too large. This lets too much oil between the guide and valve stem. Then, next time the intake valve opens, oil is sucked into the combustion chamber. Exhaust valves can pass oil the same way. But because an exhaust port is a hot, mostly high-pressure area, excessive clearance there results in blowby into the rocker cover. Air-cooled VW engines are hard on guides because of the angle with which the rocker arm contacts the top of the valve stem. So, oil loss through the guides is common.

Oil Leaks—Many air-cooled VW engines leak oil. Much of the time, the leaks are minor, and won't affect oil-consumption calculations. But, if the engine has more than one leak or one bad leak, it will affect these figures.

So you won't be fooled by an oil leak when trying to figure how much oil is being burned, let's review some of the common oil leaks. Because some oil leaks result from worn engine internals, this review should be an integral part of the engine diagnosis.

Oil-Pressure Sender—Oil-pressure senders often dribble from their plastic centers, causing a puddle right under the sender. Wiggle the plastic center. If it is loose, replace the sender. Original-equipment (Bosch) senders are best.

Oil Pump—Leaks from around the oil pump are most common in recently rebuilt engines. They result from prying the crankcase halves apart with screwdrivers, ice picks and other barbaric instruments. Tighten the oil-pump cover plate first. If that doesn't stop the leak, remove the pump and try repairing the mating area of the crankcase. RTV silicone sealer makes a good temporary fix here, but the permanent cure is disassembling the engine and adding material by welding.

Engine/Transaxle Mating Surface—Most engines leak some oil at the front, the result of an overworked crankshaft oil seal. This results in a mess on the engine's bottom, but shouldn't cause any alarm.

On the other hand, if the area is washed clean by oil flowing from the bellhousing area, the oil leak is serious. If the seal was recently replaced, it may have been installed incorrectly. Removing the engine and installing another one is the only cure.

If the engine has a lot of miles on it, and the seal has not been changed recently, then the rear main bearing may be *pounded out*—the case is actually deformed—from excessive *end play*. This is a serious problem and should be investigated right away.

The cause originates from the #1 main bearing, the one closest to the flywheel, wearing out. The wear is in two directions, one parallel to the crankshaft, called *end play,* and the other perpendicular to the crankshaft. The perpendicular force wears the main-bearing bore eggshaped, so the crankshaft is free to *wobble*.

Of course, excessive end play and wobble distort the neoprene oil seal at the bearing bore, and oil pours past it. If this problem is detected soon enough, the main-bearing bore can be machined and the case saved. If the problem is allowed to continue, machining probably won't do any good, and the case will have to be replaced—an expensive fix.

There isn't any method for detecting crankshaft wobble while the engine is in the chassis. But you can measure end play, at least in Beetles and early Buses. Mount a dial indicator to read directly off the crankshaft pulley and measure end play. It can take lots of muscle to move the crank when the engine is together, so depress the clutch and then monitor the pulley. Use the detailed directions for measuring end play on page 126, if necessary. If there is a large oil leak at the bellhousing, and the end play is or even seems to be excessive, pull the engine and rebuild it. Putting this problem off can be very expensive.

There's a chance, too, that the transaxle seal is leaking, and it also drips out the bellhousing. Get a dab of the dripping liquid on a finger tip, then smell or taste it. If you're unfamiliar with the smell and taste of gear oil used in the transaxle, open the transaxle filler hole and take a sample. Compare it to the bellhousing leak. If the transaxle is leaking, pull the engine or transaxle and replace the transaxle seal right away. This problem won't go away, and the longer it leaks the greater the chance of ruining the clutch disc.

Oil Cooler—Oil coolers leak for two reasons. Either the cooler has split apart anywhere along the tubes, or it is loose on its mountings. Both leaks are real gushers because of the large volume of oil passing through the cooler. Remove the cooler to inspect the mountings and have it pressure-checked.

Case Leaks—These can be anywhere along the case parting line. Usually these are little weeping leaks and pose no danger. When oil pours from between the case halves, however, someone has used a screwdriver to pry the case apart there; a definite mistake. The machining on VW case halves is of the highest quality, and their precision, gasketless joint is a marvel of German production technique. To stab a screwdriver into this joint is criminal, and will cause a leak. Unfortunately, there is no cheap, sure, cure. You can try RTV sealer, Devcon or some other material to fill the gap, but the only enduring cure is to weld or replace the case.

A case can also leak through a crack. The magnesium case of Type 1—3 engines will crack sooner or later from fatigue. Chapter 5 has more information on case cracking. See page 76.

POOR PERFORMANCE

Performance is best defined for our purposes as *efficiency*. This is because when an engine yields poor fuel economy and power, it is inefficient. Diagnosis should determine if the engine is using the right amount of fuel to produce the expected amount of power. Remember, performance refers to both engine power and fuel consumption. If power or fuel economy drops, engine internals may or may not be the cause. The diagnostic tests later in this chapter are designed to systematically uncover the specific problem.

To locate the source of poor performance, start with a tuneup. This means a valve adjustment, points, plugs, dwell, timing and carburetor or injection tuning. Complete the tuneup yourself before performing any diagnostic tests or have it done by a professional tuneup shop. Get a complete analysis of the engine's condition from the shop. They see many, many cars and have learned to quickly and accurately diagnose their problems.

Before attempting an accurate engine diagnosis, get engine in top-notch tune. Plugs, timing, carburetion or fuel injection, and emission controls must work correctly. A *dyno-tune* is particularly helpful before evaluating how many miles are left in your engine.

Driving without engine compartment seal may seem harmless, but lets hot, dirty, under-car air into engine bay. Cooling efficiency drops and air filter(s) must fend off constant bombardment of dirt; engine life is shortened.

Early engines have few adjustments and are great at-home tuneup projects. Later fuel-injected engines can be impossible to tune without professional equipment.

Examining engine will often reveal interesting details, especially if considering buying a car. Beetle engine here is built on a Type 3 case! Large oil passage (top photo) and U letter code (bottom) were first tip-offs. Underneath, the dipstick tube is an add-on unit, another clue.

Basic Fuel-Injection Troubleshooting—Later engines have lots of vacuum and fuel hoses that are part of the fuel-injection system. All hoses must be in perfect shape and tightly sealed, or the engine will not run right or respond to tuning. The idle may be erratic, stumbling and searching (increasing and decreasing); this is called *hunting*.

Other fuel-injection problems can convince you the engine is at fault, but can be solved with minimal work, if not expense. If an injected engine won't run, first make sure all fuel-system parts are correctly installed. Every part must be in place, including the air filter.

Check all hoses for connection and condition. A stuck *airflow sensor*, the flap in the box next to the air filter, can cause driveability problems. If the flap won't move freely after a little fiddling, buy a new one. Finicky idle problems and weird throttle response on fuel-injected Buses and 914s are often traced to the *throttle switch* mounted right next to the throttle. Unscrew the switch from the throttle, remove the cover and note the *wiper contacts* that signal throttle position. Bending the arms so they wipe a new area often helps. Fuel-injected Type 3 throttle switches often need adjustment, too. Clouds of black smoke out the exhaust and poor running below full throttle indicate *electronic control unit* (ECU) failure or a *cold-start valve* that has stuck open.

If the engine responds to the tuneup with renewed performance, you've probably cured any problems and a rebuild isn't necessary. But, go ahead and do diagnostic tests as a double-check.

If a tuneup doesn't restore lost performance, do the following diagnostic tests. Tuning the engine will have helped in two ways. One, by showing there is something internally wrong and two, eliminating external variables from the diagnostic tests. Don't skip the tuneup, or some of the diagnostic tests will be inaccurate. For example, a cylinder won't have full compression if the valves are incorrectly adjusted. Vacuum test results will also be affected.

CAUSES OF POOR ENGINE PERFORMANCE

A quick look at the most likely internal engine problems will help put them in perspective before you start testing for them individually. Keep in mind that internal-combustion engines are nothing more than air pumps. They perform work by inhaling air, compressing it and expanding it, harnessing the expansion and exhaling the byproducts. Burning fuel only makes the air expand. So, anything that hinders an engine's breathing reduces its efficiency—both power and fuel economy.

The key to an engine's pumping efficiency is the tightness of the combustion chamber: the area formed by the piston top, rings, cylinder wall, head, valves and sparkplug. If any of these parts allow air to escape from the combustion chamber, engine performance will drop. Additionally, engine breathing will suffer if the valves and valve train are in poor shape.

A worn-out engine will generally perform poorly and use a lot of oil. This is most likely caused by worn rings and cylinders. But it is also possible for engine performance to be low and oil consumption to be normal. In this case, the valves, camshaft, or valve springs could be at fault.

Burned Exhaust Valve—When a mechanic says a valve is *burned*, this means that some of the valve's *face* (sealing surface) has been eroded away or cracked by the blast of hot combustion gases. Think of combustion gases as an inefficient cutting torch and you'll understand why valves burn.

A burned exhaust-valve face can't make a gas-tight seal against its seat, which causes a large drop in power and compression. As this condition worsens, a chunk may be burned from the valve head, allowing all compression to escape out the exhaust port. The engine then runs on only three cylinders. When a chunk is missing, it's indicated immediately during a compression test because engine cranking speed doesn't change on that cylinder's compression stroke and the gage reads very little or no compression.

Even worse than a burned exhaust valve is a *dropped exhaust valve*. A valve drops when the head breaks off from the stem. Because many Type 1—3 exhaust valves are made with heads and stems joined together, they sometimes separate at the joint. When this happens, the valve head destroys the piston crown, cylinder wall and cylinder head as it gets slammed around by the piston.

The valve stem, spring, retainer and keepers usually separate also, causing all sorts of havoc in the rocker cover. Metal particles circulate with the oil and score the crankshaft, camshaft lifters and oil pump. Luckily, the exhaust valve will burn badly before it drops, so this is some warning before it destroys the entire engine.

It's painfully obvious when an engine drops a valve. It will immediately begin running on three cylinders accompanied by a lot of horrible (and expensive) rattling. Instantly shut off the engine to help minimize damage.

One primary cause of burned valves is incorrect valve adjustment. This isn't a factor on '78 and later Buses because they have *hydraulic valve lifters*. Hydraulic valve lifters adjust the valves automatically and eliminate the need for periodic valve adjustments. All other VW air-cooled engines use solid valve lifters and need periodic valve adjustments. Skipping this service or maintaining insufficient valve clearance can easily lead to burned valves.

A valve cools best only when it is fully seated. Some cooling takes place through the guide, and some at the seat when the valve is closed. If a valve stays open longer due to tight clearances or whatever, it has both less time to cool and absorbs even more combustion heat. The burning process has begun.

Even well-maintained valves can burn if a piece of carbon gets caught between the valve head and seat as the valve closes. This holds the valve partially open and can start the gas-erosion process. Once it starts, the valve-burning process is rapid. Usually a valve will be noticeably burned in 2,000 miles or less.

The exhaust valves used in 914 engines are less prone to burning than other Type 4 engines because they are *sodium-filled*. This type of valve has a hollow stem, partially filled with sodium. Sodium melts well below the operating temperature of the valve, so it is a liquid when the engine is running. Reciprocating valve motion throws the sodium back and forth in the hollow stem.

When the sodium is at the hot head-end of the stem, it absorbs heat. Then it gets tossed to the cooler stem-end, where it passes the heat to the stem. The stem is in constant contact with the valve guide, so it can cool well if heat is brought to it by sodium or some other mechanism. Thus, heat is transferred by the sodium from the head to the stem, which sheds the heat quickly to the valve guide. From the guide, the heat passes to the head and ultimately to the cooling air. The net result is a cooler-running, longer-lasting exhaust valve.

Exhaust valves are much more prone to burning than intakes. This is because exhaust valves are exposed to combustion heat on both the combustion-chamber and port sides. Intake valves, on the other hand, are cooled by each passing intake charge and are heated only on the combustion-chamber side. Because intakes run so much cooler than exhausts, they are much less apt to burn.

Another factor affecting valve burning on engines with upright cooling-fan mounting is *oil-cooler placement*. Before '71, the oil cooler was mounted inside the fan shroud, in the path of cooling air going to the #3 cylinder. This preheated cylinder #3's cooling air, causing high cylinder temperatures and prematurely burned exhaust valves.

In '71, the oil cooler was moved out of #3's airstream and an offset section was added to the fan housing to enclose the relocated cooler. Such *doghouse* fan shrouds stop #3's exhaust valve from burning any sooner than the rest. Type 3 and 4 engines never had this problem because the flat mounting of the cooling fan has always positioned the oil cooler away from any one particular cylinder.

Loose Cylinder Heads—Type 1—3 VW engines don't use head gaskets. Instead, they rely on a metal-to-metal seal between the top of the cylinder and the cylinder head to contain combustion gases. If these parts loosen for any reason, combustion pressure and gases will be lost through the gap. Type 4 engines have a thin metal gasket for better sealing. Still, Type 4 engines get combustion leaks just like Types 1—3.

What typically happens is the cylinder head studs pull out of the case, letting the cylinder head and cylinders bang back and forth with piston motion. The problem is the threads in the case, not the studs. The studs are steel, and their threads are strong. The threads in the case are magnesium, which is no match for steel when it comes to mating threads. Under normal conditions, the studs won't pull. But after 100,000 miles, the magnesium can fatique and the threads weaken. Then the weak threads are ripped right out of the case by cylinder-head torque and combustion pressure.

Excessive cylinder-head torque will also destroy these threads in short order. Some people may look at the low torque specifications given for these cylinder heads and figure they aren't enough. So, when they assemble the engine, they add ten pounds torque to the cylinder-head nuts. Or, perhaps they are having cylinder-head sealing problems. So they whip out the breaker bar and crank the head nuts down another turn.

What they don't understand, is the torque applied to the cylinder-head nuts is not the same amount of torque that *seals* the heads. When the engine is cold, there is only 18 or 23 ft-lb of torque on the studs. But when the engine warms up and expands like a balloon, the aluminum cylinder heads and cast-iron barrels grow a lot longer than the steel studs. The studs are strained and being pulled from the case. Now the effective torque on the studs is nearer 55 ft-lb. If the at-rest torque is misapplied, for example, to 40 ft-lb, then it will reach over 70 ft-lb at operating temperature. No wonder the studs pull out of the case!

Overheating the engine has the same effect as overtorquing the cylinder-head nuts. The engine expands oversize when it is overheated, putting more strain on the cylinder studs. The first point to give is the cylinder stud threads in the case.

When the studs do pull, it leaves the cylinder free to hammer the crankcase and cylinder heads. If the problem is caught soon enough, the heads and case can be machined back to service, but don't count on this remedy all the time. The hammering ruins the engine and it's not even a worthwhile core. Because once the cylinder heads, barrels and case halves are replaced, you've just about bought a new engine.

Pulled studs are a very common problem up through the '70 Type 1—3 engines. Type 1—3s from '71 have steel thread inserts installed in the case at the factory. The steel thread inserts are commonly called *case savers*. They can be added when rebuilding to earlier cases that don't have them.

With case savers installed, pulled threads are no longer a common problem. If an engine with case savers is overheated, however, the cylinder heads can warp, letting combustion pressure escape between the cylinders and cylinder heads. Hammering of the case and heads by the cylinders is not a problem with warped heads.

Carbon Deposits—Although carbon deposits don't fall under the category of engine damage, and a rebuild is not necessary to remove them, a few words about carbon will help you with engine diagnosis.

Carbon is a solid byproduct of incomplete combustion, some of which sticks to the combustion-chamber surfaces. Both gasoline and motor oil are hydrocarbons, so burning them in the combustion chamber in the wrong amounts causes excess carbon deposits.

The most common source of harmful carbon deposits is excessive oil consumption, although a rich air/fuel mixture can be just as bad. Prolonged idling and slow driving can also cause carbon buildup. So, carbon deposits are a symptom of a problem, not the source. Merely ridding the engine of carbon won't cure the problem, only delay the symptoms. Therefore, while there may be ways to get rid of carbon buildup without overhauling an engine, curing excessive oil consumption may mean an engine overhaul.

Carbon deposits cause trouble in two ways. First, they may *shroud* the valves. Carbon deposits build up on the backside of a valve and restrict air/fuel mixture flow into the cylinder.

Carbon deposits in the combustion chambers can also cause damage. Carbon easily heats to incandescence, causing *preignition* and *detonation*. These types of abnormal combustion can damage an engine by placing a heavy load on engine internals.

Imagine red-hot carbon in the combustion chamber. When a fresh intake charge is compressed on the compression stroke, the hot carbon *preignites* the mixture. A moment later, the sparkplug fires and the mixture also starts burning near the plug. The two *flame fronts* collide, sometimes producing an explosion—detonation—rather than even burning. The resulting sudden pressure and temperature rise is more than the engine was designed for. Piston, valve and ring damage can result if preignition or detonation is prolonged. Although not as severe, preignition *without* detonation causes excess combustion-chamber pressure and temperature, but without any accompanying pinging or knocking.

Detonation is very similar to preignition, but the second ignition source, the glowing carbon, lights the mixture *after* the sparkplug has fired. Again, combustion-chamber temperature and pressure exceeds engine design limits and damage occurs. Audible signs of detonation are pinging or knocking, sounds akin to colliding billiard balls.

Admittedly, substantial engine damage from preignition or detonation isn't prevalent, but severe cases can burn or blast holes in pistons, break rings, and deform the main-bearing bores in the crankcase. Also, long-term light detonation will wear the rings, pistons and cylinders more quickly. So, prompt attention to the causes of abnormal combustion is wise. They are usually associated with low-octane gasoline or over-advanced timing.

Recent research indicates a small amount of knocking or pinging is not harmful to an engine, but does reduce fuel economy and power. This is sometimes called *light pinging*. Nevertheless, be concerned if the engine is knocking heavily. Besides carbon buildup, detonation can be caused by stale or low-octane gasoline, over-advanced ignition timing and engine overheating. Check for these problems if the engine detonates.

Pay special attention to the ignition timing of a VW engine. As an air-cooled engine, it is very susceptible to overheating and preignition caused by too-advanced timing. If the engine pings at the slightest load, retard the timing a degree at a time until it doesn't ping. This timing setting may be retarded from the specified stock setting, but with today's fuel it might

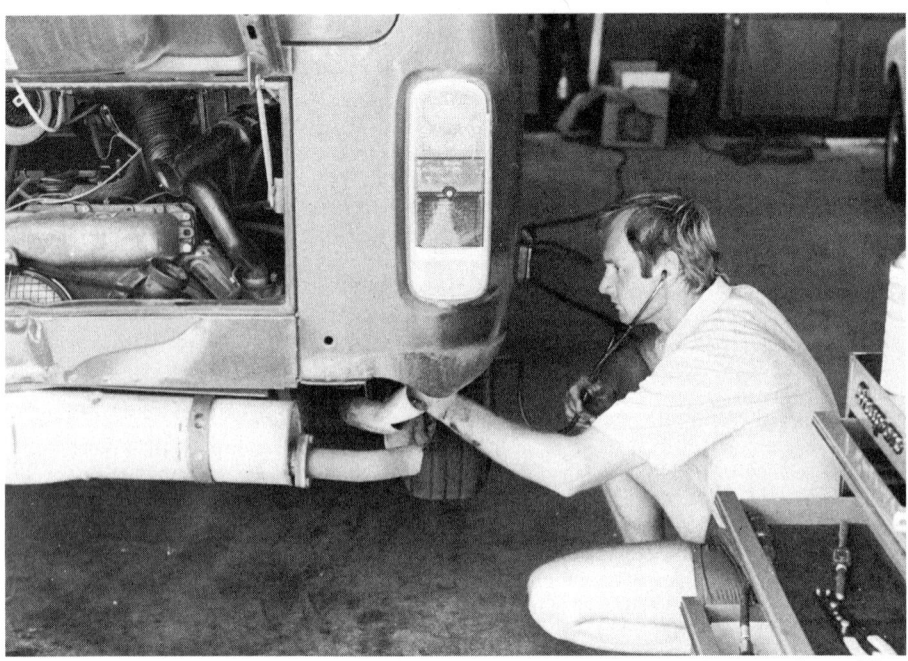

Stethoscope is preferred tool for pin-pointing internal engine noises because it amplifies sound coming through probe and reduces surrounding noise with earplugs.

be necessary.

Ignition timing is also commonly over-advanced by owners looking for more power. It's no secret that advancing the spark in air-cooled VW engines increases their power, throttle response and improves engine acceleration. But the penalty for too much total advance is severe detonation. If you advance the ignition past specification, you may pay for it with an engine overhaul.

Heed another warning: These engines self-destruct rather quickly when the cooling system fails. If the cooling flaps remain shut from a broken or missing spring, stuck thermostat, or foreign objects in the fan housing, cylinder temperatures will quickly go sky-high. The excess heat will cause severe detonation, hole a piston and spew metal throughout the lubrication system. This devastation can happen in less than one minute if the engine has been running for 10 minutes and is fully warm.

Loose carbon deposits can also lodge between the electrodes of a sparkplug, or get between a valve head and its seat, as mentioned earlier. If a piece of carbon sticks between the plug electrodes, the sparkplug will short out and the cylinder will misfire or go totally dead. Plug replacement or cleaning usually cures these problems.

A carbon-aggravated problem most people are familiar with is *dieseling*—the engine *runs-on* after the key is turned off. A hot piece of carbon acts like a diesel-engine glow plug by supplying an ignition source other than the sparkplug. Ridding the engine of carbon, slow-

ing the idle and reducing spark advance a few degrees will help reduce dieseling.

As a final note on carbon, consider vehicle operation. If you drive a delivery route, or do a lot of in-town, slow-speed, short-trip driving, carbon will build up because of low cylinder temperatures. You can easily *burn-out* excess carbon accumulated this way by taking the car for a long trip. Drive it a half hour or more at freeway speeds. This will heat the combustion chamber and burn away excess carbon. If that doesn't help, the engine may need a professional tuneup or carburetor overhaul. Worn rings and valve guides will also cause excessive carbon buildup from incomplete oil burning. They contribute excess oil to the combustion chamber and it can't be completely burned.

Fuel Shut-Off Solenoid—Type 1 and 2 carbureted engines since '70 have an electric *fuel shut-off solenoid* attached to the idle circuit of the carburetor. When the ignition is turned off, the solenoid is deactivated and a spring-loaded plunger closes the idle circuit. This should stop the dieseling mentioned above. So, if there is a problem with it, the fuel shut-off solenoid may be faulty.

To test the solenoid, remove its electrical lead. Look for the small can on the side of the carburetor with the wire leading to it. The solenoid is on the left side on '71 and later carbs (34mm) and on the right side on earlier carbs. Turn on the ignition without starting the engine. Now touch the lead to the solenoid connection. Each time you touch the lead, the solenoid should click (the plunger is moving inside). If you don't hear a click, check if the wire is supplying electricity with a test light. If the wire is "hot" (has voltage), the solenoid is bad and needs to be replaced. If the wire is "dead" (no voltage), trace and repair the wiring fault, and then recheck the solenoid operation.

Someone may have replaced the solenoid with a standard idle screw. If so, there's no problem, unless the engine diesels. Then reinstall the fuel shut-off solenoid. Carburetors not originally equipped with the solenoid can't have it added. Consequently, timing and idle-speed adjustments are required in these cases to stop dieseling.

On pre-'71 carburetors, the fuel shut-off solenoid's end doubles as the idle metering jet. The orifice is very small and even the tiniest dirt particle can clog it. A shot of compressed air usually clears the orifice.

On Type 3s with dual carburetors (and two solenoids), determine which one is defective by disconnecting first one solenoid and then the other. The engine will die when you unplug the solenoid that's working and show little change when you unplug the one that's not.

Fuel-injection systems stop fuel delivery when the engine is shut-off, so dieseling shouldn't be a problem with them.

DIAGNOSIS

Now that we've examined some engine problems, let's start in on how to find them—without taking the engine apart. The engine may or may not be exhibiting problems, but do the tests anyway. If it has a problem, you'll find it. If not, you'll have established a baseline of the engine's condition. From there, you can decide whether to rebuild now or later.

NOISE DIAGNOSIS

Internal Noises—Diagnosing engine noises is a difficult and imprecise art. Many factors influence the way sounds are perceived, not the least being the human factor. When investigating an automotive sound, try different spots. Open the hood, close the hood, sit inside the car, stand to one side, stand in back, lay down in back and to the side. You won't hear all noises from each spot. And, those that you can hear will sound different from each spot.

People are biased toward perceiving via their eyesight, so close your eyes to help focus attention to the sounds. Cupping your hands around your ears may look funny, but it helps mask sounds from the sides and amplifies those in front. It's a great way to pinpoint a noise.

Finally, learn to mentally dissect what you are hearing. Upon first hearing a running engine, the initial impression is a big jumble of sounds. By critically identifying each sound, you can more easily block out the unimportant sounds while concentrating on those you want to hear.

Aids for locating noises are a stethoscope, length of heater hose or a wooden dowel. Unlike the stethoscope a doctor uses, an automotive stethoscope has a solid metal probe at the business end. It works best when held against a solid part—head, case, manifold, bolt head or the like. If you suspect a noise from beneath a cover, place the stethoscope against a nearby bolt head or solid rail. For example, a noisy valve can best be heard by listening at the edge of the rocker cover or cylinder head, not at the middle of the rocker cover.

As a second choice, a length of hose or dowel can be used instead of a stethoscope. With a hose, hold one end firmly against the engine and the other end to your ear. When using a dowel, position the receiving end of the dowel against your skull, just forward of your ear, so engine vibrations don't bounce the dowel into your ear.

Engine noises can be lumped into three categories: intermittent ones, those occurring each crankshaft revolution, and those occurring at every other crank revolution. First, the intermittent sounds—the oddballs. These are *external* sounds coming from loose brackets, rubbing hoses, items stuck in the fan and so forth. By poking around the engine compartment, you can single out and stop these noises.

Noises that occur at every other turn of the crank—*at camshaft speed*—are most likely coming from the *valve train:* valves, rocker arms and lifters. There is one bottom-end noise that can happen at every other revolution—*piston slap*. Piston slap is the sound produced by the piston slamming against the cylinder as that cylinder fires at the top of the power stroke.

Piston slap is audible when piston-to-bore clearance is excessive. And, because there's only one power stroke for every two crank revolutions, it occurs on every other crank revolution. Piston slap is easiest to detect on a cold engine, before the pistons have expanded, reducing piston-to-bore clearance.

Noises occurring at *every* turn of the crankshaft come from the *bottom end:* worn piston pins, broken rings, worn rod bearings and main bearings.

If you have trouble telling whether a noise is at one-half or at crankshaft speed, hook up a timing light and see if the noise coincides with flashes of the light. If it does, the noise is at one-half crankshaft speed—a top-end problem or piston slap. If the noise occurs twice for every flash, it's at crankshaft speed—a bottom-end problem.

Isolating Normal Noises—Now for the hard part: What do these problems sound like? Let's start with normal engine sounds. As you listen to an air-cooled VW, the dominant sound will be the exhaust. Leaning forward into the engine compartment will help mask the exhaust, so you can more easily hear internal engine noises.

Once past the exhaust, the main noise of an idling engine should be the soft ticking of the valve train. Raise the rpm past idle and the ticking should turn into a whirr. If you hear one valve over the rest, or if all the valves are making more of a harsh clacking sound, adjust the valves.

A proper valve adjustment can only be done while the engine is cold. But, if you are checking for one loose valve, you can locate it on a warm engine. Just remember to properly adjust the valves after the engine cools. All air-cooled VWs use 0.006-in. valve clearance, unless it is a Type 4 engine with hydraulic lifters. There have been other clearances specified by VW in the past, but all have been superseded by the 0.006-in. measurement. Valve adjusting procedures are on page 136—138.

If valve adjustment doesn't cure a valve-train noise, it's possible there is a worn camshaft lobe or lifter. Then all the valve-adjustment tightening in the world won't quiet the engine; it will only burn the valve. Never close a valve adjustment tighter than 0.006 in., or the valve will burn. If the valve noise remains after adjustment, look elsewhere for the source. Try the rocker arms, lifters or camshaft.

Rocker arms can be checked by moving each by hand with the valve completely closed. If there is any appreciable movement other than 90° to the shaft, the rocker-arm bushing or shaft is worn. This, in effect, increases the valve clearance, which increases noise. Cam and lifter inspection require engine disassembly.

The other major noise in a air-cooled VW engine compartment comes from the fan. It can make several different sounds, ranging from the low-pitch noise associated with a house fan, and a high-pitch whistle. Type 3s and flat-engine Type 2s project more fan noise than Type 1s, early Type 2s and 914s because the fan is right in front of you when looking into the engine compartment.

Abnormal Fan Noises—A lot of abnormal noises come from outside the engine. For example, a whining or screaming from the fan area usually means something is caught in the fan. *Stop the engine,* and probe the fan area with your hand. Remove any debris. If you don't find anything, remove the fan belt and run the engine for a few seconds. If the noise disappears, you've got a cracked fan or bad generator or alternator bearings.

This test will isolate only the generator or alternator on Type 3 and 4 engines because their fans are driven directly off the crankshaft. Therefore, if you still have the noise with their fan belts removed, the problem is in the fan.

A sharp, intermittent rattling noise from the sheet-metal fan shroud of Type 1 and 2 engines may be a loose or broken fan. Check the tightness of the large fan retaining nut, and try to wiggle the fan on its hub. It may be necessary to remove the fan and generator assembly to investigate this noise.

11

Rhythmic scraping sounds are likely to be a bent crankshaft pulley or the cooling fan rubbing the fan shroud. Pushing or pulling on the top of the fan shroud will probably eliminate the fan noise. Bent crank pulleys are easily seen by sighting across them while the engine idles. A few well-placed wooden block and hammer blows can straighten out a bent pulley. The sheet-metal shrouding can be bent out of the way with a wooden dowel, screwdriver or the like.

Distributor Chirping—Dry distributor-cam surfaces can cause the points to give a high-pitched chirping. Make sure the distributor cam and points rubbing block are well-lubricated and then recheck.

Intake-Air Hissing—A loud hissing accompanied by poor idling usually indicates an *intake-air leak*, commonly called a vacuum leak. Check the tightness of the carburetor-to-intake manifold connections, plus the intake manifold-to-cylinder head hardware. On dual-port and fuel-injected engines, examine the rubber hose sections of the intake manifold, plus the air intake and metering area in general. Cold-start enrichment devices normally make some sucking or hissing sound while they are operating, so don't be confused by them.

Exhaust Leaks—These are often confused with other, more serious problems, so it's a good idea to check the exhaust system before jumping to any conclusions. If the exhaust system is tight, sealing the ends of the pipes will stop the engine.

So, with the engine idling, cover the exhaust outlets with palms that are swathed in wet rags or block the pipes with your shoes, if that's easier. You'll have to apply considerable pressure to exhaust openings as a well-sealed system has a lot of pressure. If there are leaks, you'll hear a phuft, phuft, phuft, phuft sound coming from the leak. Don't leave your hands or shoes over the exhaust pipes very long, or they'll get burned. Exhaust is very hot.

Exhaust systems can leak from anywhere, but mating flanges at the cylinders, muffler and tailpipe extensions are the usual spots.

Another typical exhaust sound is a whistle as the engine is accelerated. It is caused by loose or cheap replacement tailpipe baffles on Beetles and early Buses.

Piston Slap—Piston slap has already been mentioned because of its timing. It occurs at every power stroke, so it sounds off in time with the valve train. Piston slap is a dull, hollow sound. It's difficult to hear over the normally loud VW engine mechanicals and not easy to isolate. In fact, there's a better chance of hearing it while driving than listening for it with the hood open.

If you think you hear piston slap, remove and replace the sparkplug wire to each cylinder one at a time. When you get to the affected cylinder, the noise will greatly diminish. Disabling the cylinder will reduce piston slap because combustion loads no longer exist. Reconnecting the plug lead will restore the noise. Be sure to ground the plug lead when disabling a cylinder. See the sidebar for more information on disabling the ignition.

DISABLING THE IGNITION

Many diagnostic tests call for the ignition system to be disabled. With conventional ignition systems, there are several ways to short-circuit the electrical supply to the plugs; some are better than others. The best way is to remove the high-tension lead from the center of the distributor cap and ground it.

The high-tension lead is the large, heavily insulated wire running from the coil to the center of the distributor cap. To remove it, grasp the boot around the distributor-cap terminal, twist the lead slightly, and pull out the lead. The twist helps break any corrosion that resists wire removal. Now, ground the lead to the head or engine block.

On engines with *electronic ignition,* there should be *no more than* a 1/4-in. gap between the free end of the lead and ground, or *ignition-system damage may occur.* This gives the high-voltage electricity somewhere to go, instead of continuing to build voltage and trying to arc to ground inside the coil. This can destroy an expensive electronic-ignition module. Once the test is done, simply reinsert the lead into the distributor cap.

Piston-Pin Noise—All air-cooled VWs use *full-floating* piston pins. When these pins wear, or their bushings get loose, they don't make any noise. Even when a clip is broken and the pin is free to score the cylinder, it won't raise any racket. You have to take the engine apart to detect bad piston pins. They are very rarely a problem, so don't lose any sleep over them.

Rod Knock—If there is one internal engine noise associated with air-cooled VWs, *rod knock* is it. Rod knock describes the knocking sound made by a connecting rod when there is excessive clearance between it and the crankshaft. When rod-bearing clearance is large, oil can't fill the gap between the rod bearing and crankshaft.

Metal-to-metal contact starts as the rod bearing is slammed against the crankshaft. If one rod is bad, you'll hear one steady knocking beat. If you are on a long trip and overheat the engine, you may hear the rod knock start out as a light tap, move into a medium rap and finally develop into a knock. The sooner you stop and overhaul the engine, the better.

By the time a rod is knocking, serious damage has been done to the crankshaft and rod. Continued driving may spin the bearing. A *spun bearing* is one that is rotating relative to the rod and the crankshaft. In other words, the bearing is no longer clamped by the connecting rod. When the bearing spins, it covers the oil holes in the crankshaft. Oil ceases to flow to that rod and it rapidly overheats.

Many times, the rod will weld itself to the crankshaft, seize and break. The rod, case, and camshaft are sure to be broken. Usually metal particles from the rod have been pumped through the engine with the oil, and all other precision clearances have been destroyed by the passing metal. When you hear a rod knock, STOP! Rebuild the engine while there is still enough to rebuild.

Rods also wear with accumulated mileage. Then they wear in sets, so you'll hear a castanet-like rattling with old, tired bearings. This sound is often heard during a cold start before oil pressure builds.

To test for rod knock, thoroughly warm the engine to operating temperature. With the transaxle in neutral, lightly rev the engine, say from 1000—2000 rpm, and *abruptly lift off the accelerator.* Engine rpm must drop sharply. As rpm drops, the rods should rattle, knock or pound, for an instant, depending on how you hear it. This is because the rods float on their journals as they pass through the transition of being loaded, then quickly unloaded.

Main-Bearing Knock—Sounding similar to, but deeper than, worn rod bearings are bad main bearings. Main bearings knock for the same reasons as rods—excessive oil clearance—but under different conditions. To test for main-bearing knock, put the thoroughly warmed engine *under load.*

Round can at right is *throttle dashpot.* It slows throttle closing via rod resting against throttle linkage, which is being disconnected. For quicker throttle response when testing for rod knock, disconnect and plug vacuum line leading to dashpot. This is a 411 engine, but dashpots are found on all air-cooled VWs.

On a manual-transaxle car, load the engine by selecting first gear and letting out the clutch until the engine begins to labor. Then load the engine further by putting half of your right foot on the brake and the other half on the accelerator. Use the parking brake, too. Keep engine speed at about 1000 rpm during the test and don't let the car creep forward. With a little throttle and clutch juggling, the knocking main bearings will sound off with a heavy, low-frequency pounding. You can practically feel this better than hear it. Be careful! Don't do this for more than three seconds or you'll burn out the clutch. You can also hear bad main bearings while going uphill, accelerating or during other periods of high engine load.

With an Auto-Stick or full automatic transaxle, the test for bad main bearings is easier. Place the transaxle in gear, hold the brake firmly with your left foot and depress the accelerator slowly with your right. Don't overdo it; just load the engine so the car tries to creep. If the main bearings are going to knock, they'll do so right away. There is no need to keep the engine and transaxle straining—this is very tough on the transaxle. A few seconds (5—10) and the test is over.

VW main bearings are quite large and strong for the horsepower of the engines. Therefore, they rarely fail, wear out, or make noise. This is especially true of Type 4 engines. The rod bearings are much more suspect. If you hear knocking or pounding from the engine, chances are a rod is about to fail.

DIAGNOSTIC TESTS

Cranking Vacuum—The next diagnostic step is an engine-cranking vacuum test. This checks the pumping ability of the engine. By measuring the vacuum an engine produces while cranking, you are really testing how well-sealed the cylinders are. If all internal parts are in good shape, the engine will produce a lot of vacuum—if not, vacuum will be proportionately lower.

Begin by warming the engine to operating temperature—ten minutes idling or a five minute drive. Check for heat out of the heater, too. There's no need to overheat the engine, just warm it up.

Once the engine is warm, shut it off, disable the ignition so the engine can't start, and connect your vacuum gage to a *full manifold-vacuum source*. Any vacuum nipple on the intake manifold will do. Just make sure the vacuum source isn't *ported vacuum*—from one of the small diameter nipples on later carburetors or fuel-injection systems.

Ported vacuum exists in the carburetor primary venturi, just above the throttle plates. It creates a vacuum signal used for operating various emission-control switches. But, it's a vacuum signal that reads low on part-throttle applications—opposite of the high manifold-vacuum readings under the same conditions.

A note about altitude and how it affects vacuum readings. Because atmospheric pressure drops as altitude increases, cranking vacuum will drop about 1 inch of mercury (in.Hg) for each 1000-ft increase in altitude. So, at 5000 ft, for example in Denver, Colorado, cranking vacuum values will be 5 in.Hg *below* a reading taken at sea level. The vacuum values given below are for measurements at sea level.

Prop the vacuum gage so you can see it through the rear window, or have a friend crank the engine. An engine in good condition will pull a steady vacuum of about 10 in.Hg. (This same engine in Denver is registering 5 in.Hg.) A worn engine with no major problems will have a steady, but lower reading. Don't be alarmed if the needle swings about 2 in.Hg—it's normal on a four-cylinder engine, especially slow-cranking ones with 6-volt starting systems.

If the needle regularly drops to near 0 in.Hg, then there is a problem. Such a vacuum drop can have numerous causes: poorly adjusted valves, burned valves, worn cam lobes, pulled head studs, warped cylinder heads and worn cylinders, pistons or rings. To pinpoint the cylinder at fault, you'll have to perform more tests.

Power-Balance Test—This test shows how much each cylinder contributes to the power output of an engine. Thus, it also isolates which cylinders contribute little to manifold vacuum. You can perform a power-balance test at home on any air-cooled VW except for '79 and later Buses. They are equipped with electronic ignition that can't be *open-fired*.

To do a power-balance test on them, the lead has to be grounded without open-firing it first. A professional oscilloscope/diagnostic tester easily does this. You can do the same by inserting a metal spring between the sparkplug and lead. Then to ground the lead, touch a grounded wire to the spring. This is easier said than done in the confines of a Bus engine compartment, so a professional test is the best method.

To perform a power-balance test on other engines, pull all sparkplug leads off the sparkplugs, then set the leads lightly back on the tops. You're going to lift a lead off its plug without a lot of tugging. The idea is to pull the lead away from the plug and ground it against the head and stop that plug from firing—this is called open-firing. The engine will then be running on three cylinders. By comparing the resulting rpm drop for each disabled cylinder, you can determine which cylinder is at fault.

If you suspect a burned valve or other major problem, a quick, *ear-calibrated* power-balance test will tell what you want to know—which cylinder is it? Because VWs have only four cylinders, a bad one shows right away. If you are looking for a more subtle problem, however, use a dwell/tachometer to measure rpm drop for each cylinder. The car's tach is not

Manifold vacuum is excellent indicator of overall engine condition. You can detect major problems in seconds with a vacuum gage. This engine isn't drawing much vacuum during running idle test. Judging from grit and lack of air filter, rings are probably shot.

13

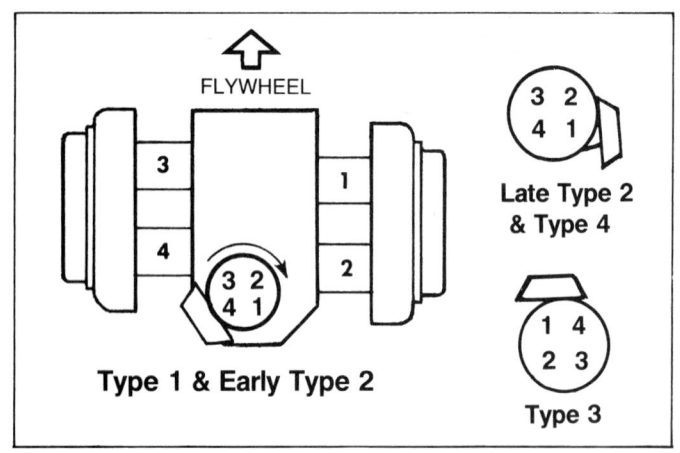

Cylinder layout, distributor position at TDC (Top Dead Center) and firing order. All distributor rotors, crank pulleys and fans rotate clockwise. Note vacuum advance can position and rotor tip points at #1 when installed.

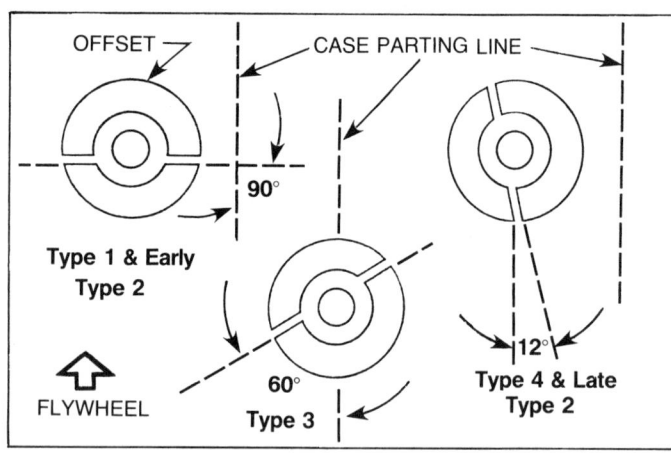

Distributor driveshaft slot position at TDC, cylinder #1. Note offset of slot: thicker arc faces different position depending on engine.

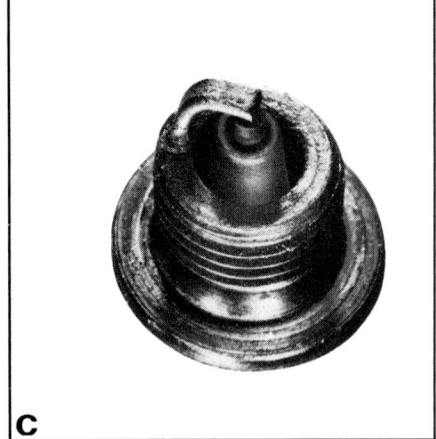

Reading sparkplugs can yield important troubleshooting clues. Plug A suffers from heavily rounded electrodes and pitted insulator. It's worn out. Replace such plugs and engine performance will improve. Plug B is oil-fouled. Shiny-black coating indicates excess oil consumption, possibly from worn rings and valve guides. Plug C is carbon-fouled; don't confuse it with oil-fouling. A carbon-fouled plug's dry, flat-black coating comes from excessively rich air/fuel mixtures, stop-and-go driving or a too-cold plug heat range. Plug D is normal. Electrode rounding is moderate and insulator is even tan or gray, indicating all is well in the combustion chamber. Photos courtesy Champion Spark Plug Company.

accurate enough for this test.

With the tachometer connected, ground the first plug lead and wait for engine rpm to stabilize. Now, write down the reading and reconnect the plug lead. Go to the next plug and do the same until you've done all four.

It doesn't matter so much how far rpm drops as how close the readings are to each other. Don't expect the readings to be any closer than 20 rpm. But when these readings start varying by more than 40 or 50 rpm, take notice. Remember, the cylinders with the least drop are the bad ones. Therefore, a really bad cylinder may not drop in rpm at all. Of course, if all cylinders are bad, none will drop very much. Good VW cylinders usually register a drop of about 200 rpm.

Reading Sparkplugs—Think of a sparkplug as a removable portion of the combustion chamber, and you'll see it has useful diagnostic potential. Because the compression test follows, which requires sparkplug removal, let's discuss sparkplug reading now.

Normally, a plug should be dry, with an even tan coating and slight rounding of the electrodes. If the fuel mixture is too rich, the plug will be coated with dry, flat-black carbon. Rub the carbon onto the palm of your hand. The black deposits should wipe off easily. If the mixture is too lean, the plug will be powdered with a white coating, and the porcelain insulator will appear burned. The insulator can also turn a pastel green or yellow in normal operation, depending on the individual fuel blend being used. Oil in the combustion chamber will leave the plug wet and shiny black. Rub that into your palm and you get an oily mess that won't rub off easily.

When reading plugs, pay more attention to the porcelain insulator around the center electrode than the metal shell. It's most sensitive to coloring and more likely to show symptoms of unusual combustion. Also be aware that sparkplugs from a street-driven engine can only show the most basic combustion conditions because of the many operating conditions a street-driven engine is subjected to.

You may have heard about the ace mechanic who read the plugs, then made a one-eighth turn adjustment to the carburetor and won the race. That's on a race engine; plugs in a street engine can't be read that way. Check for the oily plug on a street engine. It reveals a problem with the rings or valve guides.

Compression Testing—The familiar compression test is a good way to measure the condition of the rings, cylinders and valves. There are two types of compression testers: a tapered rubber-cone type that is inserted into and held against the open sparkplug hole, and the screw-in type. The rubber-cone version is difficult to use on VWs because the cone and its mount are usually too short to reach through the cooling shrouding. They are also awkward to hold against cylinder pressure while you are bent into a VW engine compartment. Use a screw-in tester, if possible. If using the cone-type, you'll need a remote starter switch or a helper to crank the engine. A helper is best in any case because it leaves you free to watch the compression gage during the test.

The engine must be warmed up, ignition disabled and all sparkplugs removed. Watch out for hot parts whenever working on a warm engine. The throttle and choke plate must be fully open for an accurate test—part-throttle openings result in low readings. So, if you're using a remote starter switch, prop the throttle linkage open with a screwdriver. If a friend helps, have him fully depress the accelerator when cranking the engine. *Note: Avoid smoking and open flames because the air/fuel charge will come out of the sparkplug holes when the engine is cranked.*

Screw-in compression testers are easier to use. Just screw it in and crank the engine yourself. If you use a rubber-cone tester, you'll be busy pushing the cone tightly against the sparkplug hole.

With either method, hold the throttle open and crank the engine so the tested cylinder has about 6—8 compression strokes. You can hear cranking speed slow as the tested cylinder comes up on its compression stroke. Note how fast compression increases and jot down the highest reading. Test all four cylinders the same way. Give each cylinder the same number of compression strokes.

Like the power-balance test, even readings are desirable. Depending on the engine, pressure can range from 75 psi (pounds per square in.) to over 150 psi. So, don't be too concerned if the figures seem generally low. Trouble is much more likely if only one or two cylinders are low.

Remember, different compression testers give different readings, so allow some leeway.

Popular diagnostic check is compression test. Screw-in tester is easiest to use, although rubber-cone type like one shown is just as accurate.

Double-check your findings with the 75% rule: All cylinders must read within 75% of the highest cylinder. So, if the highest reading is 125 psi, multiply 125 by 0.75 to get 94. Therefore, if all cylinders read above 94 psi, they are acceptable. Below that, consider them faulty. Notice I said acceptable, not desirable. It is hard to set a wear limit and say anything above is good and all below are bad. In the example given, if a cylinder yielded only 97 psi and the rest were 120 or 125 psi, I would be wary of the low cylinder.

To help determine the cause of low compression, do a *wet test* by squirting a teaspoon of oil into that cylinder. SAE 30W is fine. To determine how many squirts it takes to make a teaspoon get a teaspoon and fill it while counting the squirts. Then squirt the same amount of oil into the cylinder. Just make sure the oil can is full so you don't squirt air into the cylinder.

Crank the engine two revolutions or so to spread the oil. Retest the low cylinder. If compression comes up markedly, 40 psi or more, the trouble is poor ring-to-bore sealing. A rebuild is needed to restore the lost clearances. If compression doesn't increase much, about 5 psi, then the problem is probably with the valves. It could also be pulled head studs or a warped cylinder head.

You may notice a cylinder that takes a long time to pump up. Usually, a cylinder will produce 40 psi on the first piston stroke, another 35 psi on the next and so on. Problem cylinders may have trouble reaching 40 psi and, instead, increase by 10 psi at a time. If you crank them enough, they'll come close to the other cylinders. Wet test such a cylinder, because this condition is usually caused by poor rings.

On the other hand, a cylinder suffering from excessive oiling—from bad rings even—can yield high compression-test readings because excess oil in the cylinder seals the rings. Again, if you crank this type of cylinder enough, relatively high readings can result.

There are variables that affect the readings obtained from compression testing. One is cranking speed; higher speed gives higher pressure readings and vice versa. With a small, four-cylinder engine, it isn't likely that the battery will run down during a compression tests. But if it does, jump the battery to another one to maintain cranking speed.

Altitude will affect compression readings even as it influences manifold-vacuum readings. They will register lower values the higher the altitude. Worn camshaft lobes can also cause lower-than-normal readings. High-performance camshafts, with their long-duration profiles, also give lower compression readings. This is because such cams sacrifice low-rpm breathing for improved high-rpm breathing. Compression testing takes place at cranking speed—well below idle speed.

HIGHER ALTITUDE & LOWER COMPRESSION

Compression readings are influenced by altitude and temperature. Specifications for compression values are usually based on *standard day* conditions: 14.7 psi atmospheric pressure and 59F at sea level. Atmospheric pressure and temperature decrease as altitude increases above sea level, so compensate for this when interpreting compression-test results.

The chart supplies correction factors (accounting for decreased pressure and temperature) for different altitudes. Just multiply the specification value for compression (in psi) by the factor for the engine's operating conditions.

Altitude (ft)	Factor
1000	.9711
2000	.9428
3000	.9151
4000	.8881
5000	.8617
6000	.8359
7000	.8106
8000	.7860

An acceptable compression reading of 125 psi at sea level would register less in Denver, for example. There, at about 5000 ft, the equivalent compression reading would be 125 psi × .8617 = 108 psi. The cylinders could be reading low compared with sea-level measurements, but just fine for the actual operating altitude.

Leak-Down Testing—Although it's also a measure of combustion-chamber sealing, a leak-down test is more accurate than compression testing. Accuracy is improved because variables affecting compression-test readings—those that have no bearing on the sealing capability of an engine—are eliminated.

A leak-down tester uses an external air-pressure source. Testing is done with the engine stationary. Therefore, the test is not influenced by cranking speed, valve duration, altitude or excessive oiling. If you are diagnosing a car before buying it, a leak-down test is an excellent idea.

Leak-down test equipment is expensive. So, unless you do a lot of engine diagnosis, this is one test to farm out. Many tuneup shops can do the test for you. The cost should be minimal. It is also a test that can be skipped most of the time. A compression test gives an accurate enough picture of an engine's condition 90% of the time. This is especially true if there is a burned valve, holed piston or other catastrophic cylinder damage. If the compression readings are baffling, however, a leak-down tester will definitely help you make a decision. Of course, if you have access to a leak-down tester, skip the compression test and test the cylinders with the more accurate leak-down tester.

You'll need an air compressor (a 1/2-HP model will do) and the leak-down tester if performing the test yourself. Start by reading the instructions that came with the tester. Bring the #1 cylinder to top dead center (TDC) of the its *compression stroke*. Check the engine timing marks to make sure it's exactly on TDC. If it's slightly off, the engine will kick over without warning the instant the cylinder is pressurized. A good way to check for TDC is to insert a long, thin screwdriver into the combustion chamber through the sparkplug hole. With the screwdriver contacting the piston top, you can feel when the piston is at the top of its stroke.

During the test, have a helper hold the crank with a socket on the crank-pulley nut, generator, alternator, or fan. This will keep the engine for turning over.

Next, install the hose adapter in the sparkplug hole, then connect the tester to the adapter and the air compressor. Compressed air is pumped to the cylinder while the tester monitors how much air it takes to make up for cylinder leakage. The readout is in percent leakage.

Remember, the piston must be at TDC of its compression stroke so both valves are closed. Otherwise, leakage will approach 100% as all the compressed air blows by an open valve, or the engine will turn over. Once finished with the first cylinder, disconnect the tester, rotate the engine 180° to cylinder #4 and test it. Then test cylinders #3 and #2, by again rotating the crank 180° each time.

Leakage for an engine in good condition is 10% or less. The higher the leakage rate, the worse the problem. A 20% leakage can indicate a high-mileage engine, but doesn't normally warrant a rebuild. A 30% leakage is serious enough for an engine overhaul or valve job. And, a 90% leakage indicates serious damage, such as a badly burned valve, holed piston or the like.

You can usually tell what's leaking by listening to the engine with the tester attached. If the exhaust valve is leaking, you can hear the hiss of escaping air in the tailpipe. Leakage past intake valves can be heard at the carburetor or intake-air sensor. Bad-sealing rings and cylinders can be detected at the oil-breather or dipstick holes. Pulled head studs may cause hissing leaks between the cylinder heads and cylinders. A length of hose can aid listening in some chassis by holding one end at your ear and the other where you suspect leakage: carburetor, intake-air sensor or breather. Sometimes leakage is evenly divided and hard to attribute to one source, both the rings and valves may be leaking. If so, it's a sure sign of multiple problems.

One way to spot a suspected leaky exhaust valve is to hook an HC/CO meter to the tailpipe and squirt some carburetor cleaner into the cylinder. Reconnect the leak-down tester and watch the meter. It takes a minute for the leakage to reach the meter, but if the exhaust valve is leaking, even just a little, the HC portion of the meter will peg instantaneously! If you use a tuneup shop for the leak-down test, they should have an HC/CO meter available. This test actually doesn't require a leak-down tester, just an adapter for the sparkplug hole and a compressed-air source.

VALVE LIFT

Valve lift is the distance the valve is moved off its seat by the camshaft. As the cam lobes wear, valve lift decreases and the engine doesn't breath as well. Power is reduced and valve train noise increases because the valve clearance increases with cam and lifter wear.

If the diagnostic tests thus far indicate worn engine internals, there is little need to consider valve lift. You might as well get on with rebuilding the engine, and inspect the camshaft and lifters directly.

On the other hand, if the engine seems well-sealed at the valves and cylinders, but lacks power and has noisy valves that won't stay in adjustment, valve lift is likely your problem.

To understand the wear cycle, first consider how camshafts and lifters are made. At first glance, the working surface of a lifter and the top edge of a camshaft lobe look flat. They aren't. Instead, the lifter's bottom is convex, so its center protrudes more than the edges. The top of the lobe is cut at a slight angle, so that when the lifter rests against it, the contact point is off-center. This makes the lifter rotate with each valve opening and spreads wear over the surface of the lifter.

Problems start when the lifter wears a groove into its concave surface, or the cam lobe wears flat. Then the lifter tends not to rotate, and wear concentrates in one spot. This wears the lifter and especially the lobe very rapidly, until the lobe is considerably shorter than when new. The shorter the lobe, the less valve lift and horsepower.

Increased noise is part of the wear process because valve clearance increases as the lobe and lifter grind down. Frequent valve adjustments become necessary to keep noise down and the valves adjusted. However, even frequent valve adjustments don't stop the wear once it is started. They only reduce the valve clearance for the short time it takes the lobe to wear down some more.

It is possible to disable one cylinder by adjusting its valves. That will provoke some mystery! What happens is the lifter rotates right after a valve adjustment, so the worn spot is no longer over the cam lobe. Instead, a part of the lifter which is close to original thickness is now against the lobe. But because the valve was

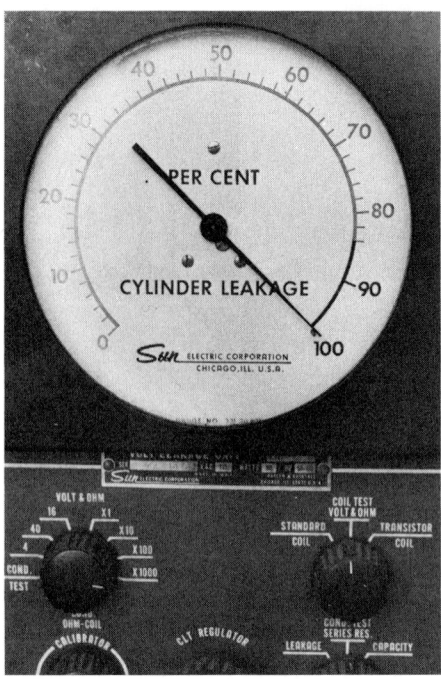

Leak-down testing can indicate more about engine condition than any other test. This unit is part of professional diagnostic tester; others are available as separate tools. Need for air compressor, cost and more involved test procedure usually prohibit home-mechanic use. Farm this test out to get accurate evaluation of engine condition.

adjusted for the worn section, *actual* valve clearance is zero and the lobe is holding the valve open *all the time*. When the engine is started, compression can't build and the engine runs on three cylinders.

If you recheck the valve clearance, you'll find it tight on that cylinder, which was just adjusted correctly. So, a bit puzzled, you loosen the adjustment to specifications and the engine runs fine. Then five minutes later it starts clacking away as the dished lifter rotates its low spot over the lobe again. No puzzle now, the engine's cam and lifters are worn out and need replacing.

If the engine has 50,000 or more miles and needs a valve adjustment every 500 miles, the cam and lifters are worn out and need to be replaced. Complete engine disassembly is required to service the cam and lifters, so it's timely to rebuild the rest of the engine, as well. Chances are the valves and cylinders are worn anyway, so endless valve adjustments are usually another clue that the engine needs an overhaul.

Measuring Valve Lift—Some VW specialists, like drag racers, use a dial indicator to measure valve lift at the valve-spring retainer. They can then determine if the cam lobe is wearing down. Racer's can't hear noisy valves over open exhaust, and don't have time to split the case to look at the camshaft between races. Measuring valve lift lets them determine there is no camshaft wear; consequently, they don't have to split the cases to determine the cam's condition.

If measuring valve lift, you'll need a dial indicator, some way to mount it near the valve springs and enough room to fit the instrument. On some chassis, like 914s, this measurement is out of the question unless the engine is out of the car. Note the space available for a dial indicator before considering measuring valve lift. Indicator magnetic bases won't attach to aluminum cylinder heads, but might mount on the cooling shroud or exhaust.

Bring the indicator's plunger to bear on the valve-spring retainer. Rotate the engine until that valve is completely closed. Zero the dial indicator, then rotate the crank pulley until the valve is completely open. Read valve lift directly on the dial.

Remember, because you are measuring valve lift, allow for the *rocker-arm ratio* and valve clearance. Rocker-arm ratio will *add* to lift measured at the cam lobe, valve clearance will *subtract*. For a test of this kind, though, don't be concerned about the absolute valve lift, but valve lift relative to the other valves. In other words, look for a valve that is lifting considerably less than its neighbors. If a valve is lifting less than the others, the cam is worn and will need to be replaced or reground.

CHAPTER 2
Engine Removal

Power train on 914 is somewhat heavy, but imagine pulling equivalent package (engine with all accessories, clutch, transmission, differential and cooling system) out of a Camaro with only a floorjack!

Use old flywheel to make wider floorjack saddle. Weld short length of pipe over gland-nut bore to secure flywheel in jack.

Engine removal and installation are important steps in any overhaul. Haphazardly removing an engine guarantees headaches during installation. It's also dangerous. A little preparation and caution before and during engine removal will reward you when installing your rebuilt engine.

PREPARATION
Air-cooled Volkswagen engines are found in many different types of chassis, but the same special tools are needed for all cars: a *floor jack*, *jack stands*, and a *piece of plywood*. Additionally, clean the engine before working on it and decide where to pull it. Containers for hardware must also be readied.
Engine Cleaning—A dirty engine is miserable to work on. Wrenches slip, fasteners hide under the goo and grime gets under your fingernails. Avoid these problems by cleaning the engine before removing it. Three methods are generally available: *steam cleaning*, *solvent blasting* and *spray degreasing*.
Steam cleaning is for truly filthy engines—such as an oil leaker driven on dirt roads. The cost is reasonable and the job takes about a half hour. Most service stations have the equipment. If not, check with a tractor or heavy-equipment shop.

Consider having only the bottom of the engine steam cleaned. Steam cleaning the top is too messy as the hot solution is reflected back at the mechanic and is trapped atop the engine by the sheet metal.
Solvent blasting uses compressed air and solvent to blow off the dirt. It works well 90% of the time. Cost is comparable to steam cleaning and practically any shop can do it. A thorough solvent blasting takes about as long as steam cleaning.
Spray degreaser can be used at home if you have a garden hose. Typically, you warm the engine, spray it with degreaser, let it set so the degreaser can penetrate, then hose off the crud. Problem is, the stinky mess ends up on your driveway. Solve that by doing the job at a car wash. Use the high-pressure water/detergent spray and leave the mess there.
Remember to cover the distributor, coil and carburetor with plastic bags or aluminum foil shaped to fit. These parts must be kept dry or you'll have a hard time restarting the engine. With patience, this method works as well as steam cleaning or solvent blasting. Don't forget to remove this waterproofing before driving. (Yes, it happens!)
Lifting & Lowering Tools—To raise and support the car during engine removal and installation, use a *floor jack* and *jack stands*. The floor jack is a hydraulic jack in a wheeled frame. For VWs a 1-ton version is adequate, but a 1-1/2-ton jack is sturdier and usually will lift higher. Besides, if you're planning to buy a floor jack, you'll need a 1-1/2-ton version for lifting most other cars.
Once the car is up, *you must support it with jack stands*. Never use any jack (bumper, scissors, screw or otherwise) as a stand. Jacks are for raising and lowering, *not for supporting* a car while you are underneath.
A jack can fail, and if you are under the car when it does, it could be fatal. Use two jack stands to hold up the rear of the car. Both the floor jack and jack stands can be rented. Look in the phone book under Rentals.
A VW air-cooled engine is *lowered* from its raised chassis with a floor jack. A balanced jacking point for lowering the engine is right under the oil strainer. You'll need a sizeable piece of 1/2-in. or thicker plywood to place between the engine and floor jack.
Besides protecting the engine from gouges, the soft plywood gives the hard engine and jack surfaces something to dig into. This makes the engine less likely to slip off the jack's pad, making engine removal safer and a lot easier. It's frustrating to pull on the engine, hoping to roll the engine/jack combination toward you, only to have the jack stop and the engine slip off.
For the same reason, try not to remove a VW engine on a dirt surface. The floor jack resists rolling; it will sink into the dirt instead. Even on very hard packed dirt, little pebbles can chock

18

Support car with jackstands any time it is raised. Place stands forward of rear wheels on sturdy chassis component; set them on concrete for best foundation. On soft asphalt or dirt, place plywood between stand and ground to prevent settling and tipping.

Label all wire and hose connections with tape and permanent marker. These labels are your insurance for correctly installing connections during engine installation, so *don't* rely on memory alone!

the jack wheels, making engine movement a jerking series of barely controlled, backbreaking grunts. If a dirt floor is all you have, lay a full sheet of plywood down to roll the floor jack on. It will also keep you cleaner as you work under the car.

Consider what you'll do with the chassis when you remove the engine. Once it's out, moving the chassis means pushing or towing. The chassis will be immobile for awhile, depending on how fast you work. Three weeks is about average. Stationary cars attract vandals, angry landlords, even the authorities in some cities.

If you don't have a dedicated working room, try renting space at a service station. Look in the phone book for a do-it-yourself auto shop or hobby shop. If you are stationed at a military base, they often have auto hobby shops available, complete with some of the larger tools.

Get Organized—It's a trying task to install an engine someone else removed. Who knows where all those nuts and bolts go? Well, pulling an engine and throwing all the hardware in one big box amounts to the same effort. When you do get around to installing the engine, you'll find your memory can't make any order from the chaos.

With the earlier Beetle, there aren't many disconnections to remember, but later engines and different chassis can definitely tax the best memory. Save yourself considerable trouble and frustration by getting several coffee cans and boxes and labeling them. Use one for bell-housing hardware, another for heater tubing and so on. Have the containers ready *before* dropping the engine, or you won't use them.

Get a roll of masking or other stout tape and a permanent, waterproof marking pen. Or use a plastic label maker. Use these for labeling the vacuum and electrical disconnections you'll make. *Labeling disconnections is a critical step, so don't skip it!*

There isn't enough room in this book to list all the hose and wire diagrams for the various chassis. Draw your own schematics of the various connections to help at reassembly. It's up to you to label and keep track of the electrical, vacuum, mechanical and fuel lines, hoses and cables. You'll thank yourself at installation.

ENGINE REMOVAL
We'll examine engine removal chassis by chassis because of the different chassis air-cooled VWs are mounted in. Yet, the first few disconnections of components about the engine are similar in all models.

Battery—On all chassis, disconnect the battery *negative* cable first, then remove the positive one. The battery is under the rear seat in the Beetle, Squareback and Fastback. On the 411 and 412 look under the driver's seat. On the Bus, Karmann Ghia and 914, the battery is on the right side of the engine compartment.

It's not essential, but now is a good time to completely remove the battery for cleaning and charging. It's aggravating to try and start your fresh engine only to find the battery dead.

Drain Oil—Now drain the oil. On Type 1—3 engines, the drain plug is the large bolt in the center of the oil strainer. On the Type 4, the drain plug is separate from the strainer. Let the oil drain while you make the various electrical and mechanical disconnections. The longer the oil drains, the less mess you'll have later when you open the engine.

BEETLES & KARMANN GHIAS (TYPE 1)
Air Filter & Housing—Open the engine cover and remove the air filter. Earlier cars had oil-bath air filters with a minimum of hoses attached to them. On the Beetle, label and disconnect any hoses, then unscrew the clamp at the air-filter housing base. Without tipping the filter, lift it off the carburetor and set it aside.

From August '67, a cable is fitted to the warm air flap on the filter inlet. The cable is connected to the engine thermostat and controls engine inlet air temperature in response to engine temperature. Disconnect it at the air filter. Remember, when handling and storing an oil-bath air filter, *keep it upright,* or oil will contaminate the upper half of the filter and spill out of the unit.

In '73 a paper-element air filter replaced the oil-bath unit. The paper element is easily identified by its rectangular shape. Its hose attachments are different, but no problem. Label them with masking tape and your permanent marker.

You don't have to know exactly what the hose does you are removing, just mark the first hose 1 and where it attaches with a 1, too. The next disconnection gets a 2 and so forth. During engine installation all you'll have to do is connect the 1s to 1s, 2s to 2s and so on.

Karmann Ghia air filters are mounted to the right of the engine. Unclamp and remove the filter-to-engine and hot-air hoses. Next, undo the warm-air control flap cable from its arm on the filter inlet. Label and remove the crankcase breather hose, then unlatch the filter assembly from its mounting bracket and remove it from the engine compartment.

Fuel-Injection Air Filter—Undo the four

19

Tipping air filter any more than this will slosh oil inside against upper section; it will also drain from under lid. Clean filter canister and change oil before filter is installed on rebuilt engine.

Needle-nose pliers hold throttle cable and linkage while cable is disconnected. Don't loosen cinch bolt against cable tension, or you'll kink cable. Always hold linkage stationary instead.

After throttle cable is removed from linkage, pull cable guide from fan housing. Sometimes a hose clamp is placed on guide in front of fan housing as a retainer. If so, just leave guide in place.

Generator disconnections are normally three wires right on top of the generator. However, slip-on connectors are used on generator-mounted voltage regulators and alternators.

clips around the air filter housing. This releases the air filter housing cover and the paper air filter element. Now disconnect the multiple wire plug on the *intake air sensor*, which is the cast-aluminum box. Be careful when pulling back the rubber boot and tugging on the connector. Never pull by the wires.

Next unscrew the rubber boot's clamp at the other end of the air sensor. Remove the two air cleaner mounting nuts on either side of the hood hinge and lift the air sensor and filter housing off as a unit. With the sensor out of the way, unclamp and remove the rubber air duct. There are several breather lines intersecting with this duct. Mark and remove them.

Carburetor—Mark and remove the automatic choke heating element and fuel-cutoff wires.

Also remove the fuel line and throttle cable. The throttle cable is removed by unscrewing or unbolting the clamping bolt and pushing the cable toward the fan housing. Pull the cable guide out of the fan housing and set it aside. Later, when the engine is partially out of the chassis, you can pull the cable out the fan housing the rest of the way.

A lot of throttle cable guides have been clamped behind the fan housing to hold a homemade grommet. In this case, you can push the throttle cable into the guide now, and pull it completely out when lowering the engine.

Electrical Connections—Look under the distributor for the oil-pressure sending unit. Disconnect and mark its single wire. Disconnect the positive coil wire. Look on the coil, near the terminals for a + sign if you don't know which is the positive wire. The negative wire also runs to the distributor, but it's not the right one.

Generator/Alternator—Mark and remove the three wires on the generator. The voltage regulator is mounted on the generator on '66 Beetles. In that case, remove the three slip-on connections.

If the car has an alternator, remove the multiple-wire connector. The voltage regulator is mounted separately on '73 and early '74 alternators. After that, the regulator is integrally mounted on top of the alternator.

Fuel Injection—Unfortunately, VW's Bosch fuel injection adds a lot of little steps to engine R&R (Removal and Replacement). You have extra marking and removing of necessary wires and hoses. Take your time when labeling these connections. You'll be reconnecting these extra fuel system wires and hoses during engine installation.

The electrical disconnections are pull-off, push-on plugs. Most plugs separate easily, but other don't. This is especially true of those brownish, rectangular connectors. You usually have no choice but to grasp the wires leading out of this type of connector and pull. Practically every time you succeed and the spade connection inside the connector separates. Sometimes the wire pulls out of its terminal end. Of course, repair the wire in that case.

The fuel-injection wiring is in a harness. Follow it around the engine, disconnecting and labeling wires at the coil, injectors, crankcase and temperature sensors. Then the harness can be pushed aside.

Make sure you follow the harness. Some wires lead from one side of the engine to the other, and there's no reason to disconnect them. Only remove a wire if it leaves the engine and attaches to the chassis.

Also remove the two fuel lines: one supply, one return. Be sure to correctly mark their flow. Get them reversed and the engine will not start.

Throttle Positioner—If the engine has a throttle positioner, you'll see an aluminum diaphragm-and-cylinder unit sticking out from under the carburetor. Most manuals say the positioner must come off for engine removal, but it isn't so. Just leave it alone. Later, when lowering the engine, you'll have to tilt the fan housing forward. This raises the positioner so it will clear the rear bodywork.

Rear Engine Cover Plate—Between the rear of the engine and the rear of the engine compartment is the rear engine cover plate. This piece of sheet metal is part of the cooling system, which works by sealing the top of the engine from air passing under it.

Because the engine must be slid to the rear to disengage it from the transaxle input shaft, the rear engine-cover plate must be removed. On early 40-HP engines, merely remove the four screws and lift out the plate.

Later engines have two large hoses leading from the fan housing to the heat exchangers. Completely remove these hoses and their rubber gaskets at the cover plate end. Then unscrew and remove the small separate shroud over the crankshaft pully. (Unless you have a fuel-injected engine. They don't have this small plate.) Finally, remove the two covers around the heat-riser tubes leading to the intake manifold.

The heat-riser tube covers are at the outboard ends of the rear engine cover plate. Four screws attach each one. With those parts gone, you can unscrew the rear engine cover-plate attaching screws and pull the plate out of the car.

Raise Car—Use the floor jack to raise the car until the engine is about a yard in the air. The rear bodywork must be high enough to clear the top of the fan housing. Put the jack under the frame just forward of the transaxle, *never* under the engine or transaxle. You can crack the case by jacking under the engine.

Immediately place the jack stands to support the chassis, and slowly lower the car onto them. Check the stability of the car on the stands by gently shaking it from side to side. VWs have to be raised a lot to get the engine out, which means most jack stands are raised to their highest, *and least stable,* position. Be sure your stands are stout and stable before getting under the car.

Heater Cables—At the front and sides of the engine you'll find the two heater-control valves. Remove the heater-control cables from their levers on the control valves. A bolt passes through the lever and cable end and is nutted on the other side.

Use two wrenches to remove the bolt and nut, then the cable will pull free. There is also a small cylinder in the lever which the cable passes through. When you remove the cable, the cylinder should fall free, so be ready for it.

Alternately, you have extra steps if the heater controls have rusted shut and then been peened over by rocks. Freeing the cable end requires rust penetrant, pliers to grip the lever, the usual two wrenches and inventive language.

These are *clean air* hoses connecting fan housing and heat exchangers. Just slip them off at both ends and set aside. Replace or tape-repair damaged hoses.

Loosen small bolt on heater-control valve arm to free heater cables. When cable and lock bolt are hopelessly frozen, snip cable and buy new ones.

Remove rear sheet-metal tray so the engine can easily slide rearward. Follow tray's leading edge to find its attachment screws.

21

Heater ducts are large flexible hoses leading forward from heat exchangers. This one was clamped, but often they clip on the exchangers and easily slide off.

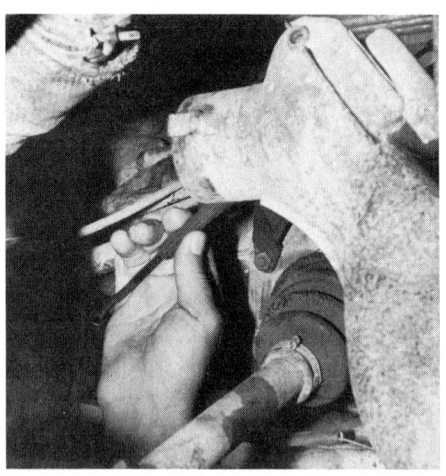

Be prepared for spillage when disconnecting fuel line. A bolt stuffed and clamped in flexible hose will stop fuel tank from siphoning dry.

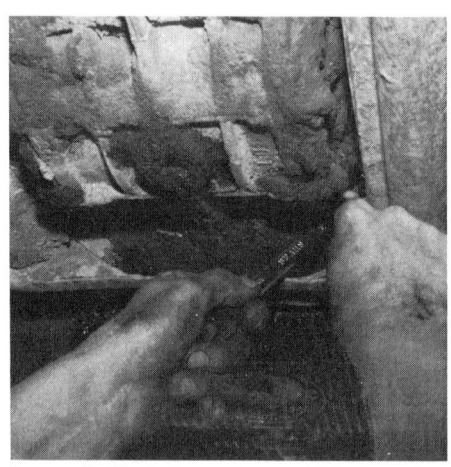

Lower bellhousing nuts are higher up between transaxle and engine than you might think. Still, they are easy to get at. This drivetrain is a candidate for steam cleaning.

Once you have the cables free, pull the large flexible hoses off the heater-control valves. Push the hoses away from the engine so they won't get torn as the engine is lowered.

Fuel Line—Above the left heater-control valve is the fuel line connection from the fuel tank. Slip off the flexible line and use a pencil or bolt to plug it. If you use a bolt, make sure it has an unthreaded shoulder. A fully threaded bolt can let gasoline leak past through the threads. Some mechanics pinch the line shut with locking pliers, then pull it off. It's fast and clean, but I don't like squeezing fuel hose that hard.

Bellhousing Nuts—Remove the two 17mm hex nuts and washers at the lower corners of the bellhousing. These nuts are threaded onto a pair of *studs* that fit into the transaxle. Don't worry about the engine falling. There are two more bolts still attached on top of the engine.

WARNING: If the lower bellhousing fasteners are nuts and bolts, remove them *only* after checking that the upper bolts are still in place and the floor jack is set up to support the engine. The engine can fall on you if the upper bolts aren't in place.

The lower bellhousing nuts may be very tight on their studs because of rust or impact damage. If so, the stud may unthread from the transaxle, not the nut from the stud. There's no problem with this, so don't worry about it. You can separate the stud and nut later and reinstall the stud.

Automatic Stick Shift (Auto-Stick)—A few extra disconnections are necessary on cars with the Auto-Stick. Two ATF (Automatic Transmission Fluid) lines need disconnecting. One line runs to the ATF tank, the other to the oil pump.

The lines are steel braided and use high pressure hydraulic fittings, like on brake lines. Use two *tubing wrenches,* also called *flare-nut wrenches,* on the fittings. If you don't have tubing wrenches, a regular open end will do, but take extra care to not round off the hex. Be ready for ATF to pour out of the line from the tank. Have a pan underneath and work fast.

Plug the disconnected fittings so they won't leak, and so dirt can't enter the transmission system. The best plug is a pipe fitting that has been soldered, brazed or welded shut, but you probably don't have one laying around. Those small plastic caps new brake master cylinders are shipped with work well, if you have the right size. Aluminum foil wrapped several times around the fitting and secured with a hose clamp works, too.

Because Auto-Stick transaxles have a torque converter between the engine and clutch, there are four driveplate bolts to remove. The driveplate is bolted at its center to the crankshaft, like a flywheel. At its outer edge it is bolted to the torque converter. These bolts are accessible through a hole in the bottom backside of the bellhousing. Fuel-injected engines have a rubber plug in the access hole; carbureted engines have an open hole.

Some of these bolts are 8mm, 6- or 12-point. Make sure your socket is clean, not rounded off and lined up straight with the bolts. These bolts are small and will break or round off if not treated with care. Have a helper rotate the engine with the crankshaft pulley while you watch the access hole. Stop rotation when the bolt is squarely centered in the hole. Remove the bolt, then have your helper rotate the engine 90° where another bolt will appear in the hole. Continue until you have removed all four bolts.

If you don't remove the driveplate bolts, the torque converter will slide out of the transaxle with the engine. That's fine if the engine is seized and you can't rotate it to gain access to the bolts. But the transaxle oil seal will be ruined if it's necessary to pull the torque converter with the engine. Replace the oil seal if that's the case.

More Auto-Stick disconnections are necessary inside the engine compartment. Look on the firewall, to the left of the ignition coil, to find the control valve. Mark and disconnect the electrical leads. Investigate the vacuum hoses to see which ones must come off. Those that don't go to the engine can be left alone. The rest need to be labeled and disconnected.

Support Engine—Get the piece of plywood and set it on the jack saddle. Then roll the jack under the engine and raise the saddle until it is just carrying the engine weight. Don't lift it too much or you'll bind the engine on the bellhousing studs and have trouble sliding it off them.

Upper Bellhousing Bolts—Slide out from under the car and turn your attention to removing the upper engine-to-transaxle fasteners. All manual transaxle cars through '70 have bolts and nuts at the upper bellhousing. Starting in '71, all cars use two bolts but only one nut. At the right side is the usual nut and bolt assembly, but the left side uses only a bolt. It threads into a special round nut pressed into the engine case.

This is necessary because the offset oil cooler used from '71 on doesn't leave enough room to get at a nut from the engine side. Auto-Stick cars in '70 use nuts on studs, so their bellhousings have four studs: two at bottom, two at top. In '71 the Auto-Sticks went with the two bolt, one nut fastening.

Whatever the attachment method, you need

22

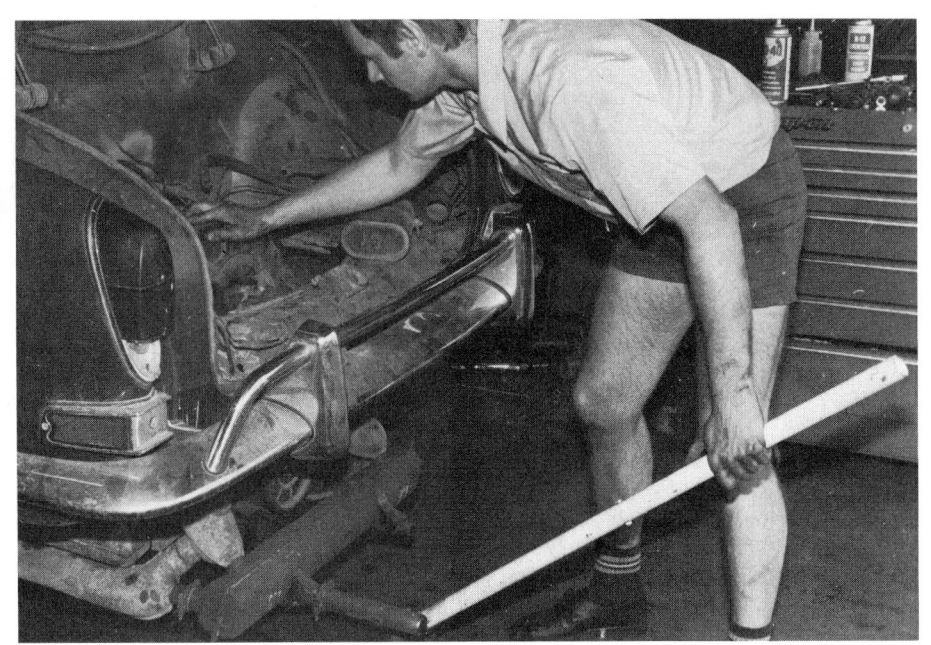

At first, engine must come back and down in small, quickly alternating steps. Once you get this far, a slow, smooth lowering is all that's needed. Keep an eye on hoses, wires and engine compartment seal.

to remove the upper bellhousing fasteners. Remember, the nuts are in the engine compartment and the bolt heads are on the transmission side, accessible only from under the car. With luck, both nuts will come off without anyone holding the bolts from under the car. More typically though, the bolts will turn.

Have your helper get under the car to hold the bolts. If no helper is handy, try pulling the engine away from the transaxle. That might bind the bolts so they won't turn. If that doesn't work, you'll have to attach a box end wrench or Vise-Grip pliers to the bolt heads, then let the wrench turn against the body.

Lower Engine—Once the upper fasteners are out, the engine is ready to lower. It's best to have two people for this job: one to manage the floor jack and another to eyeball the engine compartment to watch for hangups.

Start by pulling back on the floor jack until the input shaft clears the clutch. If you are working on an Auto-Stick, the driveplate will clear the torque converter right away. All that's left to do is clear the bellhousing studs.

Once the engine has disengaged, lower the jack. Continue to pull the engine and jack rearward while slowly lowering the jack. Constantly monitor the engine so it won't snag a wire or cable on the way down. If the engine catches on something, stop the jack, clear the snag and continue down.

Watch for the throttle cable as it pulls from the fan shroud. Steady the engine with one hand on the fan shroud and the other on the muffler. This is another good reason to have a helper move the jack, you'll have your hands full with the engine.

On '70 and later engines with a throttle positioner, you must tip the engine to clear the body. The top of the fan housing needs to be tipped toward the firewall so the throttle positioner can get past the rear body panel. Alternately, you could remove the carburetor and throttle positioner. Removal requires a very thin, specially bent box-end wrench if you do this with the engine in the car.

With Auto-Stick transaxles, once the engine is out, run a brace across the bellhousing to hold the torque converter. A simple piece of flat metal with a hole in it will do. Use one of the transaxle studs and nuts to secure the brace. Without this brace, the torque converter can slip out of the transaxle, be damaged and its oil seal ruined.

BUS & TRANSPORTER (TYPE 2 & 4)

Two sections are necessary to examine removing Bus engines because '71 and earlier Buses use the upright-fan Beetle engine and '72 and later Buses use the Type 4 engine. Early style Bus engines are one of the easiest VW powerplants to drop: a couple of disconnections and it practically falls out of the chassis. The later engine is more difficult, but not overly so.

Early Bus (Pre-'72)—Start with the air filter. Up to '68, simply remove the crankcase breather, hot-air hoses and unclamp the filter housing at the carburetor. On '68—70 models, disconnect the hot-air flap cable, but the '71 version has no cable. Don't tip the oil-bath air filters when removing them.

Disconnect Wires, Cables, Hoses—Disconnect the distributor, coil, generator, oil-pressure sender and carburetor wires. Read the Type 1 section if you need more help with these. *Label the disconnections.* Undo the throttle cable at the carburetor and push it forward through the fan housing as far as it will go. On '70—71 Buses, disconnect the vacuum hoses at the throttle positioner.

Engine-Plate Screws—Unscrew the 10 rear engine-plate screws and lift out the plate. Six are in the left- and right-forward corners of the plate. The other four are in the rear corners, mounted vertically on the plate's rear face.

An optional method is to continue rearward and remove the rear bumper and body panel. With those out of the way the engine can be slid straight back out of the chassis.

Bellhousing Bolts—Now reach way to the front of the engine compartment and remove the two upper bellhousing bolts. You'll be looking at the nutted end of the bolts from inside the engine compartment. On a '71 Bus, only the right upper bolt is accessible from the top. The other upper bolt must be removed from underneath.

Heater-Control Cables—It's time to go underneath anyway, so get under the engine and disconnect the heater-control cables and hoses. You shouldn't have to raise the chassis very high, as Buses stand pretty tall. In fact, you might not want to raise the chassis now, but wait until the engine is ready to come out. Then the chassis has to go up so the fan housing will slide under the rear bodywork.

More Cables & Fuel Hose—Pull the throttle cable all the way free of the fan housing, then loop it out of the way. Slip off the starter solenoid connections and remove and plug the flexible fuel hose. The fuel line is on the left, and it is not clamped. It's just a slip joint.

Transaxle—Support the transaxle with a second floor jack or prop it up with wood blocks. The Bus chassis is so tall, you might have to add a wood block to the supporting jack. With the transaxle supported, remove the two lower bellhousing nuts. On '71 Buses, now remove the left upper bellhousing bolt, which doubles as a bolt for the starter mounting.

Lower Engine—Place the floor jack and plywood under the engine. Just barely take up some engine weight with the jack, then disconnect the rear crossmember. There is one vertical bolt at each end of the crossmember. Remove them and the engine is ready to come out.

Pull the engine and floor jack rearward until the clutch is clear of the transaxle input shaft and the bellhousing studs are clear of the engine. Then lower the engine while guiding it by the generator and exhaust. Have your helper watch for wire and hose snags on the way down.

23

Small silver tube sticking up from Bus exhaust pipe is *oxygen sensor*. Avoid knocking this somewhat delicate and expensive part when working around engine.

Type 3 air filter removal starts with hoses. It's usually best to disconnect hoses so they lift off with filter.

Late Bus (Post-'72)—Removing this engine is basically the same as dropping the early Bus engine, but there are more disconnections to make on the Type 4 engine.

Air Filter—Remove the air filter. On dual-carbureted engines, flip open the clips at the carburetors and at the filter's center section. Lift off the top half of the filter and set it aside. Keep it upright. If not stored upright, residual oil will drain into the upper half of the filter and contaminate it. Unclamp and remove the fresh and hot air hoses from the filter bottom half. Then unclip it at its bottom edge and lift it out.

A paper-element filter is used on '73—74 carbureted engines, and it is removed like the oil-bath type.

Fuel-injected engines have a different paper element filter. It is best removed as a unit with its intake air sensor. Start by disconnecting all hoses, then unclipping the cover. Remove the cover and paper element.

Locate the intake air sensor at the left. It is the aluminum box with the cast-in grid work. Locate the electric connector plug, slip off its protective boot, and carefully pull out the connector. Unclamp the large S-shaped rubber hose from the air sensor and remove the air filter body and intake sensor together.

Electrical Connections—With labeling materials in hand, disconnect the electrical leads from the distributor, alternator regulator, oil-pressure sender, and fuel injectors or carburetors, as the case may be. Follow the wiring harnesses over the engine to find all the disconnections. At the distributor, remove the lead to the fuel-injection *triggering contacts*. It's attached near the bottom of the distributor.

Throttle Cable & Vacuum Hoses—Undo the throttle cable from the crossbar or at the throttle body. Get the cable started through its guide in the front engine plate. Later, you can completely pull it through from underneath. If the car is an automatic transmission model, remove the vacuum hose from the intake manifold. On carbureted models, the hose is attached to the balance tube. Fuel-injected models have the hoses attached to the *intake air distributor*. The intake air distributor is the black, sheet-metal section in the center of the intake system.

Coil & More Hoses—Look near the coil for an inline fuse holder. Disconnect this wire (it's for the backup lights) at the fuse holder. The engine may also have a temperature sensor mounted in the upper right engine compartment. If so, disconnect its wire.

Take out the ignition coil and remove the hose leading to the charcoal filter. The charcoal filter is the can suspended from the upper right ceiling of the engine bay. On carbureted engines there is a hose mounted to the top of the left carburetor; remove it.

All '72—73 engines, plus '74s with automatic transmissions have a vacuum advance cutoff valve mounted near the blower motor. Disconnect the electrical and vacuum leads from it. On all chassis, remove the two large diameter blower hoses.

Oil Filler—As on Type 3s, remove the oil filler bellows and dipstick. Then set about removing the rear engine plate, which is in two pieces on the late Bus. Take off the right rear plate first, then the left rear. The left plate wraps around the engine side and runs forward a little, so it isn't a mirror image of the right rear plate.

Automatic Transmission—This causes some extra work. First, remove the ATF filler pipe nuts, rotate the pipe counterclockwise and pull it out. Then remove the driveplate-to-converter bolts. On this chassis, these bolts are accessible through a hole in the bellhousing from inside the engine compartment. Look under the plastic plug in the upper left mounting flange area of the case.

Gravel Guard—Before getting all the way under the engine, remove the gravel guard from under the rear bumper. Take out four bolts and this thin bent strip will come off.

More Cables & Fuel Hoses—Now slide up to the front of the engine and disconnect the heater cables and hoses at the heater-control valves. Pull the accelerator cable all the way through the front engine plate and put it out of the way. Disconnect and plug the fuel lines from the fuel pump on carbureted engines, or pressure regulator of the fuel-injected model. Look on the front right of the engine for the fuel lines.

Transaxle—Support the transaxle with a jack or wood blocks. Now raise the jack and plywood assembly against the engine. Set the jack snug against the engine, and remove the two lower bellhousing nuts. Then go to the rear and remove the crossmember. Remove the three bolts at the each end of the crossmember which thread directly into the frame. When those bolts are out, the engine is ready to come out.

Lower Engine—Pull the jack back until the engine clears the transaxle, then lower it out of the engine compartment. Watch for hangups and guide the engine so it won't fall off the jack. You definitely need a helper with a heavy Type 4 engine.

Fit a brace across the torque converter on automatic transmission transaxles immediately after removing the engine. See page 22 in the previous Type 1 section for the reasons for this.

Once the engine is out, check the rubber transaxle mounts on Buses without rear crossmembers. Weak, mushy mounts are cited as a prime reason the cases cracked on these early Buses. If the transaxle sags almost to the ground, you know the mounts are useless. Replace worn transaxle mounts.

FASTBACK, SQUAREBACK & NOTCHBACK (TYPE 3)

Air Filter—With the battery disconnected and oil draining, remove the air cleaner. All Type 3 air filters are oil-bath units, so don't tip them during removal. On single-carbureted engines, unscrew the wing nut in the center of the filter canister, plus the air intake bellows from the hot-air control box. Disconnect the intake elbow between the filter and carburetor at the carburetor. Mark and disconnect the crankcase breather hose.

Filters on dual-carbureted engines have wing nuts over each carburetor, plus one at front-center. Unclamp the hot-air hose from the hot-air control box and remove the box with the filter. Snap the throttle linkage off the carburetors and the center-mounted bellcrank before removing the air filters. Don't fiddle with the locknuts and rod ends of the linkage, you'll only get the carburetors out of synchronization. Just pop the rod ends off the ball sockets with a screwdriver.

Air filters on fuel-injected engines require only the intake elbow and several hoses be removed. The elbow is clamped and the remaining hoses are slip ons, but remember to label them during removal. With all hoses removed, unscrew the center wing nut and remove the unit. Store in a level position.

Electrical Connections—Label and remove the electrical leads at the carburetors, oil-pressure sender, generator and coil. On fuel-injected engines, remove the wiring harness at the various connections on the engine.

The fuel injection ECU mounts inside the left-rear inner fender, and the wiring harness comes from that side. Follow the harness to the connections at the distributor, injectors, crankcase sensors and grounds, cylinder-head temperature sensors, and intake air distributor. Also remove the vacuum hose from the fuel pressure sensor on the left engine compartment wall. Label each hose and wire before removing them so you'll be able to reconnect them correctly.

Check the fuel shut-off solenoids on engines with dual carburetors. These are small cylinders with a wire mounted on the outboard side of the carburetor. They usually catch on and foul the bodywork when the engine is lowered, so unscrew them from the carburetors.

Oil Dipstick—Remove the dipstick, then unclamp and remove the oil filler rubber boot. This is the accordion piece between the dipstick tube and body. Also unclamp and remove the cooling air bellows. This is the big rubber connector between the engine and rear bodywork.

Throttle Cable—Disconnect the throttle

Intake and dipstick bellows are two Type 3 & 4 disconnections. Completely remove intake bellows at left to avoid cutting it when engine is slid back during removal.

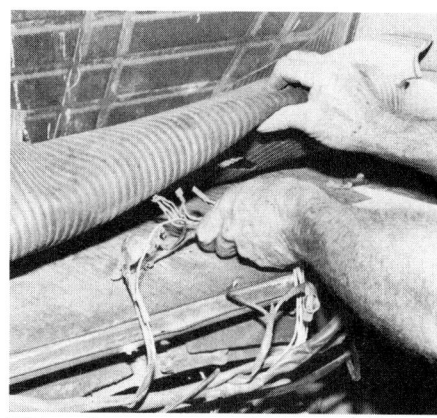

Wiring always seems to be in your way. On Type 3s, pick up rear mat and store wires underneath.

cable. With a single carburetor this is done at the carburetor. On dual-carburetor engines, disconnect the cable at the throttle linkage crossbar. Fuel injection throttle linkage is undone at the throttle body, which is part of the *intake air distributor*. That's the sheet-metal center section of the intake manifold.

Fuel Line—Remove and plug the fuel line. On fuel-injected engines, the disconnection should be made on the left and right *fuel manifolds*. The fuel manifolds are the metal sections of fuel line right above the injectors.

The rubber fuel line coming from the left front of the engine compartment and running to the left fuel manifold is the *fuel inlet line*. The fuel lines then connect to the injectors via the manifold. Another line leads off the back of the left manifold and runs to the other side of the engine. It joins the right fuel manifold, and finally runs forward, out of the engine compartment through the rear sheet metal. This last line is the *fuel return line*. It returns excess fuel to the fuel tank. The entire circuit or U of fuel lines is called the *fuel ring*.

Bellhousing Bolts—Finish the topside chores by removing the two upper bellhousing bolts. You may need a helper under the car to hold the bolt heads. On '71 and later engines, the left bolt threads into a special nut permanently attached to the case. Remove this bolt from underneath.

Another way of doing this is to wait until the engine is on its way down. Then you can get at both sides of the upper bellhousing bolts from the top. Be careful not to lower the engine too far or you'll damage the transaxle mounts.

There is supposed to be an engine mount attached to the fan housing on cars without a crossmember. It doesn't provide much support, so a lot of owners leave them off. Two bolts and it's in hand. If nothing else, it must be off to remove the engine.

Raise Car—Raise the rear of the car 3 ft and support it with jack stands. Undo the heater control-box cable connections and stow the large diameter air hoses out of the way. Pull the throttle cable through the front engine cover plate and gently loop it out of the way. Don't kink the cable or the throttle will be sticky. Disconnect and plug the fuel return line on the right side of the engine if you didn't get it from the top. On '72 Type 3s, disconnect the exhaust gas recirculation (EGR) wire from its transmission switch.

On cars with an automatic transmission, unbolt the driveplate from the torque converter. Unlike the Auto-Stick transmission, the full automatic transmission has only three driveplate-to-converter bolts. They are accessible through a hole in the front-bottom of the bellhousing. Also on the automatic transmission, slip off the vacuum hose at the balance pipe and disconnect the kickdown-switch wire.

Because the crankshaft pulley nut is inaccessible on Type 3s, use a stout screwdriver against the ring gear teeth to rotate the engine. Reach the ring gear teeth through the bolt access hole.

Support Engine—Disconnect the lower bellhousing nuts, then place the floor jack under the engine. Don't forget the plywood cushion. On double-jointed-axle cars, support the transaxle with another jack or wood blocks. Then slide rearward and unbolt the engine crossmember. Undo the two horizontal bolts at each end of the crossmember which connect the crossmember to its rubber mounts.

Don't unbolt the vertical rubber cushion-to-body bolts or the crossmember-to-engine bolts. The rubber mounts are centered by their mounting bolts. If you undo these bolts, you'll have to recenter the engine during installation.

Lower Engine—Now you are ready to pull the engine back and lower it. Watch for hangups on

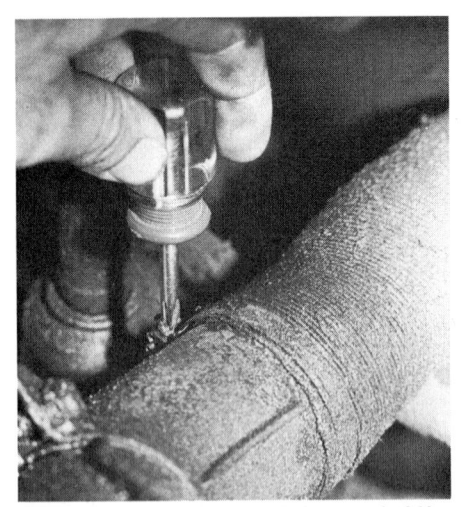

If heater ducts won't pull off, they are held by clamps. Oil and dirt can combine to camouflage these connections.

Automatic transmissions have a vacuum disconnection under left axle flange. It's probably just as easy to pull off this hose at transmission as at engine.

Bellhousing and driveplate hardware are accessible through windows in automatic transmission bellhousing.

the way down, and steady the engine so it doesn't fall off the jack. Brace the torque converter so it can't fall out of the transaxle and be damaged. A metal tab nutted to one of the bellhousing studs works fine. Wire up the transaxle on double-jointed-axle cars so you can move the chassis.

411/412 (TYPE 4)

Engine removal in the Type 4 is roughly similar to dropping the '72 and later Bus engine because they share the same engine, but there are several notable differences.

Air Filter—With the battery disconnected and oil draining, begin the engine bay disconnections. Start with the air filter. Remove the fresh air and crankcase breather hoses. On carbureted engines, unclip the upper filter half at the carburetors and center section, then lift off the upper half. Now unclip the bottom half and remove it. Fuel-injection air filters need the center wing nut undone and the filter removed. Before '72 all 411/412 filters were oil-bath type; post-'72s have paper elements.

Ducts, Wires & Cables—Remove the dipstick and oil filler bellows. Unclamp and remove the cooling air intake bellows. Take off any ducting for the heater blower motor. Disconnect and remove the ignition coil, then the oil-pressure sender lead and throttle cable. Push the throttle cable through the front engine panel.

Voltage Regulator—Pull the plug connector from under the voltage regulator. It is mounted on the right front side of the engine compartment and the connector comes up to the regulator from the bottom. Use a mirror to see this connection.

Fuel-Injection Connections—Label and disconnect the fuel-injection leads at the engine. Follow the wiring harness over the engine to find the connections. They are at the intake air distributor, ignition distributor, injectors, case ground, plus the temperature sensors at the case and heads. *Be sure you label all disconnections.* Reassembly will be so much easier, and the engine will operate correctly, too.

Automatic Transmission—Remove the three driveplate bolts from inside the engine compartment. Look on the left engine-case vertical flange for a round plastic plug. Pry out the plug with a screwdriver to expose the driveplate underneath. Rotate the engine to expose the driveplate bolts one at a time. Rotate the engine with a wrench on the cooling fan mounting bolts or by simply grasping the fan in your hand. This task is much easier with the sparkplugs removed.

Remove the transmission dipstick and bellows section of the filler tube. These parts protrude from the forward left sheet metal in the engine compartment. Disconnect the vacuum hose from balance pipe. Next, remove two upper engine mounting bolts from bellhousing.

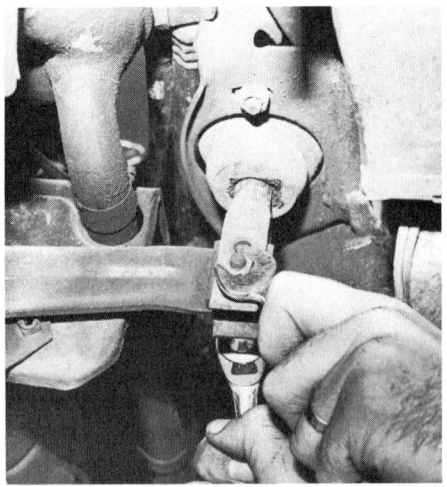

Disconnect Type 3 crossmembers where crossmember meets engine mount, not where mount meets body. If mounts are disconnected at chassis, they will have to be aligned at engine installation to keep engine straight in compartment.

Throttle linkage on Type 4 uses a clip, spring and unusually shaped cable end. Pop clip off with screwdriver (don't lose it), and pull pieces apart.

Manual Transmission—An unusual design feature of the Type 4 manual transaxle requires an extra step in engine removal. Because the differential is *between* the engine and transmission, a driveshaft runs forward from the clutch to the transmission. The front of the driveshaft looks and functions the same as an input shaft on other transmissions, except it is longer than a normal input shaft. To remove the engine the driveshaft must be unlocked and moved forward in the car 4 in. (100mm).

Moving Driveshaft—To reach the driveshaft, remove the rear seat cushion. Under the cushion is an access panel; remove it to expose the front of the transaxle. Find the round, screw-in plug, about 2-1/2-in. in diameter. Unscrew the plug. Inside the transaxle will be the end of the driveshaft with a nut threaded onto it. Unthread

Disconnect fuel-injection wiring on Bus and Type 4 engines. Set wiring loom aside once disconnections (label them!) are made. Hand cleaner gel is great hose and wire cleaner to detail wiring.

Hole in bellhousing (arrow) is for reaching automatic transmission driveplate hardware on Type 4 engines. Rotate crankshaft to bring hardware into view.

the nut, then remove the circlip from the shaft, behind the nut. You can pull the shaft forward somewhat with the circlip still attached, but to move the shaft forward the required 4 in., remove the circlip.

Lower Engine—Under the rear bumper, remove the slotted lower splash pan. Note that both the engine and transmission are supported by crossmembers. While it is possible to remove the engine with the transaxle crossmember in place, it is easier to unbolt it from the chassis. The engine/transmission unit can then tilt down at the rear, giving a clearer path for engine removal.

The transaxle crossmember mounting bolts require one person inside the car to turn the nuts and another underneath to hold the bolts. Support the transaxle with a jack. Then move forward, support the engine with a jack, and undo its crossmember at each end. Now lower the engine.

Watch the fuel injectors' clearance to the body because they are a tight fit. Monitor the transaxle jack too, as it will have to be lowered as the engine is lowered. Remove the four bellhousing nuts and the engine is ready to disengage from the transaxle. Extract the engine, then reinstall the transaxle crossmember. On automatic transmissions, secure the torque converter in place with a metal tab nutted to a bellhousing stud.

PORSCHE/VW 914

Although there is little difference among the Type 4, later Bus and 914 engines once they're on the bench, there sure is a difference in the 914 engine mounting. The 914 is a mid-engined car, so the engine is right behind the seats, and the transaxle is behind the engine. So, when you crawl forward under the back of a 914, you first see the transaxle, followed by the engine.

Because of the mid-engine mounting, the engine can only be removed with the transaxle or after the transaxle has been removed separately. If using a small floor jack, or working on dirt, it's best to remove the transaxle first, then the engine. If using a larger floor jack with a wide saddle, though, remove the engine and transaxle as a unit.

I explain removing the engine/transaxle as a unit here. If you remove the transaxle first, the steps are the same, except you must also remove the four bellhousing fasteners to free the transaxle.

Engine Cover—First step in removing a 914's engine is to take off the engine cover. Two bolts can be felt under the hinges. Remove one and have a friend hold that side up, then remove the other bolt and lift off the cover. You'll have plenty of extra room to work and no lid banging your head to get your attention.

Carburetors—Many 914s have had the Bosch fuel injection removed and dual carburetors installed. So, completely remove the carburetors, intake manifolds and throttle linkage in your installation.

Fuel Injection—If the car is equipped with fuel injection, remove the air filter and disconnect the fuel lines near the battery. There is one supply and one return line. You don't have to remove the fuel injection wiring and fuel line ring.

Instead, disconnect the connector from the fuel injection ECU located in front of the battery. First, remove the battery for clearance, then unscrew the ECU's bracket. Tip it up so the inboard end with the connector is visible. Unscrew the metal clamp around the wiring bundle, then slide off the plastic cover from the ECU.

The connector plug takes up almost the entire side of the ECU. There may be a plastic handle

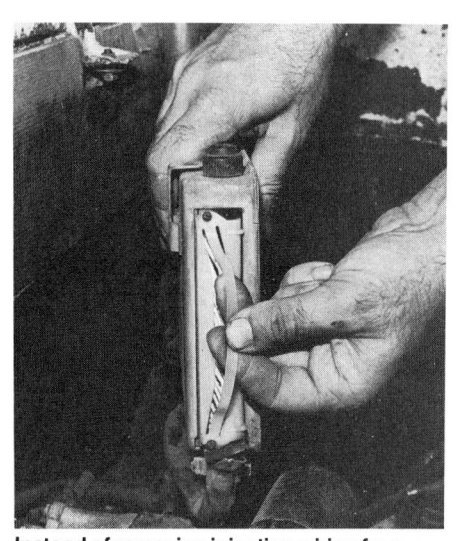

Instead of removing injection wiring from engine on 914s, take it off at ECU, relays and other connections. ECU connector is at one end; plastic handle on multi-pin connector eases disconnecting.

on the plug; if not, use a hooked instrument to carefully remove it. Reinstall the plastic cover and set the ECU in a safe place. Inside the car is a good spot.

Heater Blower Ducting—Remove the heater blower motor ducting and the rubber intake air elbow.

Voltage Regulator—Remove the voltage regulator by extracting the two screws on its mounting tang and then lifting off the unit. Now unscrew the round nut on top of the large black plastic cover next to the regulator. Lift off the cover and find several electrical connections. Label and remove all the connections, then replace the cover and regulator.

Pass 914 throttle cables through sheet metal so they completely clear engine. Otherwise, they will kink when lowering engine.

Use a firm tug to pull reverse-light wiring from transaxle-mounted switch.

Shift linkage is underneath plastic box with sardine-can band clamp. Use small Allen wrench to unlock shift rod from transmission linkage.

Air Filter & Throttle Cable—Remove the formed sheet-metal air filter support from the center of the engine. Use a long screwdriver to reach down to the case where the two front support legs attach. The other rear attachment is bolted to the case using one of the case flange bolts. With the support removed, the throttle cable can be removed. Unthread the locking nuts on the cable housing, disconnect the cable from the throttle arm and push the cable through the hole in the right side engine plate.

Some engines mount the air filter to one side. On these, leave the air filter support alone. You can remove the throttle cable without detaching the support.

Vacuum & Vapor Hoses—Label and disconnect the remaining vacuum and vapor hoses. Some common hoses go to the charcoal canister and pressure sensor. Drape all the disconnected hoses over the center of the engine so they won't be in the way when removing it.

Lift Car—Raise the car and support it on tall jack stands. Don't put them under the suspension or front engine crossmember. Instead, place them at the two small round protrusions in the body, outboard and forward of the engine compartment. Remove the rear wheels and tires. This will give a lot more room and light under the car.

Removing Cables & Exhaust—It helps to remove the bodywork panel below the rear bumper, but if the chassis is lifted fairly high, the engine can be removed with it in place. Remove the muffler from the exhaust pipes and the muffler brace from the rear of the transaxle.

Unplug the reverse-light leads from the left side of the transaxle. Unwind the sardine can clamp over the rear shifter boot and disconnect the shifter. Use a small Allen wrench to remove the set screw. Another set-screw arrangement

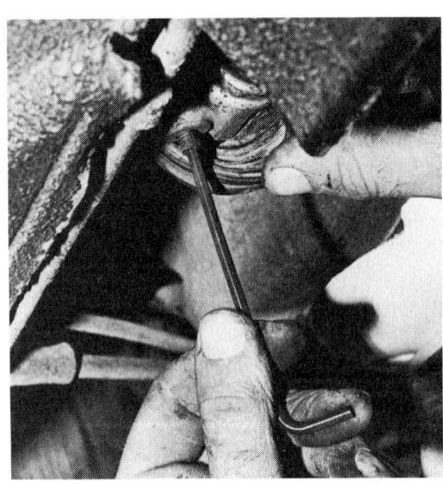

Another shift-rod set screw is located in front of engine where rod enters chassis. Pull back rubber boot to expose screw, and extract it. Once screw is out, remove shift rod.

Transaxle groundstrap bolts to chassis near rear mounts.

secures the front end of the shift rod. Look under the rubber boot where the shifter enters the bodywork at the front of the engine compartment. With the forward connection removed, the shift rod can be extracted from the car.

Unbolt the ground strap above the rear transaxle, unscrew the speedometer cable and remove the clutch cable. The clutch cable is undone by removing the self-locking nut in the center of the *cable pivot:* the round plastic wheel. Under the wheel are two nuts. Remove them, and the metal pivot bracket and cable come free. Loop the cable and bracket aside.

Return to the exhaust system. Remove each exhaust pipe/heat exchanger. Undo the flat sheet-metal shields under the heat exchangers, and disconnect the heater control valves and associated plumbing. Once the exhaust pipes are out of the way, you'll actually be able to see the engine.

Axles & CV Joints—I've left the axles until now because they are such a gooey mess. And the less time you spend pushing the disconnected axles out of the way, the better. Now's the time, so rotate the axle to get straight

28

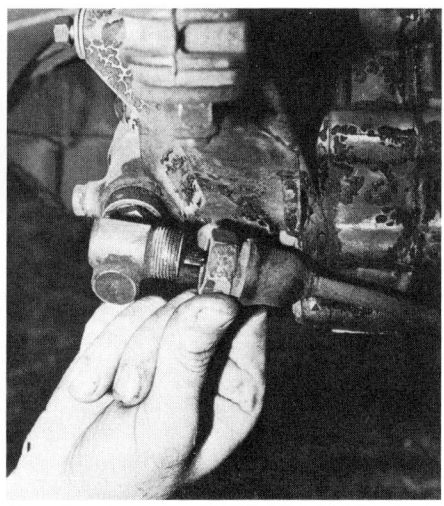

Speedometer cable is at right rear of transaxle. Unscrew large nut and pull cable out.

Clutch cable pivot uses a self-locking nut for retention. Remove nut and pulley to loosen cable and easily remove it at clutch release-bearing arm.

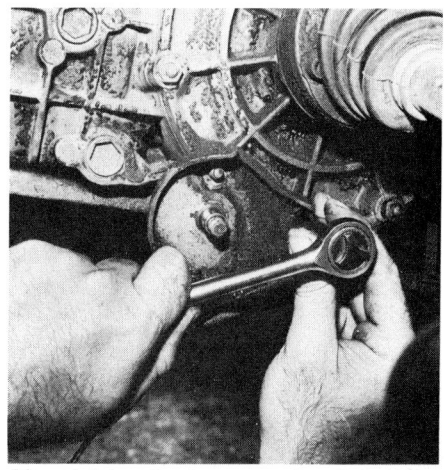

After clutch pulley is removed, unthread the two bracket nuts. This frees bracket from transaxle, leaving clutch cable-to-bracket connection undisturbed.

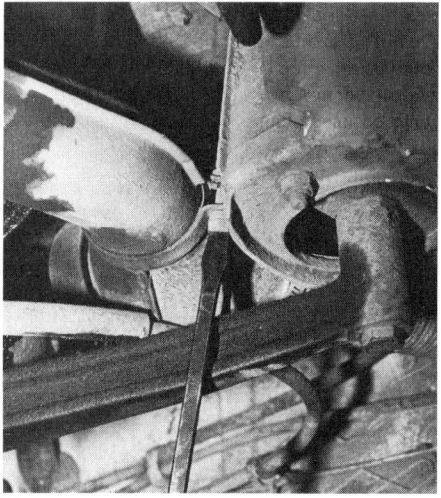

Heater ducting on Type 4 engines is often metal tubing. Examine your chassis so a minimum of heater ducting is removed with engine. On 914s, this clamp just before heater valves is best separation point.

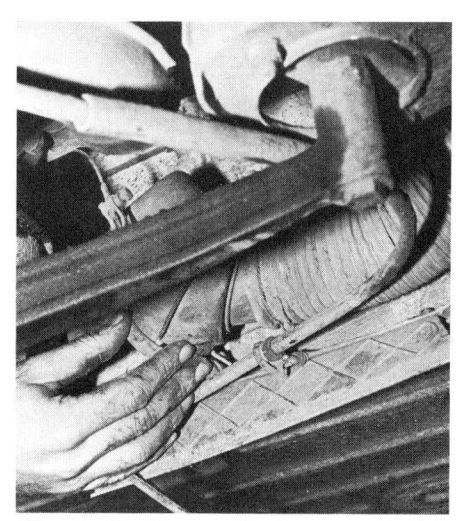

After ducting from engine is removed, push heater valves out of the way. There's no need to disconnect heater-control cables or flexible ducting.

Clean dirt from tiny splines in CV-joint bolts. These bolts are heavily torqued to prevent loosening; splines will strip if not completely clean. Short Allen head tool and wrench duo shown here is cheapest method, but special 3/8-drive socket, 12-point 6mm Allen head tool is easier to use.

access to the *constant-velocity* (CV) joint bolts. Hold the axle by inserting a breaker bar through two lug bolts, or have a helper step on the brakes. Unscrew and remove the CV bolts. They are 12-point, 6mm Allen head bolts that require a special removing tool. Rotate the axles a couple of times to reach all the bolts.

It takes quite a prying effort to free the CV joints from the *transaxle flanges*. Besides the bolts, there are dowels at the joint, so you'll have to work to free one side of it, then the other. Sacrifice the gasket between joint and flange by sticking a flat-bladed screwdriver between them. Just get a toe hold, about 1/16 in. of blade inserted, then twist the blade. Keep alternating this action from one side of the joint to the other until it separates. Don't insert more screwdriver blade than necessary or you'll scar the mating surfaces. That's why I try to put the screwdriver right through the gasket; so it will protect the metal surfaces.

As soon as you separate a CV joint, place plastic sandwich bags over the joint and its flange. Use a rubber band to secure the bags. Besides keeping the close tolerance, expensive CV joints dirt free, this maneuver will keep you a lot cleaner.

Support Engine—Place the floor jack and plywood under the bellhousing. The larger the plywood, the better. Extra length will help balance the awkward engine/transaxle unit on the jack.

Go forward and remove the two nuts from the center of the solid metal crossmember. These are the front engine mount nuts. Then remove the two large bolts from the crossmember ends. The crossmember will drop free (it's heavy) complete with the cables that pass through it. Gently set it aside without kinking

29

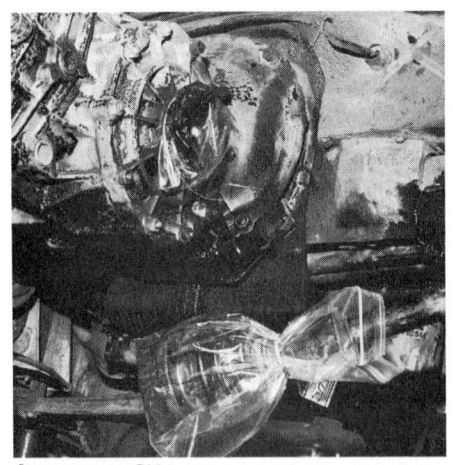

As soon as CV joints come free, wrap them and transaxle flanges with plastic bags. This will save a lot of aggravation when you drag your hair over them.

Rear engine mounts must come completely off transaxle to clear bodywork. Start with small mount-to-chassis hardware, then remove larger bolts at center.

Front crossmember attaches to engine mounts using small nuts found in recessed wells. Remove these, then large bolts at each end of crossmember.

Crossmember will drop after both sets of bolts are out. This iron piece is heavy, so be careful. Don't let it lay unsupported over clutch and speedometer cables; prop one end up with a block of wood. Once the engine is out, it can be stored by its mounts on chassis.

or smashing the cables.

Go to the rear and remove the transaxle mounts at the body. Then the entire bolt, rubber cushion and washers will come out with the unit.

Lower Engine—Gently start lowering the engine and transaxle while a helper checks the engine compartment. It's easy to bang the injectors against the bodywork or have their fuel lines hangup, so pay extra attention to them. It will take a minute or two to jockey the engine free of the chassis, so take your time.

Remove the unit and support it on the floor with wood blocks. Be careful to get the blocks bearing against the cylinder heads, not the sheet-metal pushrod tubes or another vulnerable part. Disconnect the starter motor leads, remove the four bellhousing fasteners and separate the transaxle from the engine.

CLEAN-UP (ALL MODELS)

Once the engine is clear of the chassis, get a helper and lift it off the jack. Grasp it by the ends of the cylinder heads, not the fan housing, pushrod tubes, flywheel or the like. Set the engine on the floor. If lifting a Type 4 engine be prepared for a heavy load; it weighs about twice as much as a Type 1. That's about 300 lb for a Type 4. Two strong people are needed when moving it.

Use the floor jack to get the chassis back on the ground and pushed to its storage location. Pick up all tools, rags, hardware and parts before they get scattered. Store all hardware and parts in clearly marked boxes and cans. Do this now while they're still fresh in your memory. You'll thank yourself at reassembly. Use the trunk or Bus interior for storage if garage space is tight.

CHAPTER 3
Parts Identification & Interchange

Money-saving and parts-scrounging opportunities await in VW wrecking yards. Knowing which parts fit your engine and how to identify them are keys to getting bargains.

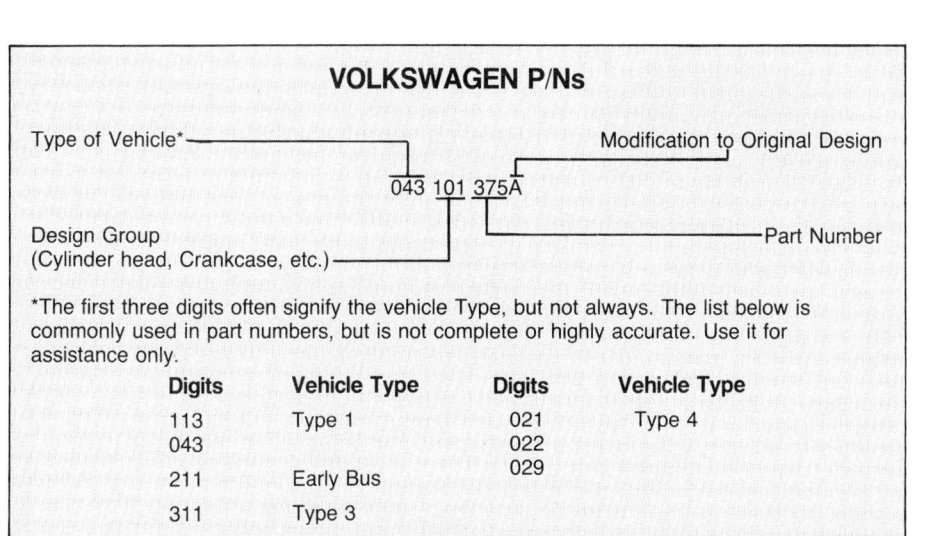

VOLKSWAGEN P/Ns

Type of Vehicle* ——— Modification to Original Design

043 101 375A

Design Group (Cylinder head, Crankcase, etc.) ——— Part Number

*The first three digits often signify the vehicle Type, but not always. The list below is commonly used in part numbers, but is not complete or highly accurate. Use it for assistance only.

Digits	Vehicle Type	Digits	Vehicle Type
113	Type 1	021	Type 4
043		022	
		029	
211	Early Bus		
311	Type 3		

If there ever was an engine with a stupefying number of variations, the air-cooled VW is it. First, it's been in production an incredibly long time, from before WWII until the '80s, and all the while being updated to meet market needs. Multiply that by several variations, including one completely different engine, and there are quite a few engine combinations.

And some confusion is caused by the wonderful interchangeability of the engine's layout. Cases, crankshafts, rods, cylinders and cylinder heads are all easily mixed, resulting in a deluge of possible combinations. Finally, add in a strong aftermarket industry busily building custom engine parts. VW engine versatility is so great, you can easily build a "Volkswagen" engine without one factory part in it.

All these options can be confusing, but also very helpful, to the engine rebuilder. This

31

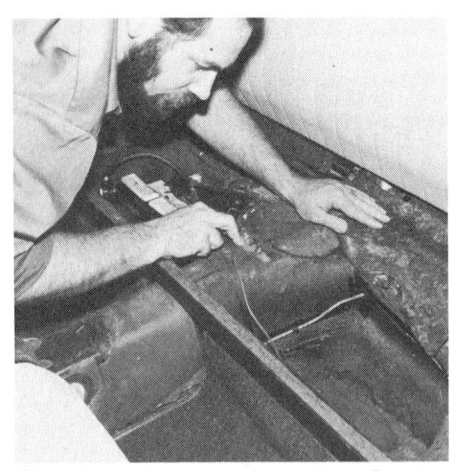

Chassis numbers are located on floorpan tunnel in front of rear seats; it's most accurate place to read chassis number. Look immediately behind driver's seat on Buses.

Code letters and engine number are located just under generator pedestal on Type 1 and pre-'72 Type 2 engines.

Type 3 engine number is at top of case just left of generator pedestal. Look for it at the case parting line (arrow).

chapter will sort the different parts so you can identify them and know which ones are interchangeable. Stock parts (as opposed to competition ones) are covered here because this is a book for rebuilding an engine to stock specifications. I realize a lot of displacement increases take place during otherwise stock rebuilds, so I've touched on big-bore kits and selected aftermarket parts. Nevertheless, there isn't enough room to include all aftermarket pieces. That's for another book.

IDENTIFICATION

The best method to identify parts is with numbers. There are four types of numbers used to identify VWs: Chassis Number, Engine Number, Part Number, and Casting Number.

Chassis Number—The chassis number is found on all VWs on the floor behind the driver's seat. Peel up the carpet and look for stamped numbers. On Beetles, the number is on the transaxle-mount access plate, which is centered on the rear floor. On Type 2s, it is behind the front passenger seat. Type 3s and 4s have the number centered on the floor underneath the rear seat. It can also be found under the front hood and on the dashboard on most models and years.

Dashboard numbers weren't used until padded dashes were installed, so early chassis won't have the number there. The dashboard number is visible through the windshield on the driver's side. The most reliable location for noting the chassis number is on the floor stamping. The dashboard numbers can change when the dash padding is changed, and the underhood tags can change when major body repair work is done, like a front clip.

The chassis number can specify which engine is supposed to be in a particular chassis. This is helpful when shopping in wrecking

Casting number and p/n are cast in most VW parts. This cylinder head is p/n 043 101 375H and casting number 152. P/N is critical number to rebuilders.

yards or when contemplating buying a car. In fact, on some Buses, the chassis number is also stamped on the right cylinder cover plate. This will quickly show if the engine has been changed, if that's important to you. For most engine work, however, the engine number is the one to work with.

Engine Number—VW stamps all engines with an engine number. On Type 1s and pre-'72 Type 2s, it is on the right case half, just below the generator pedestal. On Type 3s, it is on top of the right case, near the fan housing and case parting line intersection. Type 4s, including '72 and later Bus, Vanagon, 914 and 411/412, have the number in two spots. Look for a pad on top of the left case half, running diagonally from the bellhousing flange, or on top of the right case half near the fan housing.

The first engine number digit and sometimes the second, is always a letter on post-40-HP engines. This is called the *Code Letter*. Code letters denote the basic engine group, the remaining digits are the serial number.

The 40-HP engine uses a single number to denote the year of manufacture. It is the first numeral in the engine number.

The code letter at the beginning of the engine number is usually the primary reference to use. By checking it against the chart in this book, you can identify the engine. For example, an HO denotes a 1500 used in a '66 Bus or '67 Beetle or Karmann Ghia.

This is not to say the letter will always lead you to the original application. As in the last example, the same engine, with the same code letter, was used in three different chassis. So if you find an HO engine in the wrecking yard, you only know it originally came from one of those three cars. Who knows what happened to it in the meantime?

The serial-number portion of the engine number is ignored for engine rebuilding. But it is not useless. Sometimes changes were made during a model run, starting with a certain serial number. Then the engine number shows if the change has been made to that engine or not. This is helpful during routine service, but doesn't answer any rebuilding questions.

Casting Numbers—When some parts are made at the factory, they receive a number. The number is part of the mold the parts are cast from, so they are called *casting numbers*. Because casting numbers come from the mold, every part cast in that mold has the same number cast on it. The factory does this for internal identification and inventory purposes. Once the part leaves the factory, the casting number is no longer referred to. The parts department doesn't use them, nor do most mechanics pay

32

ENGINE LETTER CODES

TYPE 1

Code	Year	Engine	Remarks
4	'60	1200	40 HP
5	'61	1200	40 HP
6	'62	1200	40 HP
7	'63	1200	40 HP
8	'64	1200	40 HP
9	'65	1200	40 HP
F0	'66	1300	8mm oil-pump studs
H0/T0	'67	1500	8mm oil-pump studs
H5	'68—69	1500	8mm oil-pump studs
B6	'70	1600	Dual-relief, single-port
AE	'71—72	1600	Dual-relief, dual-port
AH	'73—74	1600	8mm head studs w/case savers
AJ	'75	1600	Fuel-injected, no fuel pump mount.

Replacement Cases

All cases service parts only—sold only over the counter, not supplied in new cars. All are dual-relief and have 10mm oil passages.

Code	Year	Engine	Remarks
FI		13/1600	
F2		13/1600	
D0			40-HP part when last digit is "X".
DI			Sometimes a 40-HP part
AB	'66—79	1600	Built since '73 in Mexico, Brazil, Germany
AD		1600	
AK	'67—74	15/1600	
AM		1600	
AS			Sometimes listed as Super Beetle original case, '73-on

TYPE 2

Code	Year	Engine	Remarks
			Early Bus
Numerals	'61—63	1200	See Type 1, 40-HP listings
O	'64	1500	Based on 1200, no cam bearings
H	'65	1500	
H0	'66—67	1500	No crossmember holes
B5	'68—69	1600	8mm oil-pump stud, single-relief
B5	'70	1600	Dual-relief
AE	'71	1600	w/crossmember holes
			Late Bus
CB	'72—73	1700	Dual-carb, manual trans
CD	'73	1700	Dual-carb, auto trans
AW	'73—74	1800	Dual-carb, manual or auto
AW	'75	1800	Fuel-injected
ED	'75	2000	
GD	'76—79	2000	'79 has Vanagon heads
GE	'79	2000	Calif. only, Vanagon heads
			Vanagon
CV	'80—83	2000	Unique case and heads

Note: All Early Bus engines are similar to Type 1 and are single-carbureted. Late Bus engines are Type 4 style, and are either dual-carb or fuel injected. Late Bus engines with carbs have fuel pump hole in lower right-front case half. Fuel-injected versions don't. Dipstick and oil fill hole are on right-rear case half.

TYPE 3

Code	Year	Engine	Remarks
O	'63—65	1500	Based on 1200, no cam bearings
TO	'66—67	1500	6mm oil-pump studs
UO	'68—69	1600	Fuel injected; '69 w/xmember holes
UO	'70	1600	w/xmember holes, dual-relief
U5	'71—73	1600	like UO w/xmember holes, 7.7:1 compression
X	'72	1600	7.3:1 compression

"xmember" = crossmember

Note: All Type 3 cases have no dipstick or oil-pressure sending unit holes. These can be added by a machine shop if swapping to a Type 1 or 2.

TYPE 4

Code	Year	Engine	Remarks
W	'71	1700	
EA	'72—74	1700	
EB	'73	1700	Calif. only
EC	'74	1700	

914

Code	Year	Engine	Remarks
W	'70—71	1700	8.2:1 comp., D-Jetronic
EA	'72—73	1700	8.2:1 comp., D-Jetronic
EB	'73	1700	7.3:1 comp., D-Jetronic
EC	'74	1800	Calif. only, 7.3:1 comp. L-Jetronic
AN		1800	8.6:1 comp., D-Jetronic
GA	'73—74	2000	7.6:1 comp., 3 Intake studs, D-Jetronic
GB		2000	8.0:1 comp., 3 Intake studs, D-Jetronic
VO		1700 1800 2000	European only, magnesium, some brought into U.S. in core engines, hardens and cracks like Type 1—3 case.

ENGINE DISPLACEMENT (mm)

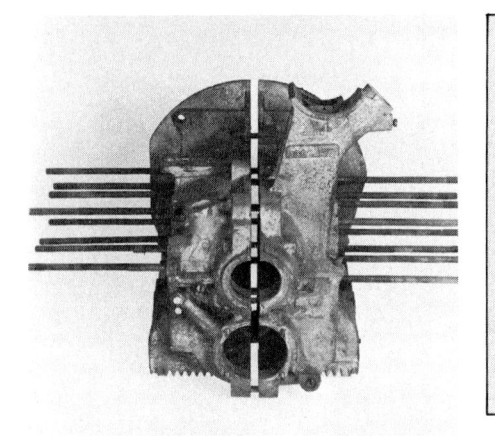

Stroke	Bore						
	77	83	85.5	87	88	90	92
64	1192	1385	1470	1522	1557	1629	1702
66	1229	1428	1516	1569	1606	1679	1755
69	1285	1493	1584	1641	1679	1756	1835
71	1322	1537	1631	1688	1727	1807	1888
74	1378	1602	1699	1760	1800	1883	1968
78	1453	1688	1791	1855	1897	1985	2074
80	1490	1731	1837	1902	1946	2035	2127
82	1527	1775	1883	1950	1995	2087	2180
84	1565	1818	1929	1997	2044	2138	2234

Dead giveaway of 36-HP and earlier cases is integral generator pedestal. Engine is smaller in all dimensions than later, more powerful versions and has almost no interchange possibilities.

any attention to them.

Instead, the *part number* (p/n) is used, which tells a lot more about the part than the casting number anyway. Luckily, VW casts the p/n onto most major pieces right along with the casting number, so it's available for reference, too.

The p/n is of great use because it will immediately tell you a lot about the part and eliminate guesswork. They are especially useful when hunting through swap meets and wrecking yards. Through a combination of visual clues and p/n, you can identify the part in question. This is a handy skill for determining if you've found the right part.

Because the p/n always stays with its part, the engine doesn't have to be intact to identify some of its separate pieces; the heads, for example. Also, on complete engines, the cylinder-head numbers can be checked against the code letter to see if they make sense. You'll often find *potluck* engines cobbled together from leftovers—which may or may not be usable to you.

A p/n isn't an infallible identification mark. Just because the number says it is such and such a part, it might not be so. Somebody may have machined it into something quite different. This is especially true of cylinder heads. Always check for *flycutting*—opening or deepening the combustion chamber in the heads.

P/N Location—On cylinder heads, look for the p/n on the floor of rocker-arm area, directly over the intake port. The nine-digit number is the p/n. A larger, three-digit number is the casting number. Cases have casting numbers found on the lower right half, but they don't tell you anything. There is no p/n *per se*, so the engine number with its letter code takes over this job.

Connecting rods have a three-digit p/n located on the main part of the rod, not the cap. Look near the parting line where the rod and cap join. The small numerals are on the bosses the bolt or stud pass through.

Crankshafts have neither a p/n nor casting number on them. Visual inspection alone can easily tell them apart, though.

ENGINE DESCRIPTIONS

In the most general sense, air-cooled VW engines can be separated into two groups: Types 1—3 and the Type 4. Type 1—3 engines are very similar, in some cases identical, underneath their different sheet metal and fan housings. The Type 4 is about as different as it can get and still be related to the other engines. There is not one major interchangeable part between the Type 4 and Type 1—3 engines, yet without sheet metal, they look similar and share a common layout.

Inside the Type 1—3 group, many different models of the same engine have been made. These are denoted by different code letters in the engine number. Different code letters are assigned to a specific engine at the factory. A change in heads, intake-port design, flywheel, induction system and so on would result in a different code letter. Type 4s follow the same procedure, but there are fewer models because it is a newer engine and used in fewer chassis.

I'll introduce the different engine models with a short description.

36 HP—Although not covered in this book because there just aren't enough of them around anymore, the 36-HP engine powered all VWs from '54—59, and the Beetle used it in '60. Compared with later engines, the 36 HP is smaller in all dimensions, and the generator pedestal is an integral part of the right case half. Some of the photos in this chapter show 36-HP parts for comparison purposes. Chassis numbers up to 5,000,000 use 36-HP engines.

40 HP—Displacement is 1192cc, the same as the earlier 36 HP. The engine is physically larger, however, with the addition of a fourth main bearing at the pulley end of the crankshaft. Look for the bolt-on generator pedestal, and intake manifold that meets the heads at a right angle. Most commonly called the *40 horse* engine, but sometimes referred to as a *twelve hundred*.

Type 1s used the 40 HP from '61—65, Type 2s, from '60 to mid-'63.

1300—An easy to remember engine is the 50-HP, 1300cc mill used only in '66 Beetles. It's full of detail improvements over the 40 HP, and is beefier, lasts longer and offers more power. The two look very similar, but the 1300 intake manifold joins the heads at an angle. An FO letter code is the best identification.

Inside are insert cam bearings, 22mm piston pins and a crank with 0.200-in.-longer stroke. Changing cylinders and cutting cylinder heads easily transforms 1300s into 1500 or 1600s, so you may find one with increased displacement.

1500—Rated at 53 HP, the 1493cc engine was another improvement. Called *fifteen hundreds*, these engines were supplied in Type 2s in '66—67. They also powered Type 1s from '67—69. Look for an H letter code and an intake manifold that joins the heads at an angle.

Another 1493cc engine was used in Type 2s from mid-'63—65 and Type 3s from '62—65. This one was variously rated at 51-, 53- and 66-HP, but was really nothing more than a bored and stroked 40 HP. Therefore, it has all typical 40-HP problems like worn cam bores. These cases are marked like 40-HP units, with numerals, not letters.

1600—The final version of Type 1—3 engines is the 1584cc engine, commonly called the *sixteen hundred*. Type 3s were the first to get them, from '66—73. Type 2s were next, '68—71 and Type 1s last, '69—79. Letter codes are the best method of identification because the 1600 and 1500 can differ only in minor details. There are a bunch of letter codes, so look at the accompanying charts.

One easy to spot difference is in the ports. Through '70, all engines have a single intake port per cylinder head. From '71 on, dual ports are the rule. Therefore, all dual-port engines are 1600s, unless you count the very different looking Type 4s.

Keep in mind that the 1500 and 1600 are the garden variety air-cooled VWs. They were imported by the millions, and are the best junkyard, swap meet and classified ad prospects. This isn't to say you won't find any 36- or 40-HP engines, not by a long shot. It's just that the chances of finding the later engines are better. They are also better engines to buy because they are stronger and make more power.

1700, 1800, 2000—The last air-cooled VWs are commonly called *Type 4s*, even if they are in a Bus or 914. The different models are called, *1.7* or *1.8* and the 2000 is simply a *2 liter*. There is little externally to tell the three apart. The 1.7 and 1.8 engines differ only in 3mm of bore, and valve size, so they are practically the same engine anyway. The 1.8 was produced in '73—74 only, so there aren't that many of them. The 2 liter used in 914s can be spotted by its *missing* intake-manifold stud. All other Type 4 engines have four intake-manifold studs per cylinder head, the 914, 2 liter has three. The single stud is between the two intake ports on these engines, where there are two on all others. Again, it's best to use engine-code letters for positive identification.

Spotting Type 4 engines among all the Type 1—3s is easy. First, all Type 4s are flat engines, that is, the cooling fan is mounted on the end of the crankshaft. So if you're looking at an engine with an upright fan, it's not a Type 4. Second, Type 4s have an aluminum case; all the rest are magnesium. Aluminum is a shiny silver; magnesium is a duller gold or gray-white. Third, all Type 4s have a spin-on oil filter; Type 1—3s don't. Fourth, all Type 4s have alternators; all Type 3s, the only other flat engine, use generators. Finally, Type 4s are larger than the 1—3s, and weigh about twice as much.

Are You Sure?—Whenever dealing with core engines, be extra careful to determine exactly what engine you're buying. Most engines are easily enlarged by installing new cylinders and flycutting the cylinder heads. The 1300 and 1700 are good examples. They may decode into these designations, but actually be 1600cc or 1800cc displacements. This isn't going to hurt anything, but you need to know the cylinder heads have been enlarged before buying new cylinders and pistons.

More radical displacement increases are possible by cutting the case to accept larger cylinders, or installing *stroker* crankshafts. The only sure way to tell is tearing down the engine and measuring. So at least disassemble the engine before buying new parts.

Finally, beware of cracked cases. All magnesium cases crack eventually. Always check Type 1—3 engines for cracks before buying. Chapter 5 has more on case cracks and pin-

Larger 8mm oil-pump studs and gallery plugs of 1600 case (arrows) at right contrast with 6mm studs and smaller gallery plugs of 40-HP.

Cylinder-head stud spacing remains same on all Type 1—3 engines starting with the 40-HP; so gap between cylinder bore and stud hole is *smaller* with increasing bores. Case at left is 40 HP, one at right 1600. Both have case savers.

points their location. You may have also heard that air-cooled VW heads crack. This is oh so true, but not so financially painful.

CASES
40 HP—These early cases don't have insert cam bearings. That is, 40-HP camshafts run directly in the case bores without any replaceable-insert cam bearing. By now, practically any 40 HP you run across will have been rebuilt, and the cases possibly modified for insert cam bearings. This is a good modification, because when the steel camshaft inevitably gores open the magnesium bearing bores, oil pressure drops. With less oil pressure, the connecting rods eventually fail.

You may come across a 40 HP with insert cam bearings originally installed at the factory. That's because it has a replacement case made after VW went to replaceable cam bearings. Replacement-case code letters are DO and DI and have an x at the end of the engine number. These code letters are also found on 1600 replacement cases.

Like all replacement cases, the 40-HP versions are *dual oil relief* with the large 10mm oil passages. In fact, the only way to tell a DO or DI 1500 replacement case from a 1600 is by the size of the cylinder openings. Take a 40-HP cylinder along when parts hunting. If it fits without looseness, its a 40-HP case. Alternately, a 1500 or 1600 cylinder won't fit into the 40-HP replacement-case openings.

The original 40-HP case isn't as heavily ribbed or beefed up as later cases. The center main-bearing saddle is not as thick and the

Early and late 40-HP cases are distinguished by breather-passage shape. Early case at left has an "8" code (for '64) and small round breather passage. Right case has the larger, "D"-shaped passage introduced in mid-'64. This example is a '65 as noted by "9" code, but could also be a '64 with "8" code.

Identify single- and dual-relief cases by checking number of relief valves at bottom of case. Extra valve is found at flywheel end as illustrated by right case (arrow).

Oiling improvements are illustrated in these SIR case cutaways. Single-relief case is at left. All replacement cases, regardless of original layout, are dual relief like one at right.

flywheel area has less material in it. This makes them weaker than later cases, but then, they were not designed to take the horsepower the larger engines produce. This is not to knock the 40 HP; it's a great engine, but it should be left close to stock. If high rpm or more power is your intention, start with a 1300-or-later case.

A good 40-HP visual clue is oil-pump stud diameter—6mm. Later engines use 8mm studs. All original 40-HP cases have a single oil-pressure relief valve, 8mm oil passages and magnesium cylinder-head stud threads with 10mm studs. Yet, by now many of these early cases have had steel thread inserts (case savers) installed.

40-HP engine numbers have no letter code. Instead, the first digit in the number denotes the manufacturing year. A 4 means '60, a 5 equals '61, 6 is '62 and so on until '65, which is a 9.

Several modifications made in mid-'64 to the Type 1 40-HP, and the 1500 Type 2 & 3 engines based on the 40-HP case, resulted in the terms "early" and "late" being applied to this engine.

Early engines were produced to mid-'64 and have an 8 number code. Then the late engines started, also with an 8 number code. You can tell which is which by looking at the oil-filler hole under the generator pedestal. Early engines have a small round hole, late engines have a larger, D-shaped hole.

1300—Look for an FO letter code on the '66 Beetle case. There will also be 8mm oil-pump studs and larger oil passages inside the case. Nevertheless, these are still *single oil-pressure relief* cases; 1500 or 1600 cylinders can be fitted to this case without machining.

1300 replacement cases are F1 and F2 letter codes. They are dual relief, 10mm oil passage, and generally strengthened.

1500—There are several 1500 cases, depending on the year and chassis. Most have all the 1300 improvements and are difficult to differentiate from them, except by letter code. Up to and including '67, cases for 1500s have the small 6mm oil-pump studs and letter codes TO and HO. In '68 they have 8mm pump studs. All are single-relief cases with 10mm head studs.

Don't forget the 1493cc engines that are more a bored-out 40 HP than anything else. They have the small oil passages and 6mm pump studs. These cases have a O ("oh") number code and beefier webbing in the bellhousing area at #3 cylinder. Well, they're beefier than a standard 40-HP case, but not more robust than a regular 1500 case. Another tip-off is the lack of insert cam bearings. Some sources claim these cases have letter codes G and K, not the letter O. If so, Type 2s had G cases, Type 3s, K cases.

1600—Detail improvements mark the development of the 1600 cases. All have 8mm oil-pump studs and large (10mm) oil passages. In '70, another oil-pressure valve was added to the case. These are called *dual-relief* cases. See the sidebar for dual-relief functioning.

Another oiling modification to all cases started in '70 was increasing the size of the oil passages leading to the oil cooler from 8mm to 10mm. Therefore, all dual-relief cases have large oil passages at both the main gallery and cooler. Upgraded oil coolers with larger passages are used on these cases. The next year, '71, the improved *doghouse* cooling system was introduced on Type 1s. It uses a third-style cooler.

In '73, the factory began installing case savers, on the Type 1—3 magnesium cases. When they added the case savers they went back to

36

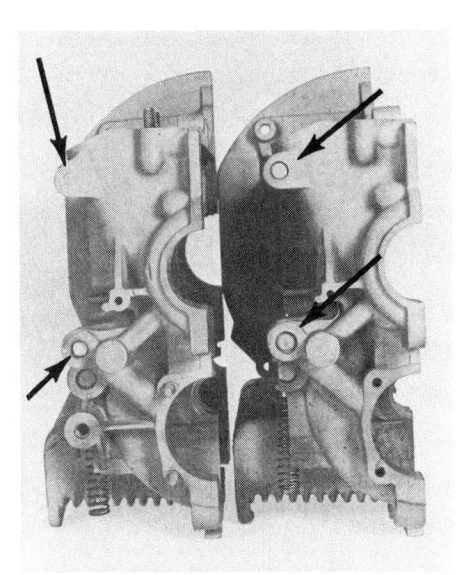

10mm oil galleries of dual-relief case (right) are another advantage of '70 and later Type 1—3 engines. Earlier single-relief case (left) has 8mm galleries.

Mexico and Brazil are only current sources for major replacement parts. Except during transition period discussed in text, all parts' origins are correctly labeled.

VW's Brazilian factory has been operating since the '50s. While its parts quality has been spotty on some items, Brazilian cases are perfectly acceptable replacements. New cases like this one are expensive, but have a full service life. Price and availability fluctuate.

8mm cylinder-head studs. You'll see a lot of cases with 10mm head studs with case savers. These case savers were added by a machine shop at sometime, not the factory.

Also in '73, Wolfsburg started to *deep-stud* all 1600 cases. Deep stud refers to the upper rear cylinder-head stud on #3 cylinder. The boss for this stud's threads is thought to contribute to a case cracking problem in the bellhousing. Some theorize that letting the stud pass deep into the case and thread into the main-bearing bore area stops this cracking. This is what the factory did in '73. Whether it works or not is debatable. Some machinists think it works, others aren't so sure.

Added Material—Cracking is a problem on all Type 1—3 cases, especially around cylinder 3. Here, the typical crack is seen from the bellhousing end of the case, running vertically down to the oil-gallery plug. To combat this crack, the factory has added more and more material to this area over the years. Early cases have plain ribbing here, followed by cases with more material added in the area between the ribs. Finally, it got to the point where replacement cases had no ribbing, just a smooth slab.

After cases got the solid slab, the Wolfsburg factory stopped all case production. The Mexico factory took over, supplying all factory cases. For awhile, the Mexican cases had the solid slab, and still said *Germany* on the lower right case side. Then some of the cases added *Hecho en Mexico* (Made in Mexico) to the upper left case half, but retained *Germany* on the lower right case. Ultimately, the Germany designation was dropped and Mexico added in its place. About this time, less material was added to the slab area, and ribbing started to show again. These cases seem to be as strong as the slabbed cases, so there's no problem in running them. Changes in alloy content probably account for the improved strength.

Cracking is more of a problem with later, higher-horsepower engines. This is because the 40-HP case is fairly strong for the modest power the engine produces, but 1500 and 1600 engines produce about 20 HP more in what is essentially the same case. An intensive strengthening program or a new aluminum case like the one used in the Type 4 would be necessary to stop this cracking.

Rebuilders currently report one of every three 1500 or 1600 cases is a fatality, where most 40-HP cases seem rebuildable four or five times. Who knows, it might be profitable to squirrel away lightly cracked 1500 or 1600 cases. Someday their value may exceed the repair cost.

Changing Cases—While I've treated all Type 1—3 cases as the same part in the preceding section, keep in mind there are really three different cases for each displacement—one each for Type 1, 2, or 3.

Type 1 cases have a vertically-mounted dipstick, no bosses, drillings or tappings for crossmember mounting-bolt holes and no cutout for the oil filler tube used on Type 2 and 3 cases. Additionally, Type 1 cases through '74 have a fuel-pump mount on the upper right case half. Starting in '75, fuel injection was used and the fuel-pump mount was not needed or provided.

Type 2 cases through '71 don't have the oil

37

Type 4 case is robust and features crack-resistant aluminum construction. Non-enclosed breather area and open oil-filler hole at lower right-rear corner on left engine immediately identify this one as a 411/412 or pre-'80 Bus case.

filler tube hole in the lower right case rear, but use the Type 1 style generator pedestal/breather/oil filler. The dipstick is like the Type 1 also (vertical). From '68, Bus cases had drillings for a crossmember.

Type 3 cases have no dipstick hole and no oil-pressure sending unit hole, although the boss is cast into the case. A long stud is provided for oil cooler mounting. Before '69, there were no crossmember holes.

There are some possible interchanges. First off, the Type 1 and 2 cases are good only for Type 1 and 2 interchanges as the oil filler tube cannot be added for Type 3 use. Type 3 cases can be used as Type 1s or 2s by blocking over the oil filler tube hole with commercially available plates, and drilling the dipstick and oil-pressure sending unit holes as required. VW machine shops can handle all drilling, tapping and plugging operations. If your Type 2 or 3 has a crossmember, make sure you get a '68—72 Type 2 or a '69—73 Type 3 case with crossmember mounting holes because these holes can't be added.

Type 4—These engines use two different cases: one for every engine up through '79 and the second for the '80—83 Vanagons. There is no interchange between the two styles only because the crankcase-breather area is different. Type 4 cases are practically bulletproof—unless a rod lets go—so you shouldn't need a new case.

All Type 4 cases are aluminum, dual-relief, with 8mm oil-pump studs and replaceable cam bearings. A spin-on oil filter and 8mm head studs are standard. All cases have the same size cylinder openings, so any stock cylinder will fit without machining.

Because the case is aluminum, it is not subject to cracking like the Type 3. But be aware there is a magnesium Type 4 case used in Europe. Some have made their way to the U.S. in core engines, but aren't used in rebuilds because the aluminum case is so much stronger. In the odd chance you run across one here, it has a VO letter code. Magnesium cases can also be spotted by their yellow-white color and distinctive sound when scratched.

Don't be confused by Type 4 engines that have been rebuilt. Often they are painted a gold that looks a lot like magnesium at first glance. The scratch test can be done with a pocket knife scraped sideways over the metal. Magnesium makes a harder, higher-pitched screech than aluminum. Scratch any Type 1—3 case to calibrate your ear.

Like Types 1—3, the Type 4 cases differ somewhat according to application. All carbureted engines, for example, '72—74 Buses, have a fuel-pump mount on the lower right front case with a passage for the fuel-pump pushrod. Fuel-injected engine cases don't have the fuel-pump mount, and it can't be added to them. But, the mount can be blocked off for use on an injected engine.

CRANKSHAFTS

40 Horse—Stroke changes and oiling improvements mark Type 1—3 crankshaft evolution. The 40-HP crank was the first VW crankshaft with four main bearings and looks more like a 1500 crank than a 36-HP piece. Like all Type 1—3 crankshafts after the 36-HP one, these two are the same overall length, but when viewed on end the difference is apparent in their counterweight shape. The 40-HP crank is simply rounded on the counterweights, the 1500 has a bump on one side. Note the accompanying photos.

Journal diameters are the same for all Type 1—3s. The larger mains, #1—#3 are 55mm. The smaller #4 is 40mm. Rods measure 55mm also. But, 40-HP rod journal width is wider than all later crankshafts by 1mm. Another

Vanagon Type 4 engines date from '80—83 and have a different case casting. Of interest here is cast-over breather area at left. Separate breather box is replaced by closed casting and sheet-metal hose nipple.

dimension the 40 HP doesn't share is stroke. It is 64mm.

The 40-HP crankshaft is known to break, especially in engines with big-bore kits. Fitting larger cylinders to the 1200 makes extra horsepower you can feel and many owners over-rev the engine during acceleration because the engine feels powerful. Unfortunately, this revving causes the crank to flex, pounding out #2 main bearing and then fatiguing the crankshaft. It will finally break because it has lost its support from this bearing. Crank breakage is directly related to engine rpm in this case, so not over-revving is the best insurance.

Short crank at left is 36-HP part. Next is 40-HP shaft, followed by 1600 and 1500 parts. Dowels make 1500 crank at far right seem taller than 1600 and 40-HP, but all are same size.

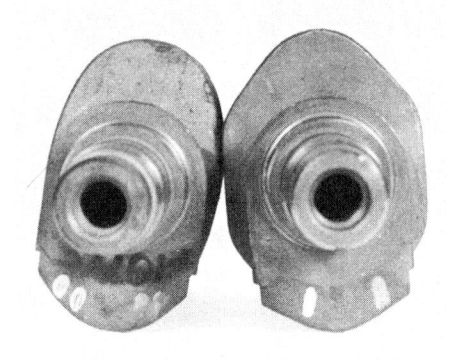

Rounded flanges identify 40-HP crankshaft (left). Pointed, larger flanges belong to 1500 crank (right).

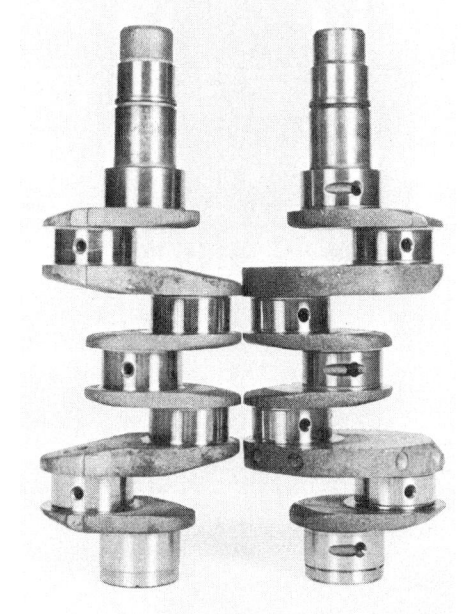

Another crank identification clue is angled flanges of 40-HP crank (left). Constant-thickness flanges are used on 1500 and 1600s. Oil troughs are part of cross-drilling; identify 1500 crank as a '67 or later part.

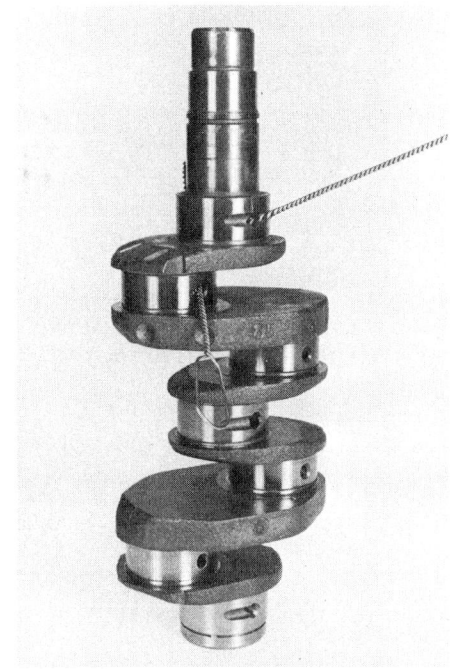

Wire-handled brushes passing through this 1600 cross-drilled crank illustrate how rod journals are bored completely through when cross-drilled.

1300, 1500, 1600—The major difference between the 40 HP and the later Type 1—3 crankshaft is the extra 5mm in the stroke, bringing it to 69mm. Also, the journals are 1mm narrower and the flanges thicker, with a generally more robust construction. This is best seen on adjacent throws. The connecting flange is not knife-edged like the 1200, but is a constant thickness all the way to the edge as shown in the photos.

Beginning in '67, a major visual and performance difference is the oil holes. This is when VW started *crossdrilling* crankshafts. Crossdrilling means an extra oil passage is bored through the rod and main journals, ensuring that oil reaches all parts of the rod bearing at all times. To help get more oil out to the rods, a trough is cut into the main-bearing journals leading to the oil hole. The net effect is that an oil passage is open to the rod bearing through more degrees of crankshaft revolution with crossdrilling.

In a nutshell, with Type 1—3 engines, there are only two strokes: the 40-HP at 64mm and the 1300/1500/1600 at 69mm. All journals are the same size, and all crankshafts will physically fit in all cases. Since '68, all cranks have been crossdrilled and are visually identified by oil troughs.

So, the fundamental question is, will the 69mm-stroke crankshaft fit in the 1200cc case? Well, it fits right in with no modification to the case or crankshaft, but that's not the end of it.

First, the 40-HP engine is about 3/4-in. *narrower* than the later engines. That means the connecting rods, pushrods, pushrod tubes, studs, intake manifold, exhaust and sheet metal are all shorter on the 40-HP engine. It's a waste of time and effort to fit later cylinders and pistons because you'll then have later engine internals inside the weaker 40-HP case.

Stick with the 40-HP rods, pistons, 77mm cylinders and so on. This will yield 1285cc, or 1300 if you prefer. To make the combination work, grind 0.020 in. off each side of the 40-HP rods. A machine shop can do this and have them set up the rod side clearance (0.008—0.012 in.) at the same time.

Cylinder spacers are also required. Use 0.100-in.-thick spacers between the cylinders and crankcase. Check deck height as explained in Chapter 7, page 131. It should be correct with 0.100 in. spacers, but adjustments may be necessary.

All cross-drilled crankshafts have oil troughs on main-bearing oil holes. Generous chamfer around oil hole helps scoop oil and strengthens drilling.

Oil-Control Valves

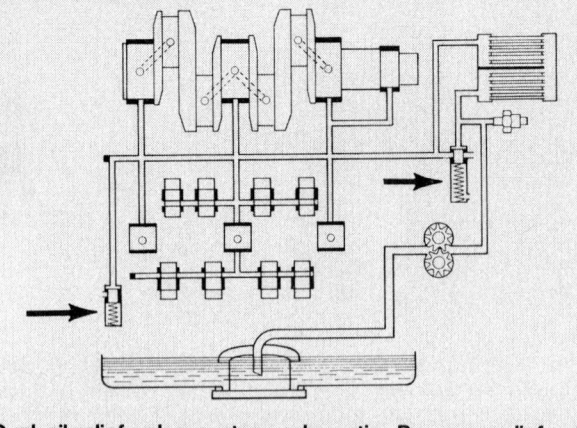

Single oil-relief valve (arrow) system schematic.

Dual oil-relief valve system schematic. Pressure-relief valve is at right, pressure-control valve at left.

OIL-CONTROL VALVES

Air-cooled VWs have valves that regulate oil pressure, routing and supply. Up to 1970, one valve was used, after that two were fitted. Understanding the operation of these valves is important in diagnosing and servicing the oil system.

The first valve is the *oil-pressure-relief valve*. It is located at the oil-pump end of the left case half and controls passages leading from the oil pump to the oil cooler, oil gallery and sump return.

When the engine is cold and the oil thick, oil pressure is high. Such high pressure can rupture the oil cooler, and when oil is cold, no cooling is desired. Therefore, the relief-valve spring is designed to yield under excess oil pressure; it releases some oil through the sump return and lets the rest lubricate the engine via the oil gallery. The cooler is completely bypassed when the valve is fully open.

As oil temperature rises, oil pressure drops. The valve begins to close from its spring tension. The first passage closed is the sump return. Now all oil must pass through the oil gallery to the bearings, or through the cooler to the bearings. Most oil passes through the gallery because it's a less restrictive path than the cooler. If the oil continues to heat, the relief valve closes farther. Now it must flow through the cooler on its way to the bearings.

What isn't controlled with this design is ultimate oil pressure when the oil is warm. With cold oil, the relief valve lowers pressure by venting the excess through the sump return. But when the oil is hot, the only path is through the gallery or cooler. At high engine speeds too much oil is supplied by the pump. This causes increased oil consumption due to overloaded piston rings, horsepower losses to the pump, unnecessary oil heating and bearing wear.

These problems were eliminated with the addition of the *oil-pressure-control valve* in '70. The valve is located at the flywheel-end of the left case half—at the far end of the oil

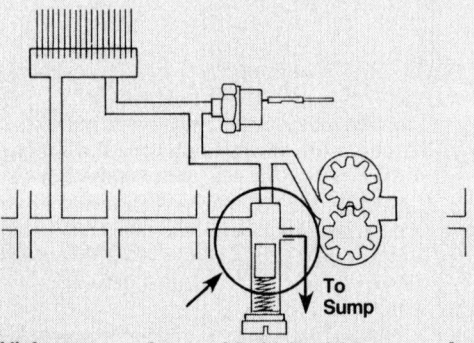

High pressure from cold, thick oil forces relief valve fully open, bypassing oil cooler.

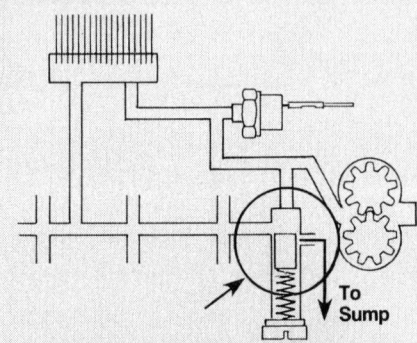

As oil warms, relief valve partially closes. Some oil flows through cooler, some directly to bearings through main oil gallery.

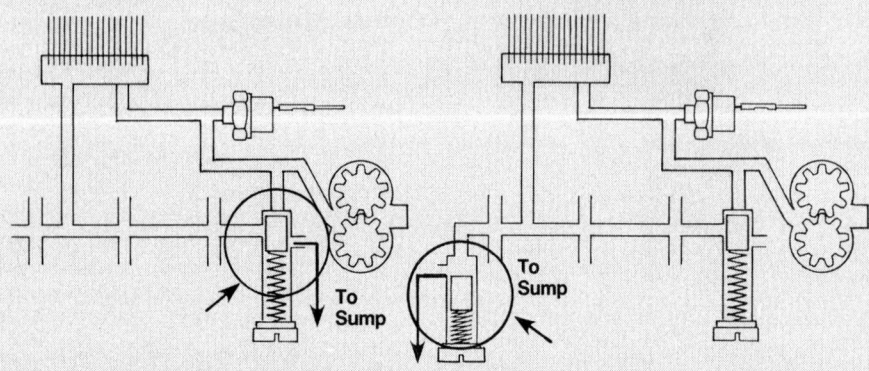

Hot, thin oil lets relief valve close completely. All oil must pass through cooler before reaching main oil gallery.

Dual oil-relief valve system functions at high engine speeds with hot oil. Oil-control valve at left opens to vent oil to sump at pressures above 42 psi.

gallery. Thus, when pressure rises too high, the control valve opens and vents the excess to the sump.

Spring pressures are set for 42 psi operation when using 30W oil at 158F (70C) at 2,500 rpm. A wear limit of 28 psi is specified, but you have to hook up a gage at the oil-pressure sending unit hole to accurately read it. The sending unit is calibrated to warn you of oil pressure at 2.1—6.4 psi (0.15—0.45 kg/cm sq.), so when its warning light comes on, it's all over but the check writing.

During rebuilding, check the relief- and control-valve spring specifications. Checking tension is not the easiest job as the loads are so light. Read the valve spring section on page 112 for tips on checking springs. Don't swap the springs or use sealing compound on the threaded valve covers. They are tight enough as is.

CHASSIS SERIAL NUMBERS

Year	Type	Model	Starting Chassis #
'59	1	Beetle	2 226 206
	2	Bus	374 811
'60	1	Beetle	2 528 668
	2	Bus	469 506
'61	1	Beetle	3 192 507
	2	Bus	614 456
'62	1	Beetle	4 010 995
	2	Bus	802 286
'63	1	Beetle	4 846 836
	2	Bus	971 550
'64	1	Beetle	5 677 119
	2	Bus	1 144 282
'65	1	Beetle	115 000 001
	2	Panel Truck	215 000 001
'66	1	Beetle	116 000 001
	2	Panel Truck	216 000 001
	3	Fastback	316 000 001
'67	1	Beetle	117 000 001
	2	Panel Truck	217 000 001
	3	Fastback	317 000 001
'68	1	Beetle	118 000 001
	2	Station Wagon	228 000 001
	2	Panel Truck	218 000 001
	2	Kombi and Campmobile	238 000 001
	2	Pickup Truck	268 000 001
	3	Fastback	318 000 001
	3	Squareback	368 000 001
'69	1	Beetle	119 000 001
	2	Station Wagon	229 000 001
	2	Panel Truck	219 000 001
	2	Kombi and Campmobile	239 000 001
	2	Pickup Truck	269 000 001
	3	Fastback	319 000 001
	3	Squareback	369 000 001
'70	1	Beetle	110 2000 001
	1	Karmann Ghia/Convt.	140 2000 001
	1	VW Convertible	150 2000 001
	2	Station Wagon	220 2000 001
	2	Panel Truck	210 2000 001
	2	Kombi and Convertible	230 2000 001
	2	Pickup Truck	260 2000 001
	3	Fastback	310 2000 001
	3	Squareback	360 2000 001
	4	914	4702900001
'71	1	Beetle	111 2000 001
	1	Super Beetle	131 2000 001
	1	Karmann Ghia/Convt.	141 2000 001
	1	VW Convertible	151 2000 001
	2	Station Wagon	221 2000 001
	2	Panel Truck	211 2000 001
	2	Kombi and Campmobile	231 2000 001
	2	Pickup Truck	261 2000 001
	3	Fastback	311 2000 001
	3	Squareback	361 2000 001
	4	411 2 Door	411 2000 001
	4	411 4 Door	421 2000 001
	4	411 Wagon	461 2000 001
	4	914	4712900001
'72	1	Beetle	112 2000 001
	1	Super Beetle	132 2000 001
	1	Karmann Ghia/Convt.	142 2000 001
	1	VW Convertible	152 2000 001

Year	Type	Model	Starting Chassis #
'72	2	Station Wagon	222 2000 001
	2	Panel Truck	212 2000 001
	2	Kombi and Campmobile	232 2000 001
	2	Pickup Truck	262 2000 001
	3	Fastback	312 2000 001
	3	Squareback	362 2000 001
	4	411 2 Door	412 2000 001
	4	411 4 Door	422 2000 001
	4	411 Wagon	462 2000 001
	4	914	4722900001
'73	1	Beetle	113 2000 001
	1	Super Beetle	133 2000 001
	1	Karmann Ghia/Convt.	143 2000 001
	1	VW Convertible	153 2000 001
	2	Station Wagon	223 2000 001
	2	Panel Truck	213 2000 001
	2	Kombi and Campmobile	233 2000 001
	2	Pickup Truck	263 2000 001
	3	Fastback	313 2000 001
	3	Squareback	363 2000 001
	4	412 2 Door	413 2000 001
	4	412 4 Door	423 2000 001
	4	412 Wagon	463 2000 001
	4	914	4732900001
'74	1	Beetle	114 2000 001
	1	Super Beetle	134 2000 001
	1	Karmann Ghia/Convt.	144 2000 001
	1	VW Convertible	154 2000 001
	2	Station Wagon	224 2000 001
	2	Panel Truck	214 2000 001
	2	Kombi and Campmobile	234 2000 001
	2	Pickup Truck	264 2000 001
	4	412 2 Door	414 2000 001
	4	412 4 Door	424 2000 001
	4	411 Wagon	464 2000 001
	4	914	C 4742900001
'75	1	Beetle	115 2000 001
	1	Super Beetle	135 2000 001
	1	VW Convertible	155 2000 001
	2	Station Wagon	225 2000 001
	2	Panel Truck	215 2000 001
	2	Kombi and Campmobile	235 2000 001
	2	Pickup Truck	265 2000 001
	4	914	C 4752900001
'76	1	Beetle	116 2000 001
	1	VW Convertible	156 2000 001
	2	Station Wagon	226 2000 001
	2	Panel Truck	216 2000 001
	2	Kombi and Campmobile	236 2000 001
	2	Pickup Truck	266 2000 001
	4	914	C 4762900001
'77	1	Beetle	117 2000 001
	1	VW Convertible	157 2000 001
	2	Station Wagon	227 2000 001
	2	Panel Truck	217 2000 001
	2	Kombi and Campmobile	237 2000 001
	2	Pickup Truck	267 2000 001
'78	1	VW Convertible	158 2000 001
	2	Station Wagon	228 2000 001
	2	Panel Truck	218 2000 001
	2	Kombi and Campmobile	238 2000 001
	2	Pickup Truck	268 2000 001
'79	1	VW Convertible	159 2000 001
	2	Station Wagon	229 2000 001
	2	Kombi and Campmobile	239 2000 001

Tapered snout is mount for fan hub and instantly identifies Type 4 cranks. All Type 4 cranks are cross-drilled and noticeably longer than Type 1—3 parts.

VW achieved longer stroke in 2.0 liter engine by *offset grinding* rod journals. Besides being narrower, note how rod and main-bearing journals overlap less on 2.0 liter crank (right). This 2.0 liter crank has been polished, so it's brighter than 1.7/1.8 crank (left).

Counterweighting adds visual complexity and 5,000-plus rpm stability. Stock crankshaft is OK for normal use, but popularity of high-rpm VWs makes counterweighted cranks a common parts-house piece.

The same procedure can be used with 82mm or 83mm aftermarket cylinders to get 1457cc and 1493cc displacement, respectively. It's easier and cheaper to simply install the big-bore kits without buying the 69mm crankshaft, so most 40-HP owners looking for a power boost simply bolt on the bigger barrels. They supply all the increased power a 40-HP case should handle anyway. If looking for more power, get a 1300-or-later engine and enlarge it.

Type 4: 1700, 1800, 2000—Type 4 and Type 1—3 crankshafts don't interchange. To identify a Type 4 crankshaft, look for its tapered snout. The taper supports the fan hub, and also gives this crank a distinctive profile that can be seen across a dark garage.

There are only two Type 4 crankshafts, one for the 1700/1800 and one for the 2 liter. Any Type 4 crank will work in any Type 4 case, but you may have to change rods and pistons, as well. Again, the stroke difference is why. The 1700/1800 crank has 55mm rod journals and 60mm main-bearing journals. The extra 5mm stroke of the 2-liter crank comes from centering the rod journal farther away from the main-bearing journals than on the 1700/1800 crank. To keep overall crank dimensions the same, so the rods and crank will not hit the case, the rod journals are made 5mm smaller. Therefore, the 2-liter crankshaft's 50mm rod journals are noticeably narrower than 1700/1800 rod journals. Rods are made smaller to match; consequently, a special case isn't necessary for 2-liter engines.

A 1700/1800 crank can be stroked to 2-liter specifications by *offset grinding*. Offset grinding is exactly what it sounds like: more metal is ground from one side of the rod journal than the other, resulting in the 2-liter journal diameter and center. Matching 2-liter rods must be used with an offset-ground crank.

All Type 4 cranks are crossdrilled and have an oil trough on the main-bearing journals.

Counterweights—There is no such part as a counterweighted air-cooled VW crankshaft, at least, not from Wolfsburg. Those parts of the crank which connect the journals are correctly called *flanges*, not *counterweights* as they are commonly misnamed. Counterweights are material added to the flanges to counteract connecting-rod and piston weight. Such counterweights are commonly welded onto stock VW cranks and sold by machine shops, rebuilders and parts houses.

Counterweights help offset the strain caused by piston and rod motion. Thus *crank flex* is reduced, especially at high rpm. With less crank flex, the #2 main bearing doesn't get pounded into an egg shape quite so fast. The next question is: Do you need a counterweighted crank?

For Type 4 owners, the answer is no. Bulletproof best describes the Type 4 bottom end. If you can ruin it with spirited street driving, you could probably destroy an anvil with a rubber mallet.

Type 1—3 owners aren't so lucky. If you use your Beetle as efficient, inexpensive transportation, then you probably don't need a counterweighted crank. You don't rev your engine to a high enough rpm to rapidly wear the center main bearing. Installing a counterweighted crank would cost a lot of money for very little return in bearing life.

TYPE 1—3

Oil-Pressure-Relief Valve Spring

	Loaded Length	Load @ Length
'68—69	1-11/16-in. (43.20mm)	5.5—9.2 lb. (2.5—4.2 kg)
'70—79	1-3/4-in. (44.10mm)	12.3—16.0 lb. (5.6—7.3kg)

Oil-Control-Valve Spring

'70	13/16-in. (20.20mm)	6.3—7.9 lb. (2.9—3.6kg)
'71—79	13/16-in. (20.20mm)	6.8—8.4 lb. (3.1—3.8kg)

TYPE 4

Oil-Pressure-Relief Valve Spring

All	1-9/16-in. (39.00mm)	15.0—19.4 lb. (6.8—8.8kg)

Oil-Control-Valve Spring

All	1-1/16-in. (26.00mm)	3.8—4.4 lb. (1.7—2.0kg)

On the other hand, if you think of VWs as inexpensive performance, the extra money is worth it. The limit is 5000 rpm. If you drive at this region or higher, the counterweighted crank will pay off in longer-lasting main bearings. Don't confuse counterweights with balancing. Balancing brings parts to the same weight (pistons), and equalizes weight side-to-side and end-to-end (crankshaft, rods). Balance an engine for maximum longevity.

FLYWHEELS

On Type 1—3 engines, several flywheels have been used. Diameter, number of ring-gear teeth and O-ring installation are the variables that have changed on the different flywheels.

Diameter is measured in millimeters, and specifies the working surface the clutch disc bears against. The overall diameter increases sometimes with increased clutch surface, too. Any variation in flywheel outside diameter (OD) is slight, and can be compensated for by grinding the bellhousing inside diameter (ID), if swapping.

The number of teeth on the ring gear is either 109 or 132. Six-volt starters mesh with 109-tooth flywheels, 12-volt starters need the one with 132 teeth. Therefore, a flywheel is called a *6 volt* or *12 volt* depending on how many teeth it has and what starter it is used with.

Finally, later flywheels (post-'65) have an *O-ring* on the crankshaft mating-area ID; earlier ones don't. This O-ring seals oil in the case. Crankshafts are designed to accept O-ringed flywheels or they aren't. You can tell by looking at the flywheel end of the crank, right in front of #1 main bearing. There is a shoulder machined into the bearing OD.

If a non-O-ringed flywheel was intended for the crank, the machined shoulder on the crank will be quite short, about 1/8-in. long. Cranks designed for an O-ring have a wider shoulder, about 1/2-in. long. If the flywheel doesn't have an O-ring or groove, it mates to any crankshaft. But O-ringed flywheels *must be used* on O-ringed cranks. They just won't slide completely onto a non-O-ring crank.

The nearby chart explains Type 1—3 flywheels specifications and combinations.

The 1300 European flywheel wasn't imported to the U.S. on any cars by VW, but some have made their way here anyway, so it's possible you could come across one. It simplifies converting early 6-volt cars to 12 volts because it bolts right on and requires no bellhousing grinding.

All '66 Beetles sold in the U.S. were 6-volt cars and had 6-volt flywheels. Another flywheel that merits special mention is the '65 Type 2 or 3 because it is the later diameter (200mm) but fits 6-volt starters with its 109 teeth. If you want a larger clutch in your 6-volt car, it's the one to get.

Type 4 flywheels are less confusing. Their

FLYWHEELS

Engine	Diameter (mm)	Ring Gear (teeth)	O-Ring
TYPE 1 & 3			
40-HP	180	109	No
1300 U.S.	180	109	Yes
1300 Euro	180	132	Yes
1500, Type 1	200	132	Yes
1500 Type 2, 3—'65	200	109	No
1500 Type 2, 3—'66	200	109	Yes
1500 Type 2, 3 '67—71	200	132	Yes
1600	200	132	Yes

Note: Non-O-ring flywheel will work on any crank. But, O-ring flywheel cannot be used on non-O-ring cranks because O-ring flywheels have a longer step to accommodate the O-ring. Flywheels with 109 teeth are 6-volt; 132 teeth are 12-volt.

Engine	Diameter (mm)	Ring Gear (teeth)	O-Ring
TYPE 2			
17/1800 '72—74 (partial)	210	132	Yes
1800 '74(partial)—75	215	132	Yes
2000 '76—83	228	132	Yes
TYPE 4			
1700, '69	200	132	Yes
1700 '70—73	215	132	Yes
914			
All	215	132	Yes

Note: 914 flywheels will *not interchange* with any VW flywheel due to 0.125-in. difference in ring-gear placement. 914 flywheels have timing marks on their circumference; others don't.

CONNECTING RODS

P/N	Engine	Remarks
113 401	1200	Symmetrical small-end, early rod
113 401A	1200	Offset small-end, w/bump on shank
311 401A	1500	Bolts, non-self-aligning
311 401B	1600	Studs, self-aligning
Type 4		
021 401A	1700/1800	1.7 marked w/"021", 1.8 marked with "022".
039 401A	2000	

primary difference is their diameter. All are 12-volt (132 ring-gear teeth) and O-ringed.

The 914 flywheel is unique to the Porsche transaxle in that car and won't work with VW transaxles. The ring-gear teeth are set about 0.125-in. closer to the transaxle on the 914 part, and won't engage properly with starters mounted in VW transaxles. Identify 914 flywheels by the timing marks on their OD, which are on the flat area next to the ring-gear teeth.

CONNECTING RODS

For all the air-cooled engine variations VW has sold, there are only six connecting rods to keep track of.

40 HP—The 40 horse has two connecting rods: early (to mid-'64) and late (post-'64). Early rods have a *symmetrical* small end. That is, the pin bore is evenly centered on the shank. They are numbered 401. The later rod, 401A, has an *offset* small end, which can be seen if the rod is held on edge. The offset centers the rod in the piston. This is because the rod and rod journals are not quite centered with the cylinders.

Another 40-HP early and late rod variation is hardware. The early rods use capscrews, late rods use bolts and nuts. Early rods with capscrews are not *self-aligning*. That is, during assembly the cap's edges must be visually lined up with the rod. Adjustments are made by tapping with a hammer. Late rods with bolts are self-aligning, because the bolts are securely anchored in the caps.

While it's desirable to keep a matched set of rods, the early and late rods will interchange. Both have big ends sized for 55mm rod journals and 20mm piston-pin bores. Center-to-center length is 130mm.

1300/1500—The main difference between the 1500 rod and the 40-HP one is length. The later rod is 7mm longer (137mm center-to-center), effectively ending any interchange possibilities with 40-HP engines. The 1500 rod also has a 22mm piston-pin bore. Big ends remain sized for 55mm journals.

Additionally, the 1300/1500/1600 rods are narrower at the big end than the 40-HP piece. The 40-HP rods can be ground 0.020 in. on each side to fit the later cranks, but the later rods can't be used in the 1200 engine. Curiously, with all these changes, 1300/1500 rods are still marked 401A. You'll have to physically compare the big-end thickness to distinguish them from 40-HP rods.

1600—The 1600 rod looks just like a 1500 at first glance. It is the same length, has the same big end, small end, and width. The difference is in its hardware. All 1300/1500 rods use bolts secured by nuts on the rod end. The 1600 uses studs anchored in the rod which pass through the cap and secure by nuts at the rod end. These are called *self-aligning rods* and are marked 401B.

Type 4—If the increased strength and durability of the Type 4 engine needs graphic explanation, look at its rods. Massive is the word.

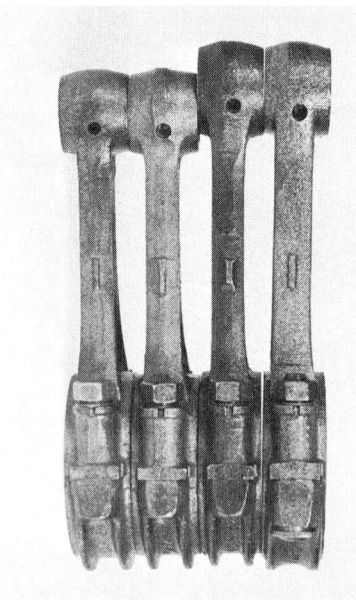

Basically, there are two Type 4 rods, one for the 1700/1800 and another for the 2 liter. All have a center-to-center length of 127mm, but they don't interchange because of the difference in big-end size. Rods for the 1700/1800 are sized for VW's standard 55mm rod journals, but the 2-liter rods are made for the 50mm offset journals of the 2-liter crank. Both rods use studs and are self-aligning.

Here are all Type 1—3 connecting rods. From left to right: early 40 HP, late 40 HP, 1500 and 1600. Visual differences are length, small-end offset and rod bolt or stud configuration. Extra 1mm width of 40-HP big end doesn't show.

If small end is offset, it looks like this late 40-HP example. Offset centers rod small end between piston-pin bosses because bore and rod journal center lines aren't quite lined up. Bump on shank points up when rod is correctly installed.

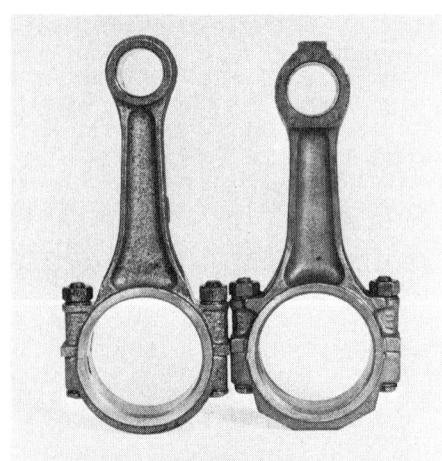

1600 Type 1 rod (left) looks tall and skinny compared to stout 1.7/1.8 liter Type 4 rod. Short balancing pad atop small end and barely visible "022" by left bolt show this particular Type 4 rod came from 1.8.

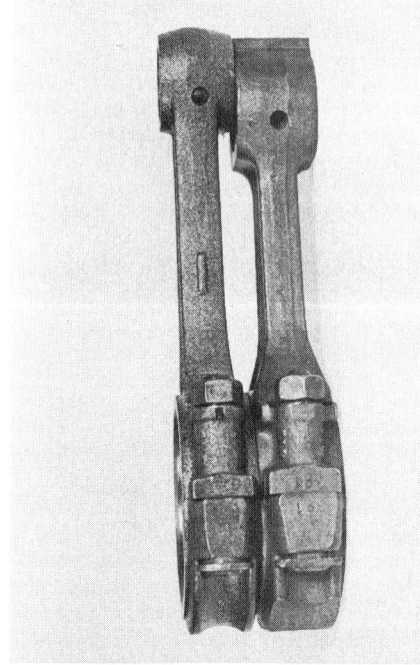

Large difference in overall mass and center-to-center length show when 1600 (left) and Type 4 rod (right) are compared. Type 4 rods do not have offset small ends or bumps on shanks for rod orientation.

The rods are easy to tell apart. The 1700/1800 rod has a *balancing pad* on top of the small end; the 2 liter does not. The big-end diameter is noticeably larger on the 1700/1800

Flat top piston (left) is Type 1 part, while dished one (right) was used in otherwise same engine installed in Type 2s. Dish lowers compression and reduces chances of detonation in heavier, hotter-running Bus.

Both pistons fit 1.7 liter Type 4 engine, but are used in different chassis. Slight compression ratio differences from such modest domes and dishes allow fine tuning compression, but wholesale changes require head work or special aftermarket pistons.

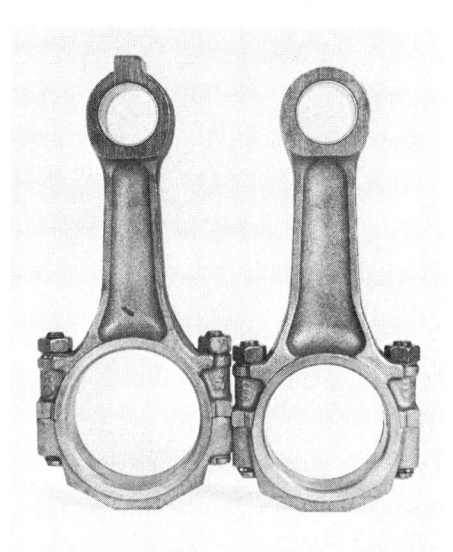

No balancing pad, thick shank and smaller bottom-end diameter are all trademarks of 2.0 liter rod (right). Compare tall balancing pad of 1.7 rod at left with shorter pad on 1.8 rod in nearby picture.

rod, too. You won't have to tell the two rods apart very often because VW scrap yards and rebuilders are filled with 1700/1800 rods, but the 2-liter rods are fairly rare. I guess they are all busy in 2-liter engines because you just don't see them around.

I stated earlier there are basically two Type 4 rods. For practical purposes, that's true. Yet, there are actually three types of rods, but many people (including some rebuilders) never notice the difference. This third type of rod is used in the 1.8 liter engine. The cast-in p/n is 022, but this is the only place this number appears. The p/n is cast in the rod near the bolt hole. VW never did assign a p/n to this piece, so it doesn't appear in their parts books.

This rod is slightly different from the 1700/1800 ones I've been describing. The only visual clues are the p/n and the balancing pad directly on top of the small end. The 1.8 balancing pad is noticeably lower than the 1700/1800 pad, yet this doesn't mean the rods are necessarily different weights. Many times a 1700/1800 and 1.8 rod are the same weight, and coexist in the same engine without problems. But like any rod, have 1.8 rods being installed into a 1700/1800 engine or vice versa, checked for total weight.

Type 4 Interchange—Because the three Type 4 engines share so many dimensions, parts swapping is straightforward. If you have a 1700, you can turn it into an 1800 with 1.8-liter cylinders and by flycutting the existing cylinder heads. If building a 2 liter, use the 2-liter crankshaft, connecting rods and cylinders. Again, flycut the cylinder heads.

The problem with this approach is finding the crank and rods. Fewer 2-liter engines were built, and their internals are in demand by all the other people wanting to build a 2-liter engine. Valve sizes vary among these engines, so be sure to read the cylinder-head section, page 46, before deciding on a swap.

Swapping Rods—Any rod going in a engine must be weighed against the existing rods. Weights vary among all rods, and must be corrected by grinding off material until all rods weigh the same as the lightest rod. This is called *total weight,* and is an inherent part of engine construction.

For high rpm, longevity, smoothness or all of the above, *balancing* is needed. Balancing means bringing all internal moving parts to their design weights, with a much smaller than factory tolerance. In VWs, balancing involves the crank pulley, its bolt, crankshaft, connecting rods, pistons, gland nut and flywheel. When it comes to balancing rods, they are checked for end-to-end weight. Thus each rod will have the same total weight, and each small end and big end will weigh the same.

PISTONS & CYLINDERS

At one time it was economically sensible to hone and reuse VW cylinders if you used a set that wasn't too worn. Rising labor prices and inexpensive parts from overseas have changed that. Now, it's rare when pistons and cylinders are reused. Shops routinely discard these parts without a glance and order new ones. It's cheaper to use new parts—particularly when you know those parts have a full life ahead of them. Therefore, this section isn't intended to help resurrect old cylinders, but to show which cylinders are used in stock and modified applications.

Type 1—3—Another chart is the best way to specify Type 1—3 bore sizes:

Engine	Bore (mm)	Piston-Pin Dia.(mm)
1200	77	20
1300	77	22
1500	83	22
1600	85.5	22

1200/1300—At first glance, it would be logical to assume the 40 HP and 1300 pistons and cylinders are the same. They are the same in bore, but overall length is different. The 1300 barrels are 0.200-in. longer for that engine's longer stroke. Remember that 1300 and later engines are about 3/4-in. wider than the 40 HP and you'll see there is little sense in increasing 40-HP capacity by using stock cylinders from later engines.

45

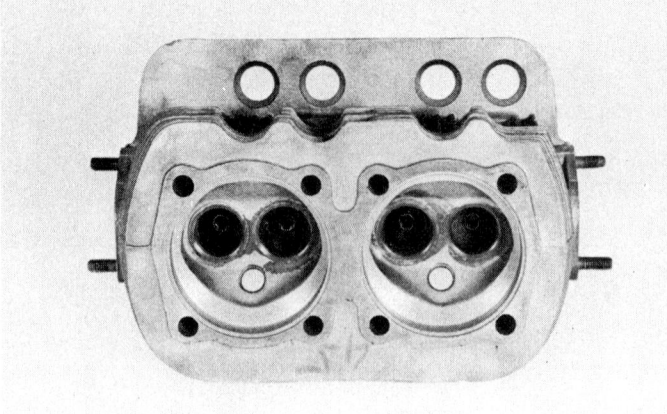

Compact overall size, small rocker area with square rocker pedestals, and deep combustion chambers housing tiny valves are all 36-HP cylinder head trademarks.

It can be done, but plan on flycutting the case and installing 1300 connecting rods, cylinders, pistons, pushrods, cylinder-head studs, heat exchangers and sheet metal. By the time this is all done, the product is a 1300 engine on top of a 1200 case. Why not start with a 1300?

Another consideration is piston pin size. The 40 HP uses 20mm pins, and later engines 22mm pins. So, you'll have to change the 1200 rods to fit the larger 1300 pins.

There are three methods of increasing 40-HP displacement: install a larger 1300/1500/1600 engine, fit aftermarket big-bore kits, or install a 1300 crankshaft with cylinder spacers as outlined in the crankshaft section. Buying the larger engine is the best option.

First, it has the stronger later case and more potential for aftermarket goodies. Fitting big-bore kits is the cheapest action, because you'll have to replace stock cylinders and pistons at rebuild time anyway. Using a 1300 crank and spacers is a good deal if you already have the crank.

Of course, another route is to combine the big-bore kits, spacers and 1300 crank. A common kit is the 83mm big-bore and 1300 crank, which yields 1493cc. Using the big-bore kit with the stock crank gives 1385cc.

The compression ratio will increase with the displacement increase in these kits. Stock 40-HP compression is 7.0:1. Adding the 82mm kit yields 7.5:1, and the 83mm kit raises it to 7.9:1. The 1300 crank with a big-bore kit achieves about the same figures, which are fine with today's pump gasoline.

1300—Probably the easiest displacement increase is using 1500/1600 barrels on the 1300 case. This gives the stock displacement for those cylinders (1493cc and 1584cc, respectively) as the crank stroke is common to all these engines. The 1300 cylinder heads will need flycutting to accept the larger cylinders, but the 1300 case will accept them as is. No spacers are needed.

Compression will rise because the 1300 cylinder-head *combustion chamber volume* is smaller. You'll have to run premium fuel and listen for detonation under high-heat, high-load conditions. A more satisfactory installation would be to fit later cylinder heads with larger valves and even dual ports. Add the cost of a new intake manifold, sheet metal and so on to complete this.

1500/1600—Interchanging 1500 and 1600 cylinders is as simple as putting them on. Because the OD of both is the same, they bolt on either engine with no modifications.

There are no stock cylinders for Type 1—3 engines larger than the 85.5mm used in 1600 engines. Use aftermarket big-bore kits for more displacement. Plenty of these kits are available, 87mm and 92mm sizes being most popular. The 87mm yeilds 1641cc and the 92mm equals 1835cc. Both require flycutting the case and cylinder heads.

Be careful about using big-bore kits that don't require flycutting the case or heads. As you can guess, these kits use cylinders thinner than stock to fit in the case and still supply a displacement increase. The bottom line is such *slip-in* cylinders are too thin for good longevity if larger than 87mm. At the first sign of heat or high mileage, these cylinders warp. Then oil control and blowby become problems. Cut the case and install thick cylinders if you want a larger displacement. You'll be better off in the long ride.

Type 4—There are three Type 4 bore sizes. The 1.7 uses 90mm cylinders, the 1.8 has a 93mm bore and the 2 liter uses 94mm. The 1.7 and 1.8 cylinders are a different OD, so installing the larger cylinders means flycutting the cylinder heads. The Type 4 case opening will accept all bores, though, so there is no need to cut it. Changing from 1.8-to 2-liter cylinders doesn't require flycutting the cylinder heads, because both the cylinder ID and cylinder opening in the head are the same, but you do need the 2-liter crank and rods to make a workable, worthwhile swap.

Interchanging 1.7 and 1.8 cylinders will give you stock displacement (1679cc and 1795cc) because the crankshaft is the same. If you have an 1800 and a set of 1700 cylinders, get the 1700 cylinder heads as well. They are needed to fit the smaller OD of the 1700 cylinders.

Most of the 2 liter's increased displacement comes from its 5mm longer-stroke crankshaft, so installing 2-liter cylinders on the smaller engines won't gain a whole bunch: 1832cc. The swap won't work without cylinder spacers, either, so a big-bore kit is the preferred method.

Big-bore kit options for the Type 4 are less numerous than the Type 1—3 variety, but sizeable engines can still be built. How about 2.4 liters on the 2000 crank or 2.7 liters on strokers? Check the popular VW magazines for information on these expensive engines.

CYLINDER HEADS

The most plentiful changes to air-cooled VWs have been on the cylinder heads. To the novice, all VW cylinder heads look alike, yet, after some inspection they show almost as much variety as snowflakes. Get familiar with those cylinder heads you might use. Chances are good you may need or want to change the heads during a rebuild. Performance could be reduced or improved, depending on which type used.

As part of VWs constant design updating, you might find the same head in several versions. Each version is denoted by a letter suffix, A is the first design, B the second and so on. The suffix is found immediately at the end of the p/n, 113 101 373B, for example. The casting

CYLINDER HEAD P/Ns

TYPE 1

P/N	Year	Engine	Remarks
113 101 371A	'61	1200	31.5mm Intake (In.) X 30mm Exhaust (Ex.), round rocker-arm boss, needs short-stud repair kit.
113 101 371B 113 101 371C 113 101 371D	'62—64	1200	31.5mm In. X 30mm Ex., round rocker-arm boss, needs short-stud repair kit.
113 101 371F		1200	31.5mm In. X 30mm Ex., replacement for above heads, round boss.
113 101 373	'65	1200	31.5mm In. X 30mm Ex., square rocker-arm boss doesn't need repair kit.
113 101 373B	'66	1300	33mm In. X 30mm Ex., single-port.
311 101 373A	'67—70	1500/1600	35.5mm In. X 32mm Ex., single-port.
113 101 375A	'71—73	1600	35.5mm In. X 32mm Ex., dual-port, w/ledge in combustion chamber.
043 101 375A	'74	1600	35.5mm In. X 32mm Ex., long upper-exhaust studs for dual pre-heaters, often called "long stud," w/ledge, dual-port.
043 101 375R or 043 101 375H	'75—76	1600	33mm In. X 30mm Ex., 9mm exhaust-valve stems, H heads have CHT sensor, R heads don't, fuel-injection heads, dual-port.
041 101 375 2			39.3mm In. X 32mm Ex., replacement head, 15% larger ports, often called "racing head," dual-port, forged-steel valves, made in Brazil.
040 101 375 2		1600	35.5mm In. X 32mm Ex., stock replacement head, dual-port, made in Brazil.
040 101 373		1600	35.5mm In. X 32mm Ex., stock replacement head, single-port, made in Brazil.

TYPE 2

P/N	Year	Engine	Remarks
113 101 371	'59—62	1200	Same heads as used on 1200 Type 1, '61—64, with rocker-stud problem.
211 101 371B	'63—65	1500	31.5mm In. X 30mm Ex., round rocker-stud pedestal w/breakage problems, angled intake-manifold mating area.
211 101 373A	'66	1500	35.5mm In. X 32mm Ex., single-port.
311 101 373A	'67—70	1500/1600	Practically identical to 211 101 373A, single port.
113 101 375A	'71	1600	Type 1 part, see Type 1 listing.

*All U.S. 914 heads are fuel-injected; 1.7 and 1.8 liter heads same as appropriate year Type 2 heads.

Type 4 Style Engine

P/N	Year	Engine	Remarks
021 101 371J/Q or 021 101 372	'72	1700	39.3mm In. X 33mm Ex., no breather pipe or CHT sensor on carbureted engine, also used on 914.
021 101 371J or 021 101 371Q	'73	1700	39.3mm In. X 33mm Ex., two EGR holes by In. ports, no pipe or sensor, carbureted engine.
021 101 371H or 021 101 371S	'74	1800	41mm In. X 34mm Ex., one EGR hole on one, two EGR holes on other, carbureted engine.
021 101 371H or 021 101 371S	'75	1800	39.3mm In. X 34mm Ex., no EGR, CHT sensor on one side, plain on other, fuel-injected head.
022 101 372G	'76—78	2000	37.5mm In. X 33mm Ex., CHT sensor on one side, other plain, sodium-filled Ex., fuel-injected head.
071 101 371A or 071 101 371B	'79—83	2000	37.5mm In. X 33mm Ex., w/different port exhaust and studs, no interchange, sodium-filled Ex.

TYPE 3

P/N	Year	Engine	Remarks
311 101 371B	'63—65	1500	Like Type 1, 40-HP w/round rocker pedestals and breakage problem, but 33mm In. X 30mm Ex., carbureted head.
311 101 373A	'66	1600	Similar to above, but w/square rocker pedestals, no breakage problems, single-port, used on Type 1—3.
311 101 375 311 101 375C	'67 & '69—70	1600	35.5mm In. X 32mm Ex., dual-port, carbureted head, no ledge, used on right side of fuel-injected engines into early '69. '69—70 versions have CHT sensor on top, used on left side.
311 101 375A	'68	1600	Same as above, but with CHT sensor below exhaust port, used on left side only.
311 101 375G	'71—73	1600	Breather pipe in rocker box, right side only.
311 101 375D	'71—73	1600	Breather pipe and CHT sensor, for left side.
311 101 375D		1600	35.5mm In. X 32mm Ex., universal replacement, both CHT sensor locations.

TYPE 4

P/N	Year	Engine	Remarks
021 101 371J, 021 101 371Q	'72—73	1700	Same p/n as Type 2 head, but has breather pipes & CHT (sensor holes.)

Note: All other Type 4 heads are identical to their Type 2 counterparts.

914*

P/N	Year	Engine	Remarks
039 101 371A		2000	41mm In. X 34mm Ex., Fuel-injected version has breather pipe & CHT sensor; carb version has two EGR holes, three intake studs, steeper sparkplug angle.

47

40-HP cylinder head evolution is represented by a 113 101 371A—D (left head in both photos) and 113 101 373 at right. Broken rocker-arm studs are common in 371 series heads. This example has a kit installed replacing long studs and round pedestals with bolts and steel thread insert. Extra cooling fins on 373 head reduce cracking.

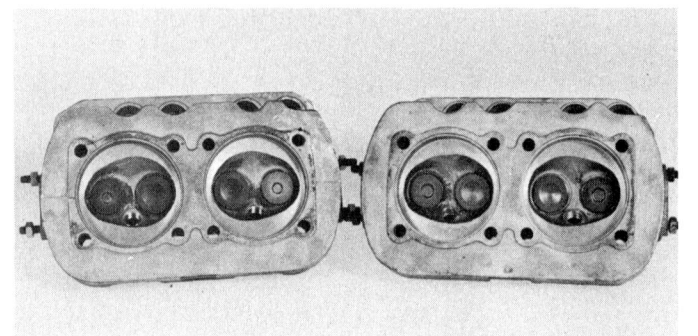

Reduced intake valve shrouding (at right) improves power on 373, too.

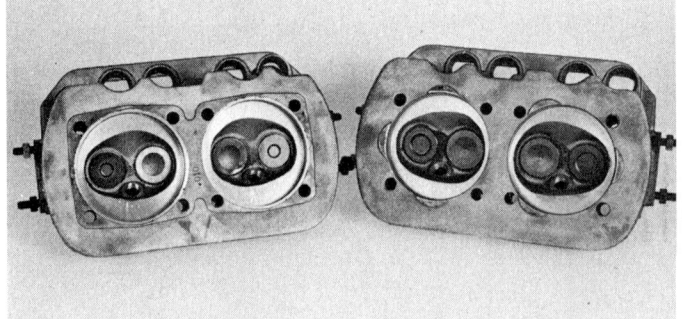

1300 head (left) and 1500 (right) differ most noticeably in combustion chamber. Flycutting has opened 373B (left) combustion chambers to 1600 size, but 33mm X 30mm valve sizes are obviously small compared to 1500's 33.5mm X 33mm dimensions.

number is located on the center of the *rocker-box floor*, the area defined by the outline of the rocker cover.

TYPE 1

36 HP—Here's a short description of this head just so you'll recognize it is when you see it. Note first how small it is. All its dimensions are much smaller than those on later engines. The rocker arm box is particularly small compared with the rest of the head. Inside the rocker box are two pedestals for rocker-arm mounting. The combustion chambers are very deep in the head and the valves are tiny. Valve-guide ID is 7mm. All 36-HP heads are single-port units.

40 HP—Two heads have been used on the 1200 engine: the 371 in A—D versions and the 373. From late '60—64 the 371A—D was used. For '65 only, the 373 was introduced. Visual differences are minimal among all versions, except for the rocker-stud supports, which I'll explain shortly. There are some cooling-fin, combustion-chamber shape and volume changes, but they are all slight.

Compared with later heads, the 40 HP's single intake port is flat against the rocker-box side. Intake valves are 31.5mm and the exhausts 30mm. This is true for both p/n ranges.

A weakness in 40-HP heads is the rocker studs. The 371A—D heads have long studs that pass through the rocker-box floor and thread into the cylinder head, right above the combustion chamber. Because the studs anchor so near the combustion chamber, they receive a lot of vibration. This vibration causes the stud to erode the rocker-box floor, creating an oil leak. It also wears the stud so it often snaps in two.

Modifications aimed at curing stud breakage resulted in the A—D versions of the '60—64 head. The 371A was used in '61 only. From '62—64 the B, C and D versions were tried, all without curing the rocker-stud problem. All of these heads have round *rocker-arm bosses*—the metal surrounding the rocker-arm stud. There are also minor changes in combustion-chamber shape and intake-port volume among these heads, but nothing substantial that affects interchange.

Various kits are available for the fixing this weakness and they do work. See page 104 for a complete explanation of these kits and their advantages and disadvantages.

Another problem with early 40-HP heads is cracking between the *sparkplug boss* and *valve seats*. These heads are quite prone to this crack, one which happens to all air-cooled VW heads.

The 373 head was released in '65. Increased surface area for better cooling was obtained by adding a cooling fin to the bottom of the head—see the photos. Material was added to the rocker-box floor and short rocker-arm studs installed. More aluminum was added around the rocker-arm studs, making them into square pedestal supports, similar to a 36-HP head. This did the trick, eliminating stud breakage and related oil leaks. Additionally, the combustion chamber was enlarged around the sparkplug and intake valve, improving breathing. The 373 is the best 40-HP head.

One idiosyncrasy of these heads is they are slightly thicker than the 371A—D versions because of the extra cooling fin. Slightly longer pushrods and cylinder-head studs (for mounting the barrels) are used on these engines to compensate.

You may hear the 351A—D heads called *long stud* and the 373s *short stud*. Long and

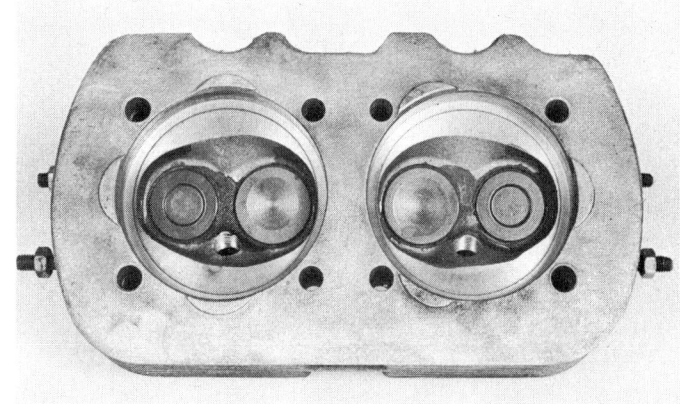

Dual-port heads started on Type 1s with 113 101 375A p/n. Dowel between ports locates intake manifold. Combustion chambers have 1.5mm ledge to reduce quench, and shouldn't be flycut out.

short refer here to the rocker-arm stud length; the short-stud style is the one to use. This is another way to tell the two basic 40-HP head designs apart.

Replacement heads are designated with a 371F p/n. These are called *universal replacement heads* because they fit any 40-HP engine. They are similar to the 373, but don't have the extra cooling fin. The rocker-box floor is not as thick, but short studs cast into pedestals are used. Cooling is supposedly not as good as the 373 version, but oil leaks and stud breakage are not a problem. This head doesn't require longer pushrods or cylinder studs when used on any 40-HP engine.

1300—Many pick the 1300 Beetle as their favorite VW. It is light, simple and has a engine with more torque. Some of the increased torque comes from the larger displacement, and some from the cylinder head. To match the increased displacement of the 1300 engine, VW gave the 113 101 373B cylinder heads 33mm intake valves and an enlarged intake port. The exhaust valves remained 30mm.

The 1300 head looks very similar to the later 1500, not the 40 HP. The 1300 head was the first to get a 20° angle on the intake port, which is a good visual landmark. Intake and exhausts valves in the 1300 are noticeably smaller than those in the 1500 and later heads, so the combustion chamber is always worth an inspection. Also beware of 1300 heads that have been flycut to 1500/1600 dimensions, which is a popular modification.

If planning to cut a 1300 cylinder head for use with 1500/1600 cylinders, be aware that this will increase compression. The 373B combustion chamber is not as large as that on later heads, so sliding on the larger cylinders means you are squeezing the same cylinder volume into a tighter chamber. Compression will be too high for street use and detonation protection will be marginal. Premium gasoline is a requirement with this combination.

1500—From '67—70, Beetles used the 311 101 373A head. This single-port casting uses 35.5mm intake and 32mm exhaust valves with enlarged intake ports. The rocker-box area is trouble free and uses reinforced short rocker-arm studs.

Over the production run the combustion chamber was slowly enlarged, but all 373A heads interchange with each other. Differences in combustion-chamber shape are noticeable if you look around the sparkplug area. This head will also bolt to any 1600 without modification, or 1300 with 1500 or 1600 cylinders. Actually, the term *1500 head* is a misnomer, as '70 was the first year of 1600 production in the Beetle. Thus, these heads were supplied stock for one year on 1600 engines.

1600 Dual Port—The most dramatic change in Beetle engines came in '71 with the 1600 dual-port engines. *Dual port* refers to a separate intake port for each cylinder. The single-runner intake manifold joins a cast-aluminum divider at the cylinder head on these engines—an instant visual clue. Dual ports give more power than the single-port design, which the 1600 really needed as emission controls strangled its output in the '70s.

Beetles started in '71 with the 113 101 375A head. Intake valves are 35.5mm and exhausts, 32mm. The revised combustion chamber has a 1.5mm ledge. This ledge runs around the circumference of the chamber, and raises the head 1.5mm away from the cylinder. This lowers compression, of course, but a smaller *bowl area* in the combustion chamber compensates for the reduction. More importantly, the ledge opens the *quench area* of the cylinder.

Quench areas are those parts of a cylinder head which are very close to the piston when it is at top dead center (TDC). This area encourages turbulence in the chamber, because the mixture in the quench area is squeezed into the more open part of the chamber when the piston nears TDC. A side effect is a pocket of rich mixture is left in the far reaches of the combustion chamber at the farthest point away from the sparkplug. This design deters detonation.

Quench areas work, but the rich pocket raises emissions, and in the early '70s engineers began removing quench areas to make engines pass smog requirements. It this head, the ledge is used to combat emissions.

The '70 head was just listed under the 1500 heading; it's the 373A and has no ledge. The '71—73 heads were the 375A with the ledge. In '74 the 043 101 375A head was used. It is exactly like the 113 101 375A version, except the upper exhaust-port studs are longer. The studs accommodate dual intake manifold-heat fittings used that year. Dual manifold heat and this head were used only in '74. Changing exhaust studs allows swapping the two 375A heads.

Don't flycut the ledge because you think it will restore the quench area and help increase power through raised compression. The compression will certainly rise but it will be way too high to run with pump gasoline.

The '71—73 heads also have a dowel to locate the intake-manifold casting, so that can help determine if the head is from these years. This head has a ledge on it where the cylinder butts against the combustion chamber. If this head has been flycut, the ledge will be gone or substantially shorter than 1.5mm high. Don't use the head if the ledge has been flycut out.

Exhaust valves with chromed stems were fitted to dual-port heads starting in '71, but these were probably replaced years ago in most engines. Most shops usually install superior chromed-stem valves as replacements.

Fuel injection was made standard in '75 on

49

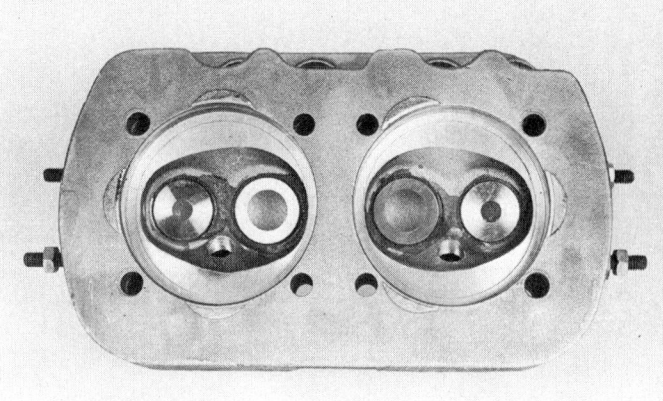

Fuel injection required modifying Type 1 head with smaller valves. A CHT sensor hole (arrow) is cast into 043 101 375H head used on driver's side. A plain version is bolted to passenger side, p/n 043 101 375R.

Long stud on exhaust port identifies this as 043 101 375A, used only on '74 Type 1s. Long stud mounts extra manifold heat fitting, and is only difference between this and 113 101 375A head.

the Beetle, and another version of the dual-port head was produced: 043 101 375R or H. The exhaust-valve stems were enlarged from the previous standard 8mm to 9mm. Also, exhaust-valve size was reduced to 30mm. This was required for correct fuel injection operation. A tapped hole for a CHT (Cylinder Head Temperature) sensor is placed near the rocker-box outer surface on the driver's-side cylinder head. Those heads with the tap are the 375H; without the sensor tap, they are numbered 375R.

1600 replacement heads use the same part numbers as the originals.

TYPE 2

40 HP—From '59—62 the Type 2 used the same 40-HP engine as the Type 1. This is the 113 101 371 series head, complete with 31.5mm intake and 30mm exhaust valves, flat intake port and rocker-stud problems.

1500—The heavier Transporter really needed the 1500 engine, used from '63—67. Three heads were used on the 1500, and the last of these overlapped onto the 1600 that followed. The first head was the 211 101 371B, used from '63—65.

In some respects, these are 40-HP heads designed to fit on 1500/1600 cylinders. They have 31.5 intake and 30mm exhaust valves and rocker-stud problems like the Type 1 40-HP parts, but don't have the flat intake port.

Next was the '66 head, 211 101 373A. It's like the earlier 1500 heads, but has the 1600 valve sizes of 35.5 intake, 32mm exhaust. Then in '67 the 311 101 373A was released. For all practical purposes, it is identical to the '66 head. The 311 101 373A head continued on in '68 with the 1600 engine. In fact, it ran through '70, until being replaced by dual-port designs.

1600 Dual Port—There was only one year for Type 2 1600 dual-port engines: '71. These engines used 113 101 375A heads, like all other Type 1—3 VWs that year. They are described in the Type 1 section.

1700/1800/2000—In '72 the Bus switched to Type 4 power, and a completely different set of heads. They are covered in the Type 4 section.

TYPE 3

As we follow the VW engine families, you see how new engines or modifications were introduced on the Type 3 or possibly the Type 2 before being applied to the Type 1. This is because the Type 1 is the smallest and lightest of the air-cooled VWs, and didn't need the extra power such improvements gave until a year or two later.

1500—The Type 3 head from '63—65 is the 311 101 371B. It is similar to the Type 1 40-HP head with the rocker-arm problem. It uses the same round rocker-arm bosses, and the studs break in the same way. However, the valve sizes are larger: 33mm intake, 30mm exhaust, like the Type 1, 1300. These heads were only used on carbureted engines.

In '66 VW applied what it learned on rocker-stud breakage with Type 1s to the Type 3 head. The result was the 311 101 373A. It is just about the same part as the 371B, but has the square rocker-stud bosses and has no stud-breakage problem.

1600—Type 3s got the 1600 in August '65, but didn't get dual ports until the '67 head: p/n 311 101 375C. This was a carbureted engine head, and has no ledge as discussed in the Type 1 dual-port section. Valve sizes are 35.5 intake; 32mm exhaust.

In '68 the Type 3 was fuel injected. The 311 101 375 continued in use, but a 375A version was added to the *left side* of the engine. The only change on the 375A head was the addition of a temperature sensor below cylinder 4's exhaust port. This sensor location is alternately called on the *side* or *end* of the head. The 375 and 375A combination was used into '69.

Starting with later '69 engines, the temperature sensor was moved on *top* of the head near cylinder 4's sparkplug. These heads are simply numbered 311 101 375, which means there are 375 heads with the sensor and those without. You've got to check a head for the sensor's tapped hole, if you need it. If the sensor isn't necessary, run the head with the hole empty. The sensor hole is just a shallow blind hole tapped for the sensor threads. Leaving it open doesn't cause any harm.

In '71—73 Type 3s, a breather pipe was added to both heads. This pipe extends out of the rocker box and has a nipple for the breather hoses. The right head has only the pipe. It's p/n 311 101 375G. Heads fitted to the left side have

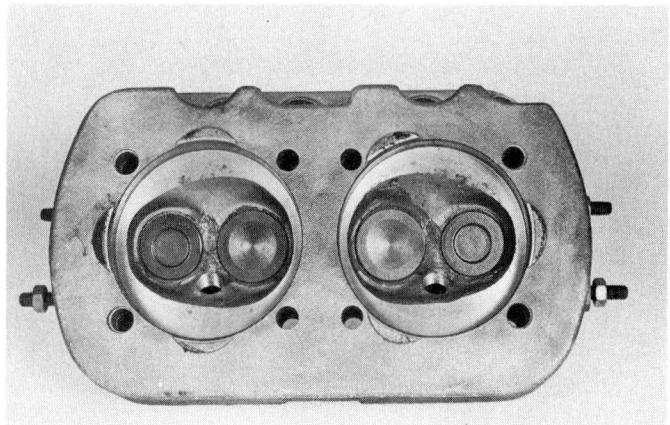

Curiously, Type 3 engines use larger valves even when fuel injected. This 311 101 375 head with 35.5mm X 32mm valves is used on carbureted engines from '67 through injected ones in '70.

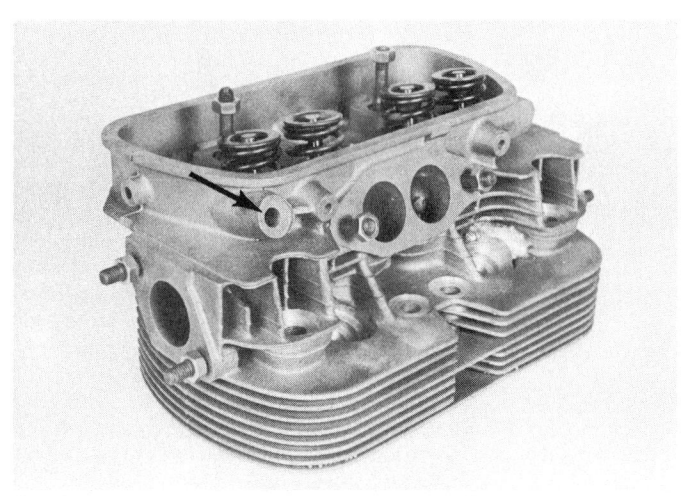

Starting in '69, some 311 101 375 heads have CHT sensor tapping in rocker box (arrow), but others don't. If your engine requires sensor hole, check for it. Replacement heads for this application have both tappings.

Another Type 3 variation is 311 101 375G head with breather pipe (arrow) jutting from right of rocker box. It's used on right side of '71—73s. If this head had pipe *and* CHT sensor, it would be a 311 101 375D and mount on left.

the pipe and temperature sensor: p/n 311 101 375D.

A universal replacement head for Type 3s is available, too. It has both side and top CHT sensor locations drilled and tapped so it can be used with early or late fuel-injection wiring. It is p/n 311 101 375D.

Aftermarket Heads—There are several aftermarket cylinder heads available for Type 1—3 VWs. The heads being built today are for racing, and aren't designed for street use. Nevertheless, in the past there were some aftermarket heads made in the U.S. and Japan strictly as stock replacements. The quality varies on these parts, from OK to not good. In fact, most of the U.S-made heads are not acceptable cores at major rebuilders, although some of the Japanese ones are. The U.S. castings are too porous, and don't wear or repair very well.

It's easy to identify aftermarket heads. They are marked with their brand name and unique p/n in the rocker-box floor just like genuine VW heads.

TYPE 4

This section includes all Type 4 heads used on the 411/412, '72—83 Type 2, and all Porsche/VW 914s with four cylinders. The Type 4 car used only the 1700 engine, but the Bus and 914 got all three versions. Cylinder heads are different for each displacement and sometimes differ inside that displacement depending on which chassis the engine was installed in.

The Type 4 cylinder head is completely different from those used on Type 1—3s. It is larger in all dimensions and has completely revised porting. The large, dual-intake ports descend straighter into the combustion chamber, so their mating flange is much closer to being parallel with the rocker box than Type 1—3 heads. The exhaust ports exit the bottom of the head, not the sides. The pushrod tube holes are much larger because the tubes are designed to pass completely through them. In short, it's no problem telling Type 4 from Type 1—3 heads.

1700—Three cylinder heads are used on '72—

Old aftermarket heads for stock applications are usually poor seconds to factory parts. Some are not even acceptable cores at volume rebuild shops. Non-VW trademarks in normal p/n area identify these fairly rare parts.

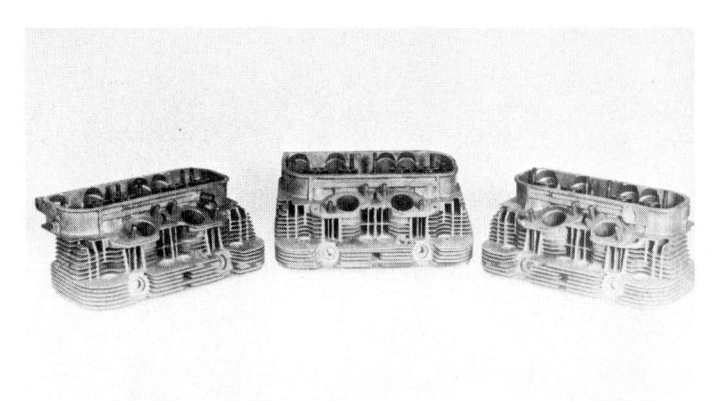

Large, widely-spaced intake ports are one quick Type 4 head recognition guide. Center head is "plain" 021 101 372; left is one with breather pipe and CHT-sensor hole, and at right a two EGR hole example. See text for more on this and its application.

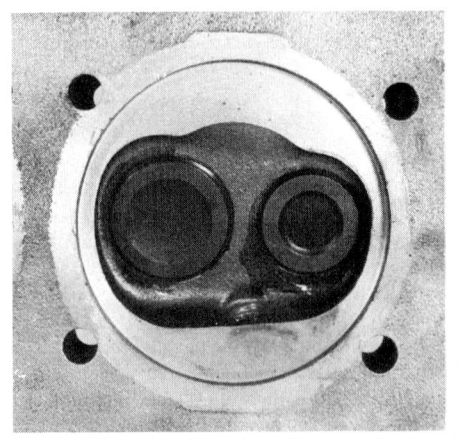

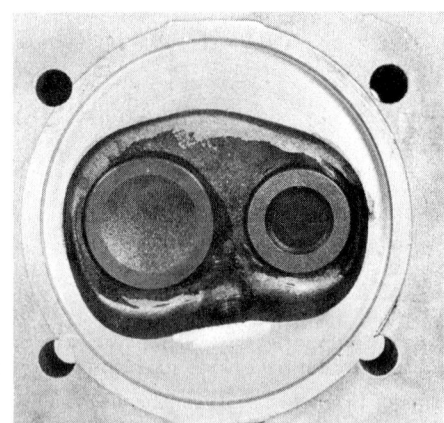

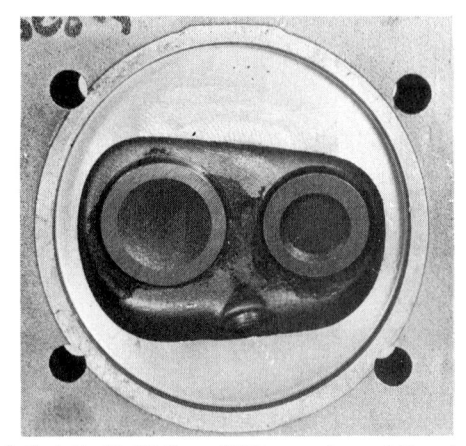

Combustion-chamber size, shape and valve sizes vary among the three Type 4 engines. From left, 1.7 with 39.3mm X 30mm valves, 1.8 and 41mm X 34mm valves and a 2.0 with its small 37.5mm X 33.4mm valves. Despite differences, heads can be interchanged as long as cylinder openings and cylinders fit.

'73 1700s used in Type 4s and Buses with dual carburetors. All these carbureted heads have 39.3mm intake and 33mm exhaust valves. In '72 the heads were numbered 021 101 372 and 021 101 371*J* or *Q*. These heads are "plain" with no breather pipe, temperature sensor or EGR (Exhaust Gas Recirculation) holes, *unless* they are 371J or Q heads on a Type 4 car. Then they have pipes and sensors. In '73 the 021 101 371J or Q head got two EGR holes, but still without pipe or sensor fittings on the Bus. The EGR holes are next to the intake ports.

1800—Carburetors were used on the '74 Bus 1800 engine. These heads are numbered 021 101 371*H* or *S*. These engines use a total of three EGR holes, two in one head and one in the other. Both these letter suffixes have been used on two- or one-EGR heads, so count EGR holes when shopping for these. Another difference is valve size. To accommodate the extra 100cc, the factory installed 41mm intake and 34mm exhaust valves in the 1800 heads.

Fuel injection replaced carburetors the following year ('75), and the 021 101 371H or S head was slightly modified to match. EGR was no longer needed, so the EGR holes were deleted. Instead, a temperature sensor was drilled and tapped on one head. The other head was left plain.

Once again, check for the sensor hole as the p/n or letter suffix doesn't make any difference. Valve size was reduced to 39.3mm intake and 34mm exhaust to work with the fuel injection. But the p/n remained the same as the '74 head, so make sure you get the right head by checking valve size, sensor hole, and EGR holes.

2000—All 2-liter engines are fuel injected. Those used in '76—78 Buses get 37.5mm intake and 33mm, sodium-filled exhaust valves. The 914 section has more on sodium-filled valves. Look for p/n 022 101 372G. It seems strange the larger engine gets smaller valves, but they were deemed necessary for the fuel-injection system.

Modified exhaust ports were incorporated into the '79—83 heads, earning them a new p/n: 071 101 371*A* or *B*. Otherwise identical to the earlier 2-liter heads, these later heads won't interchange because of the exhaust-port shape and altered stud locations. The earlier exhaust pipes won't bolt to the heads.

914—It's easy to say the 914 is the Porsche with a VW engine, but it isn't exactly true in all cases. Basically, the 1.7- and 1.8-liter 914 engines are very similar to their VW counterparts, differing only in air filter placement, sheet metal and exhaust valve material.

The cylinder head p/n is the same as the corresponding Bus engine part. The 2 liter is another story. It is called the *Porsche* engine by VW rebuilders, and does differ noticeably from the VW version.

The cylinder heads are the primary difference. Cast with p/n 039 101 371A, they have 41mm intake and 34mm exhaust valves, breather pipe and temperature sensor, three in-

52

take studs and a different plug angle than 2-liter Bus heads.

If nothing else, the three intake-manifold studs reduce this head's interchange possibilities. The intake pipes from a 914 must be used if applying these heads to a Bus, or aftermarket carburetors and manifolds will be necessary. Furthermore, there aren't many of these heads waiting around in wrecking yards because there were many fewer 914s built than Buses. To put bigger valves in your Bus engine, the best route is to have them installed in the original heads.

All 914 exhaust valves are sodium-filled from the factory. Their function is explained in Chapter 6, but for now, that makes all 914 cylinder heads different.

Many of these heads have been rebuilt and swapped since new, so you may or may not be getting sodium-filled valves. The identifying clue is a small hole in the center of the valve head. This is where the sodium is injected into the valve. If there is no hole in the valve head, it is not a sodium-filled valve. This is an important point, as these valves are expensive, but worth it. Usually a sodium valve can be reground, when no other VW exhaust valve can. This really affects the price of a valve job.

OIL PUMPS & CAMSHAFTS

Oil pumps and camshafts are like pistons and cylinders; they are not the kind of parts shopped for in wrecking yards. Because of the relationship between VW oil pumps and camshaft gears, you need to know the possible problems when mixing new parts.

VW oil pumps and camshafts are *matched parts*. When the pumps have been updated, new camshaft gears have been introduced to work with the new pump. And because air-cooled VW camshafts and gears are riveted together, changing gears means changing the camshaft, for any practical purpose. Most VW rebuilds result in a new or reground camshaft, so guarantee it will work with the oil pump being installed.

Type 1—3 Oil Pumps—As the photo shows, a lot of different length oil pumps have been used on air-cooled VWs. Changes have also been made to oil-passage size in the cases. Certain combinations of oil pump and case work, and others don't.

Case oil passages are considered *small* through '69, and measure 8mm. Starting in '70, the passages were opened to 10mm and the second oil-control valve was added. These are the *dual-relief* cases.

Oil pumps are measured by the length of their gears. The 40-HP through '67 engines had 17mm pumps. From '68—69, a 19mm pump was used. For '70 only, a 24mm pump was supplied and in August '70, ('71 model year), a 26mm pump was introduced.

The 17mm and 19mm pumps mate with the same type cam gear, known as the *flat* gear. It

Up to '78, all Type 4 heads used almost round exhaust port (left). So called *Vanagon* heads dating from '79—83 use more rectangular port (right). Port shape also changes exhaust pipes, so engine swaps should include at least heat exchangers.

Porsche 914 1.7 and 1.8 heads are exact VW parts, like one at left. But 2.0 Porsche head differs from Bus and 411/412 parts. Porsche head at right has three intake studs and different plug angle. Deep-breathing 41mm X 34mm valves don't show here. All exhaust valves used in 914s are sodium-filled.

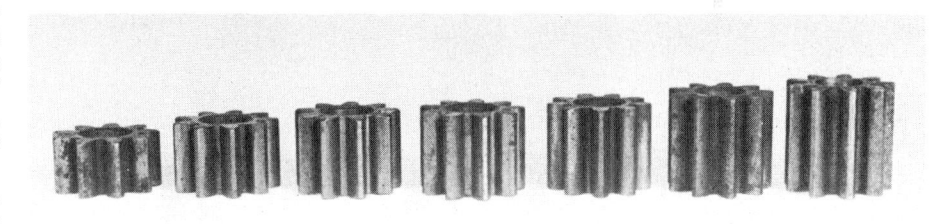

All these pump sizes have been used, but only last four since '61. Higher horsepower ratings and increased engine speeds required larger pumps.

got this name because the surface area facing the oil pump is essentially flat. Three rivets bond the gear and cam together. When the 24mm and 26mm pumps were installed, the cam gear had to be scooped out to clear the longer pump. These are called *dished* gears. The 24mm pump is held by three rivets and the 26mm by four.

Always keep oil pumps and camshaft gears matched. Later pumps and cams can be installed in earlier cases, but don't put the early pump and cam in a late case. Universal cases are dual-relief too, so they must have the 24mm or 26mm pump and matching cam gear. Pumps and gears must be matched together also. Don't install a late, four-rivet cam gear in an early case or an early three-rivet gear in a late case.

The objective is to not mismatch the hole size in the oil pump with the oil passage size in the case. The early pumps and their small outlets

There's no mistaking 36-HP lifter and integral pushrod. Best way to spot cam is its short length compared to 40 HP and later engines.

Flat or three-rivet cam gear (left) is used with 17mm and 19mm oil pumps fitted to '61—69 engines. Dished or four-rivet gear (center) mates with 24mm and 26mm pumps used from '70—79. Large gear with five rivets (right) is Type 4 part.

This is oil-pump bore of modern, Mexican-built replacement case. Note large, sleeved oil passage leading to oil-pump bore. *Never* fit a flat cam gear and oil pump combination in such a case. This hole won't be completely covered, and at low oil levels pump will suck air and not oil. Low oil levels can occur during hard cornering and braking, so this mismatch can cause engine damage even if oil level always reads properly on dipstick.

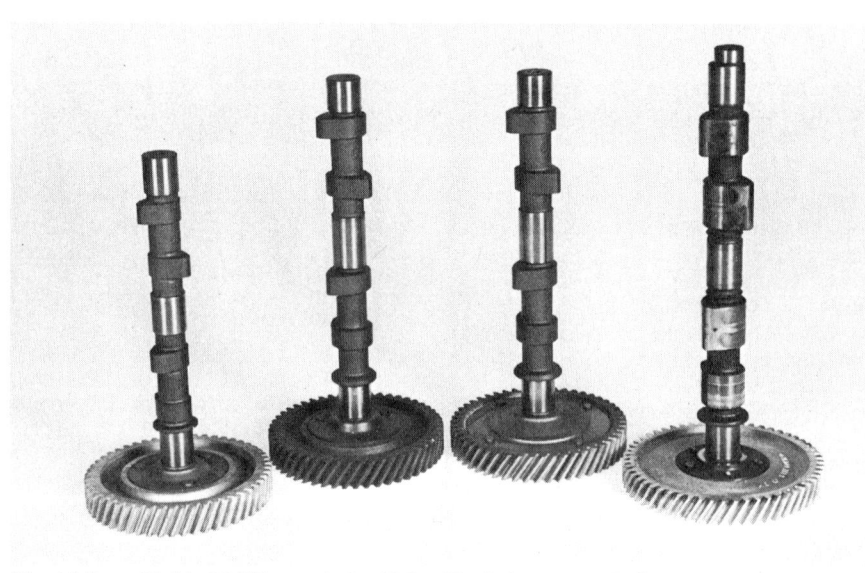

Short fellow at left is 36-HP camshaft with its 15° pitch gear teeth. Center two cams are flat and dished Type 1—3 cams. Type 4 cam is at right. All 40 HP-and-up gear teeth are 30° pitch. Also, all Type 1—3 cam profiles are identical.

will restrict oil flow in later, large-passage dual-relief cases. And the later pump, with its larger passages, will supply the maximum amount of oil possible in earlier cases.

There is no need to swap camshafts and oil pumps looking for a better grind camshaft. All Type 1—3 cams are ground alike. Type 4 cams are ground with minimal variations, but there is no discernible performance difference among cams. Nevertheless, the mechanical- and hydraulic-lifter versions are different. You won't gain any more performance from one VW cam than another.

Mechanical and hydraulic lifter camshafts can be distiguished by looking for raised rings on either side of the center-bearing journal. Mechanical lifter camshafts have one ring on either side of this journal. Hydraulic lifter camshafts have two rings or none in these positions.

Auto-Stick—Engines mated to Auto-Stick transmissions have a two-stage oil pump. The first stage supplies oil to the engine in the usual manner. The second stage is much smaller and pumps oil to the torque converter in the transmission. The stages are separated by a plate and seals. Two-stage pumps can be 24mm or 26mm, and use the regular three- or four-rivet cams. Two-stage pumps will fit in any Type 1—3 case and have large oil holes.

If thinking about using the two-stage pump in a manual-shift car for an increase in oil volume or pressure, forget it. Regular single-stage pumps supply more pressure, when carefully blueprinted, than the two-stage pump. It's a lot less work and a lot cheaper, too. See Chapter 5, page 94, for oil pump service.

High-Volume Pumps—Several aftermarket high-volume oil pumps are available for air-cooled VWs. For any sort of street driving they are not necessary or desirable. Power will be wasted driving the larger pump gears. Also, some longer pumps won't clear the fan housing on Type 3 engines. Then, a hole must be cut in the fan housing and checks made to ensure the pump doesn't hit the fan. The hole must also be

Early 40-HP lifter (right) is shorter due to thin base. Thin lifters require longer pushrods than thick-base lifter (left) used in all later engines. Thin lifters and long pushrods are history now, being culled out at rebuild time years ago. Some still remain, so note what comes out of your engine. All new and reground lifters are thick-base. One pushrod is used for all 40-HP engines, a slightly longer one for 1300/1500/1600s and a third for Type 4s with mechanical lifters. Hydraulic-lifter Type 4s have a fourth style pushrod.

Two-stage oil pump from Auto-Stick is just two pumps in common housing. Primary stage at left supplies engine, shallow right stage is for transmission. Intermediate plate locates housing halves and shaft seals.

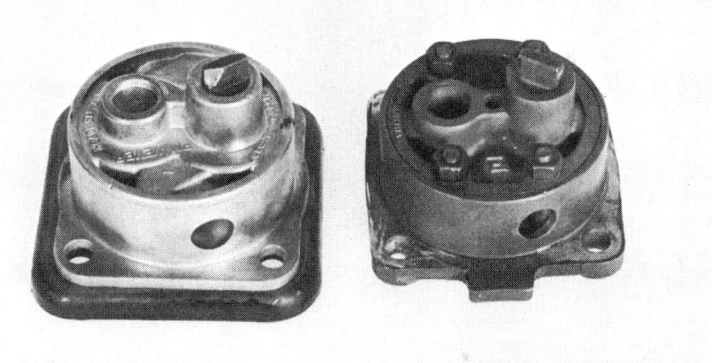

Type 1—3 (left) and Type 4 (right) oil pumps share common ancestry. Four small nuts and cast ears quickly identify Type 4 part.

sealed so cooling air doesn't escape. The work necessary to fit a larger pump isn't worth it. Put your energy into blueprinting the stock part. If you have the money and want to improve the oiling, the best modification is installing a full-flow oil filter.

Type 4 Oil Pumps—The oil pump used in Type 4 engines is completely different from any Type 1—3 pump. Only one Type 4 pump has been made by VW, and it's interchangeable with the Type 1—3 unit. The Type 4 part is more expensive than the Type 1—3 pump, and would offer no performance gain, so why spend the money? There is no two-stage pump for the Type 4 because the automatic transmissions used with Type 4 engines have their own internal pump.

All Type 4 camshafts gears are five-rivet designs. Any Type 4 camshaft, whether designed for mechanical or hydraulic lifters, will work with the Type 4 oil pump.

But don't mix a mechanical camshaft with hydraulic lifters or a hydraulic camshaft with mechanical lifters. Valve train damage can result from these combinations.

OIL COOLERS & SHEET METAL

Oil coolers differ in the size of the oil inlet and exit holes, and the way they are mounted. All Type 1—3 coolers from '61—69 have 8mm passages. In '70 the holes were enlarged to 10mm, and in '71 the cooler was relocated toward the flywheel on an adapter.

If you buy a new cooler, it will be a '70 part and will come with 10mm grommets to fit the '70 case. If you have a pre-'70 case, separately purchase two special grommets. These have tapered IDs, with one end 10mm and the other 8mm. Depending on which way you flip the grommets, they will install a 10mm cooler to an 8mm case or vice versa. From '71 the cooler looks like a Type 3 part, but there is no boss for an oil-pressure sending unit. It is used with an adapter that bolts the cooler to the case. The '71 and later coolers come with the correct gaskets when buying replacements.

In stock rebuilding, there usually isn't a lot of swapping sheet metal and oil coolers. But, if you drive in a hot climate or want to increase engine efficiency, a change to the '71 and later *doghouse* sheet metal and oil cooler on a Type 1 or pre-'72 Type 2 is beneficial.

A lot of aftermarket fan housings are available. Some are sold on their superior cooling properties, others are for applications where the oil cooler is mounted remotely. But the majority of these coolers are sold for their looks. One glance at all the polished, painted and powder-coated parts reinforces that.

The stock '71 and later fan housing with the offset oil cooler is the best cooler available, bar none. It's nicknamed the "doghouse" because it moves the oil cooler into its own little compartment. Moving the cooler forward lets cylinders 3 and 4 get their equal share of cooling air. Also, air is not heated by the oil cooler, so it is "better" (not pre-heated) air. And giving the oil cooler its own ducted cooling supply provides maximum oil cooling.

OIL PUMPS & COOLERS

PUMPS

Type 1—3

Gear Size (mm)	Year	Cam Gear
17	'60—67	Flat
19	'68—69	Flat
24	'70	Dished, 4 rivet
26	'71—79	Dished, 4 rivet

Pumps work only with specified cam gear. All camshafts have same lift and duration, so no performance gain is possible from swapping stock VW camshafts.

Type 4
All Type 4 oil pumps are identical and will work with any Type 4 camshaft.

COOLERS

Type 1—3

Year	Passage Dia. (mm)	Remarks
'61—69	8	
'70	10	Replacement part for all '61—70 Type 1—3 engines, requires special grommets to adapt to pre-'70 cases.
'71—79	10	"Doghouse" style, uses adapter, available from VW as service part.

Note: Some Type 3 service coolers supplied with 8mm passages, regardless of packaging or p/n. Always check passage diameter at parts counter. Two grommets used: 8mm before '70, 10mm for '70—73.

Type 4
All Type 4 coolers are identical, and don't interchange with Type 1—3 coolers.

Doghouse shrouds are commonly sold separately or in a kit. A good kit includes the complete shroud with all air-control flaps. An offset oil-cooler mounting flange is needed to move the cooler forward—it bolts to the stock oil-cooler mounting flange on all cases. A longer fan is used with the doghouse because it must reach forward to pump air to the cooler, so the kit should include it, too.

For normal engine operation, the stock oil cooler is fine, so don't change to another type. This is particularly true when using the doghouse fan shroud. Type 3 and 4 cooling fans, shrouds and oil coolers are also quite adequate, and no improvement is necessary. Now, running flat-out across the desert places different demands on the oil cooling methods, and requires larger, remote-mounted coolers. See HPBooks' *How To Hotrod Volkswagen Engines* for more information.

If the cooler is broken or damaged by an foreign object in the fan housing, it's best to buy a new cooler. Don't use one off a wrecking-yard engine, or any used engine unless you know that engine was well cared for with frequent oil changes. I never reuse a cooler from an engine with destroyed bearings or another disaster that caused lots of metal particles. Some mechanics do, however, and seem to have good luck doing so. They recommend an extended solvent washing at 40-psi pressure. Don't use any pressure amount higher or the cooler will rupture. Flush it in both directions through the cooler until exhausted, then clean a little more. You don't want any grit in your new engine.

CHAPTER 4
Teardown

Before disassembling the engine, have a large selection of containers on hand. Old soup and coffee cans are useful; have about 15—20 of them to separate nuts and bolts into. Get several cardboard boxes for large bulky items like sheet metal, the generator or alternator, intake manifold and so on. It may be a nuisance to gather containers right now, but separating and labeling hardware by its function as it comes off the engine pays huge dividends during reassembly. You may even have all the parts you started with.

Don't be afraid to put only two or three bolts in one can. The more you separate and distinguish hardware, the easier it is to reassemble the engine. Wrap a piece of masking tape around each can and label it.

Engine Stand—Rebuilding an engine is faster and easier when it is bolted to an engine stand. VW engine stands are available either as free-standing units or as bench-mount models. If doing even an occasional VW rebuild, a stand is a must. For a one-off rebuild, rent a stand or do without. If renting, get a stand designed for VW engines. Such a stand will have two curved arms with flat pads that bolt against the bellhousing flange. Universal stands used on in-line and V-type engines will only work with modifications.

Clutch—The first job is to remove the clutch and flywheel. This will remove a lot of weight so the engine will be easier to manipulate. It also allows the engine to be bolted to a stand.

Start with the clutch pressure plate. It is held to the flywheel by six bolts. Loosen the bolts half a turn at a time in a crisscross (diagonal) pattern. This lets the pressure plate relax evenly around its circumference. If you just remove the bolts one at a time around the plate's edge, it may bend the pressure-plate cover. After several times through the pattern, the pressure plate will relax and the bolts will be free. Unthread and remove them.

The pressure plate will now fall out of the flywheel. Catch it and the clutch disc behind it. Don't touch the friction material (surface) of the clutch disc. Grease or oil from your fingers can cause clutch grab or chatter. Even one fingerprint can cause trouble, so handle the clutch disc like a record album—by the edges.

Flywheel & Gland Nut—With the pressure plate and clutch disc removed, the flywheel is ready to come off. Type 4s have a roll pin that aligns the two parts. Type 4 engines use five bolts to hold the flywheel to the crankshaft. They are torqued 80 ft-lb; they can be removed with a breaker bar, socket, and flywheel lock.

Type 1—3 engines require more effort. Lots of it. They use a single *gland nut* to bolt the flywheel to the crankshaft. The gland nut is hollow and holds the pilot bearing, but that doesn't mean it is weak. A large diameter, 1-7/16 in. (36mm), and good steel allows for high torque. From the factory, gland nuts are tightened to 217 ft-lb, but some rebuilders use more like 300 ft-lb. Either amount is enough to defeat efforts with an adjustable wrench, ratchet or normal breaker bar.

There are two ways to remove a gland nut. The best is with a large impact wrench. When I remove a VW engine, it goes straight into the back of a truck and to a shop. There I have them take off the gland nut with an impact wrench. I also have them remove the crankshaft pulley nut while I'm there. If you have these tools, all the better.

Expect a lot of impacting before the gland nut moves. With a typical 1/2-in. air wrench and 125 psi of compressed air, it can easily take over a minute of continuous impacting. Impact in both directions to help break the gland nut loose, but favor counterclockwise, of course.

The other method is to use a long bar and a lot of muscle. Use a bar that's a minimum of 4-ft long; 6 ft is even better. You'll need as much leverage as you can muster. The bar must have a connection for a socket that fits the gland nut, or you can weld the socket to the bar. Use a 1-7/16-in. (36mm) six-point socket.

Very long torque wrenches are available through mail-order houses. Look in the popular VW magazines for ads featuring these wrenches. They are better tools to use because you'll have to torque the gland nut during installation. Obtain a flywheel lock, too. Use one of the inexpensive ones available through VW parts houses or make your own.

The flywheel lock must be secure. Use one that bolts to the case mating flange, or a 4-ft or longer bar that bolts to two pressure-plate holes. A flimsy flywheel lock is dangerous. If it gives while you are levering, you and the bar could go flying. Finally, get a helper to hold the engine down while you apply the torque. Without a helper you'll just roll the engine around on the floor.

While you've got the flywheel locked, loosen the crankshaft-pulley nut on Type 1 and pre-'72 Type 2 engines. Don't skip this step or you'll have a heck of time holding the crank stationary without the flywheel locked or removed entirely. Once the pulley nut is loose,

Most convenient way to remove flywheel gland nut is to zap it off with impact wrench. Service station can handle this job if you don't have proper tools.

57

Once flywheel is removed, pry out crankshaft oil seal. Don't let screwdriver drag across case and scratch it.

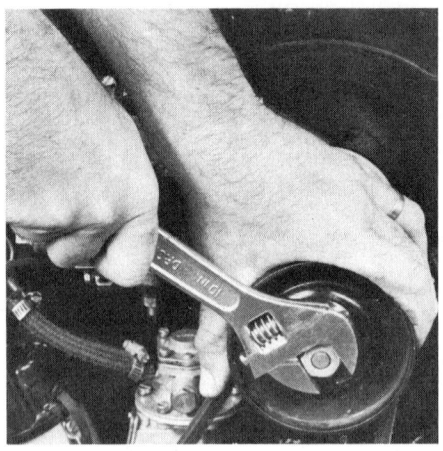

If you're lucky (and strong) you can hold generator stationary with bare hands. Otherwise, insert a screwdriver through notches in front pulley half to immobilize it.

Use extension to reach generator retaining-strap bolt (left photo). Loosen it enough to slide it forward against fan housing (right photo) and disengage strap from generator pedestal.

turn to the gland nut. Work carefully, and avoid rapid jerks, twisted positions and poor footing, all of which can lead to a serious fall when working with such high torque levels.

Once the gland nut is removed, pull the flywheel off its dowels. There may be a metal or paper gasket under the flywheel as well. Now remove the sheet-metal piece running across the engine right above the case mating flange. This piece is screwed to the side sheet metal and clips to the mating flange. On Type 4 engines, electrical wiring passes through this part. Remove the sheet metal with the wiring intact. On all engines, pry out the crankshaft oil seal that's exposed when the flywheel is removed.

The engine is now ready for mounting on a stand. VW engine stands are designed to attach to only the left case half. If using a universal stand, attach it to the left case. This allows for case separation later.

Note: If working on a 914 engine, when I refer to the front of it I mean the flywheel end.

Driveplate Removal—If the car has an automatic transmission or an Auto-Stick, there is a driveplate instead of a flywheel. It must be *match-marked* to the crankshaft before removal. Center-punch a mark on the driveplate and a corresponding mark on the crankshaft end to establish their registration.

A flywheel lock won't hold a driveplate immobile because there is no ring gear on the driveplate. The ring gear is attached to the torque converter, and that stays with the transaxle. So, install a breaker bar and socket on the crank-pulley nut, have a stout helper hold the crank stationary, and then use an impact wrench to remove the gland nut.

To remove the crank-pulley nut, put a screwdriver through one of the pulley's two slots and place it against the case so it holds the pulley immobile. Then loosen with an impact wrench or breaker bar.

Engine Accessories—When a VW engine has all its sheet metal, fan, generator, fuel pump, intake manifold, carburetor, distributor and so on installed, it's called a *complete* engine. When all accessories are removed it is called the *basic* engine. The terms *short block* and *long block*, are not descriptive or correct when applied to VW engines.

From the standpoint of a teardown, the difference between upright and flat engines is in removing the accessories. I've separated accessory removal into two sections, one for the uprights: Type 1, pre-'72 Type 2, Things and Karmann Ghias. The other is for flat engines: Type 3s, Type 4s, '72 and later Type 2s, and all Porsche 914s. Follow the accessory removal section for your engine, then continue under the "Basic Engine" section, page 66.

ACCESSORY REMOVAL—UPRIGHTS

Fan Shroud—On all upright engines, the first major component to remove is the fan shroud. Start by disconnecting the electrical leads between the coil and distributor at the coil. Remove the generator belt. Do this by removing the generator-pulley nut and pulling off the outer pulley half and spacers.

You'll have to hold the generator immobile to remove its pulley nut. Do this by inserting a screwdriver in the slot cut into the inner pulley half, and lay it against the generator-housing bolt.

Keep track of how many spacers go between the two pulley halves. They are used for spacing the pulley halves to set the generator-belt tension. After removing the belt, replace the pulley half with the same number of spacers between the halves. There will be other spacers outside of the pulley, between the nut and pulley. They are spares not needed to adjust the belt currently on the engine.

Now unclamp the generator from its pedestal. Follow the clamp around the generator until you find the clamping bolt. Loosen the clamping bolt, then slide the clamp toward the fan shroud. This disengages the clamp from the tang on the generator pedestal. Although I've specified how to remove a generator, some engines may have an alternator. The removal procedure is the same.

Underneath the right cylinders is the thermostat. It may have been removed from the engine by a previous owner, so it may not be there. It is connected to the fan shroud by rod linkage. The small rod transmits the expansion and contraction of the thermostat to the flaps inside the fan shroud; these control the amount of cooling air from the fan which reaches the cylinders.

Clean air ducts are great handles for lifting off fan housing. Watch its balance as generator will try to tip it from your hands.

Begin intake manifold removal at manifold heat connections. After loosening, completely remove bolts so they won't hang up when lifting manifold.

Box-end wrench is one way to remove single-port manifold hardware underneath manifold bend.

Leave distributor's clamp attached when removing it. Besides plug and distributor wires, only disconnections necessary are vacuum advance and clamp-to-case nut. Before removing, scribe marks in distributor body or shaft and matching mark on case. Then, just align marks at reassembly.

After locating the thermostat, note that it is mounted to the case by a small bracket. Scribe the side and bottom edge of the bracket against the case with a sharp screwdriver, pick, or scribe so you'll know exactly where the bracket fits during engine assembly. Then unscrew the thermostat from the metal rod. This will free the thermostat, leaving the bracket nutted to the case. Remove the bracket nut and pull it off its stud. Now you can lift the fan shroud off with the thermostat linkage attached to it.

Back on top of the engine, grasp the fan shroud at each end and lift it off. The hose fittings make good handles, if the fan shroud has them. Set the fan housing aside.

Distributor—Remove all the sparkplug wires from the sparkplugs. Then use a socket and extension to reach down past the distributor to the distributor hold-down nut. This nut threads onto a stud sticking up out of the case between the distributor and carburetor. It isn't the distributor-clamp nut setting sideways underneath the distributor. The clamp nut is the one loosened when adjusting the timing. Leave it alone for now; just remove the hold-down nut. Once the nut is out of the way, grasp the distributor and pull it, clamp, cap, wires and all out of the case. Set it aside.

Carburetor & Intake Manifold—The single-carburetor intake manifold attaches both at the cylinder-head intake ports and at the rear exhaust pipe. Remove the nuts and washers at the intake ports and unbolt the exhaust-heat connections. Then the carburetor and manifold combination will lift off.

Dual-port manifolds use an aluminum casting at each head joined to a tube section by rubber gaskets or bellows. To get the manifold off, remove the rubber gaskets at each end, then lift off the center tube section with the carburetor attached. Then you can go back and remove the aluminum castings at each head. Don't worry about the rubber gaskets. They are usually cracking through, and should be replaced at every rebuild.

Fuel Pump—Remove the two nuts at the fuel-pump base, then slide the pump off its studs. If the pump sticks to its mounting block, lightly tap it sideways several times. That should break the seal between pump and mounting block. Once the pump is removed, lift out the pump drive shaft and pull the mounting block out of the case. If your Beetle is fuel injected, you don't have an engine-driven fuel pump, so there is nothing to do at this step.

Exhaust—The exhaust system comes off in two parts. First the tail pipes, muffler and connecting pipes come off, then the two heat exchangers. Start by disconnecting the exhaust pipes from the cylinder heads at the muffler end of the system. Then loosen the clamps between the heat exchangers and muffler assembly. There are four of these clamps, two each per

59

Remove manifold complete with carburetor. Dual-port manifolds require a few more steps, but carb removal isn't one of them.

Remove fuel pump nuts, give pump light tap to break gasket's grip, then lift pump and extract pushrod.

Finish by lifting out plastic pushrod-guide/fuel-pump mount.

Use box-end wrench on exhaust fittings to avoid rounding off nuts. Penetrating oil helps too, but these nuts usually come loose with minimal effort.

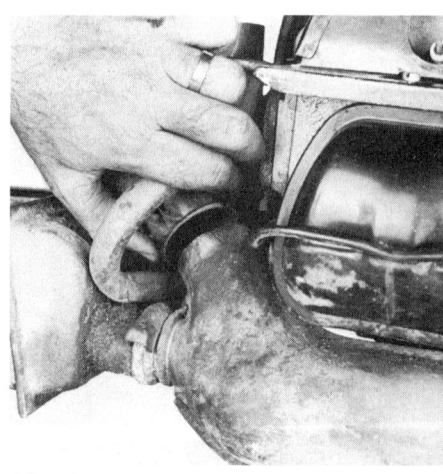

After loosening wide, flat clamp between heat exchanger segments, slide it off one end or other.

Heat exchanger-to-muffler clamp hardware is typically rusted shut. Don't worry about destroying these parts. Inexpensive replacements are available at VW parts suppliers; use them on all rebuilds.

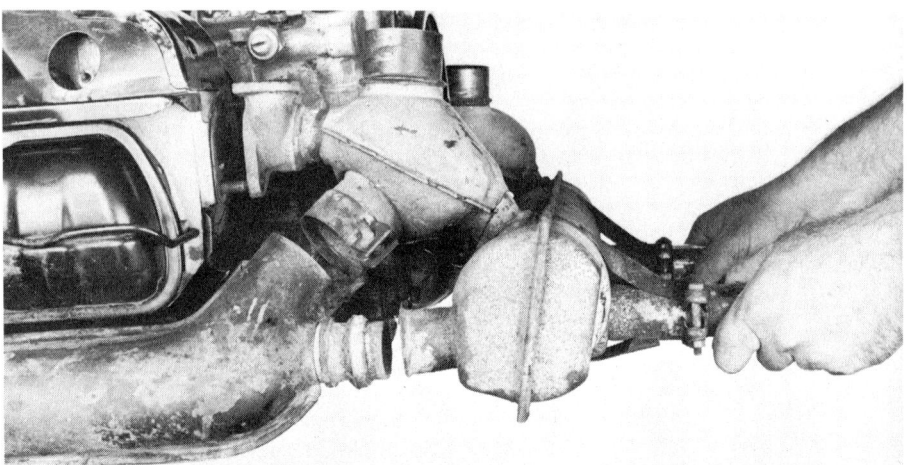

Tailpipes make handy grab points for exhaust removal. Slip joints sometimes rust shut, but persistent pulling will get the two apart.

Sheet-metal connections underneath engine must come apart before heat exchangers will pull free. Old soup or coffee cans are a must to keep small screws like these from getting lost.

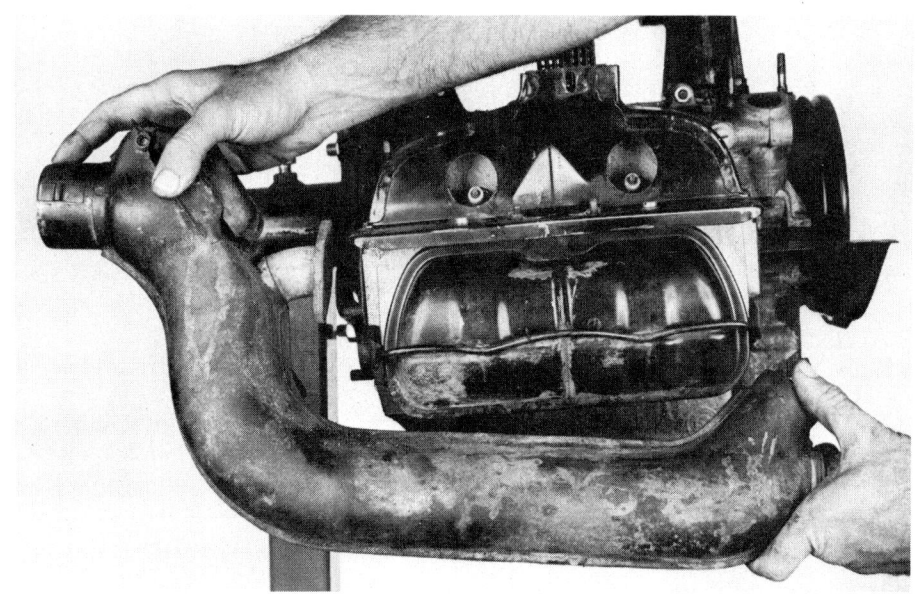

Pushing heat exchangers forward will slide them off front exhaust-port studs.

All sheet-metal screws are interchangeable, so toss all in a single can. Like this one, many screws do double-duty, securing more than one piece of shrouding.

After removing screws, sheet-metal parts simply lift off. Taking off upper cylinder air deflector uncovers a lot of engine.

side. One is a simple band clamp, sometimes called a *flat clamp*. The other is a two-piece, two-bolt clamp. Completely remove this two-bolt clamp.

Once the clamps are loosened, grasp the tailpipes and wiggle the muffler off the heat exchangers. Sometimes the slip joint between muffler and heat exchanger rusts together. If it is stuck, try twisting it as much as possible. Penetrating oil and heat from a torch will also help. Avoid using punches, screwdrivers or other sharp instruments as levers in the slip. They'll distort the mating surfaces and cause a leak at installation.

Roll the engine over and unscrew the two sheet-metal pieces between the heat exchangers and case. Then unbolt the heat exchanger at the cylinder head and remove the heat exchanger.

Sheet Metal—Get a flat-bladed screwdriver and take off the sheet-metal shrouding for cooling air. Three of the shrouds screw together in the rear corners where the cylinders meet the case by the crankshaft pulley. Remove the screws and lift off the bottom cylinder air deflector. Then remove the upper cylinder air deflector. This is the one with the sparkplug holes in it. Next is the crankshaft pulley. Simply unscrew the pulley bolt and pull off the pulley if you loosened the bolt during flywheel removal. Sometimes a light tug from a puller is necessary. Behind the pulley, unscrew and remove the curved sheet metal. Continue around the engine to get the same pieces on the other side.

Generator Pedestal—Remove the four nuts at the generator-pedestal base and lift it off. The oil filler and vent tube come off with the pedestal. Between the pedestal and base is a louvered metal plate. There is a new plate in an engine gasket set, so you can discard this one.

Oil Cooler—Three nuts hold the oil cooler in position. One is clearly visible and two are hidden underneath the cooler. The visible nut is toward the center of the case. The hidden pair are to the outside of the case, under the cooler where it overhangs the case. It takes a lot of short swings of a wrench to remove these nuts because there isn't much room for the wrench and your hand so close to the cylinders and case.

Crankshaft pulley frequently slides off by hand, but sometimes it takes a puller. Widely available harmonic-damper puller works here.

Behind pulley are two screws holding rear sheet metal. Remove screws and take away sheet metal.

Undo four nuts and generator pedestal lifts off. Underneath pedestal is louvered plate that slides off studs. Use small screwdriver to lift one corner and start plate off.

Oil cooler secures with three screws, two on bottom of case ears and one atop case. Remove all three to free cooler.

Oil-cooler grommets will stick to case or cooler. Pry them off. Save all shims found on studs or bolts. They correctly space cooler so grommets aren't crushed.

Type 1 and early Type 2 oil-pressure sending unit is mounted near top of case by distributor. Remove it now before it gets banged into destruction. If oil stains show on center section, replace sender.

After lifting off the cooler, extract the O-rings and rubber gaskets that seal it to the case. Note what style gaskets were used to seal the cooler passages to the case. During engine assembly you usually have a choice of gaskets, depending on what gasket set you get. It's useful to know what worked, or didn't, previously.

ACCESSORY REMOVAL—FLAT
Carburetors—Type 3 and 4 engines can have single or dual carburetors, or fuel injection. The 914 engine has never been offered in the U.S. with anything but fuel injection, but many 914s have been converted to dual Webers or Solexes. If you have a converted 914, you removed the carbs during engine removal.

With single-carburetor Type 3s, remove the carburetor and intake manifold as a unit. Once the fuel lines and throttle linkage are undone, all that's necessary is to unbolt the intake manifold and hot-air pipe at the cylinder heads and lift the assembly off the engine. Remove the heat riser under the one hot-air pipe-to-head connection and remove the throttle linkage arm from its case bolt mount.

Dual Carburetors—Separate the carburetors by removing the balance pipe and disconnecting the throttle linkage. The balance pipe has slip connections and is clamped down by an air-filter support. Remove the support and slip off the pipe. The linkage should be undone at the bellcrank. Use a screwdriver to pop the linkage socket off its ball. Then unbolt the intake manifolds and lift the manifold/carburetor assemblies off as a unit. Remove bellcrank, support and any linkage pieces as a unit, too.

Fuel Injection—Depending on which chassis the engine came out of, it will have different amounts of fuel injection equipment on it, so it might look different from the 914 engine in the photos. The idea is the same; remove the fuel injection in one complete step. Do this by circling the engine, labeling, and then disconnecting all fuel-injection wires, hoses and manifolds, then lifting them off together.

Start at the intake manifold-to-cylinder head connections. Remove the manifold nuts. Leave the electrical and fuel connections at the injectors alone, they'll lift off as an assembly. Note how the fuel lines connect all injectors together. This is called the *fuel ring*. Follow the fuel ring around the engine. It will take you to the *fuel-pressure regulator*, which must be dis-

62

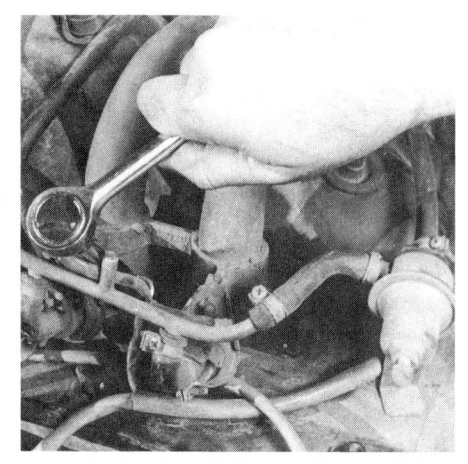

Extension and socket are only way to remove fuel-injection manifolding at cylinder heads. Fuel-pressure regulator is at right. Regulator is part of *fuel ring*, as are injectors under ratchet.

Start fuel-injection manifolding off engine by pulling at least one set of manifold runners outboard off their head studs. Runners will uncouple from intake-air distributor at center of engine.

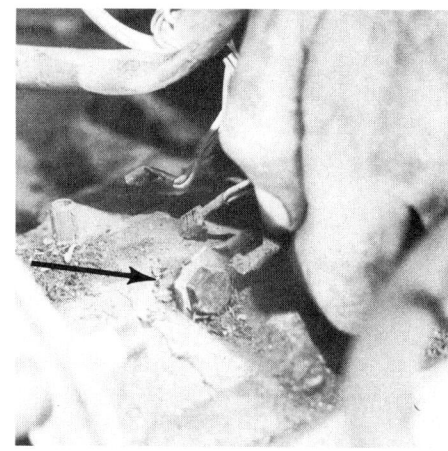

Search under intake-air distributor for multiple ground connector (arrow). Don't label these wires in such close quarters. Wait until injection system is removed as an assembly.

connected from the engine. The regulator on Type 3s is mounted above the flywheel, on Type 4s, it's usually supported by a bracket mounted to the sheet metal at the flywheel end.

Once the manifolds are loose, lift the outer ends off their studs. Because of the angle of the studs, the intake pipes will disengage from the *intake-air distributor*—the sheet-metal piece in the center.

The pipes are joined to the distributor by slip-joint hoses, so they will just slide apart. It's good the pipes come off the air distributor because there are disconnections to make under it. Look for the case bolt with the multiple electrical spade connectors attached to it. Slip off the spade connections. Clean these terminals thoroughly to remove any corrosion and maintain good ground connections. The air distributor is held to the case by a case bolt. Squeeze a wrench or socket under the air distributor to remove the bolt and free the air distributor.

Once these disconnections have been made, lift the intake-air distributor, intake pipes, injectors, hoses and wires off the engine as a unit. As always, mark all disconnections and place the hardware aside in a safe place.

Air Injection—If your Type 4 engine has air injection, remove the pump and brackets now. Loosen the adjusting and cinch bolts, push down on the pump and remove the belt. Then remove the bracket bolts and lift the pump off complete with its brackets.

Also remove the drive pulley that is connected to the engine fan. Unbolt the drive-pulley shaft from the fan, then unbolt and remove the three pulley supports. The engine fan will be loose because the drive-shaft bolts are also the fan-hub bolts.

Coil & Distributor—After removing the fuel-

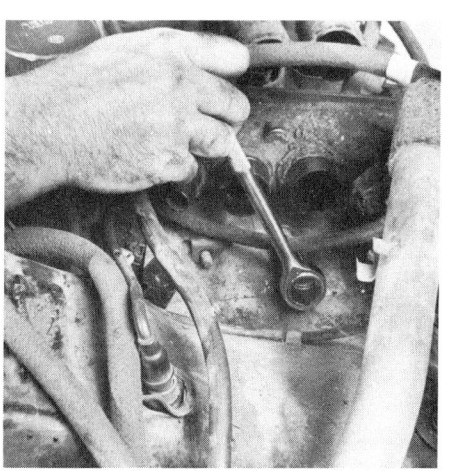

With manifold-runner slip joints pulled free of intake-air distributor, remove air distributor bolts along case parting line.

With hoses marked and dangling, lift off intake manifold and injection components as assembly.

injection system you have a lot better access to the remaining accessories on top of the engine. Begin their removal with the coil and distributor. The coil should be disconnected from the distributor, then the bracket bolt removed. Set the coil aside and remove the distributor wires at the sparkplugs. Find the nut that holds the distributor to the case. Remove this nut, then pull the distributor with cap, wires and clamp. You may have to loosen the distributor clamp and rotate the distributor to get at the hold-down nut.

Depending on the year and model of the engine, there may be some breather hoses attached to various valves or the oil breather.

Remove any hoses and valves as a unit. Remember to mark any disconnections, even if replacing the hose. Then when you install the new hose you can mark it just like the original.

Oil Breather—The oil breather and filler are removed next. On Type 4s through '79, use a screwdriver to pop off the bale wire, then lift off the breather. The 914 engine in the photos shows a breather with an oil-filler neck. Bus and 411/412 engines don't have this neck because they have a separate tube for adding oil. Late Vanagon engines ('80—83) don't have a removable breather. All that is removed on these is the breather hose, which is attached in the same area.

63

Underneath crud is distributor clamp-to-case nut. Either remove it, then pull distributor with clamp, or loosen distributor in clamp and pull it free as shown. Then go back for clamp and keep it with distributor.

Mark each connection, and then remove all breather hoses and valves.

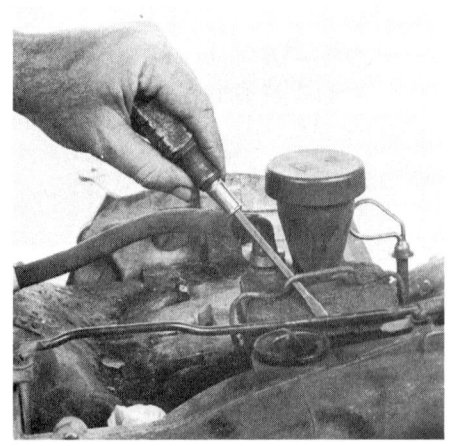

Snap over wire bale with screwdriver, then lift off oil breather box.

Removing rear sheet metal is first step in dealing with fan-end accessories.

Snap out plastic cover and loosen Allen head alternator-adjusting bolt.

Two attachment points are used for Type 3 oil fillers and breathers. The filler tube bolts near the bottom of the case, the breather bolts to the case top. The parts can be removed by unbolting them at these points. The filler can be separated from the breather for cleaning. Pry off the plug on the breather to expose a nut inside. Remove the nut to separate the filler and breather.

Type 3 Generator—Before removing the generator, the generator belt must first come off. Take the belt cover off the cooling air intake housing. Now you can get a 13/16-in. socket or wrench on the pulley nut. If the pulley turns with the nut, hold it with a 15/16-in. open-end wrench on the special washer under the pulley nut. Remove the outer pulley half, spacers and belt. Undo the generator retaining strap, then push the generator toward the flywheel and lift it off.

It's difficult to get wrenches on the Type 3 generator pulley, so here's an alternate method. Remove the retaining strap and tilt the generator so the pulley end sinks deeper into the fan housing. Now roll the belt off without disassembling the pulley. The reverse procedure can be used during engine assembly, so you don't have to take the generator pulley apart, unless the new drive belt doesn't fit properly.

Type 3 Intake Housing—With the generator out of the way, the cooling-air intake housing can be removed. This is the sheet-metal ducting the fan belt passes through. It is sometimes called the *belt cover*. Unbolt the seven fasteners around the housing perimeter, and the one inside the housing, then take off the housing. You'll find the bolt inside the housing at 10 o'clock to the fan opening. Note the condition of the housing-to-fan seal.

914 Panel—On 914 engines, unscrew and remove the sheet-metal panel that runs across the fan opening.

Fan Housing & Alternator—If interested in a quick rebuild, it's possible to remove the Type 4 fan housing intact with the alternator. I prefer to remove the alternator separately so I can clean it and the fan housing thoroughly. It's the method I explain here. With either method, there is no need to split the Type 4 fan housing into its two halves. The only time a Type 4 fan housing needs separating is for a fanatical cleaning session.

Type 3 Fan Housing—This fan housing must separate into two halves to come off the engine. Unlike the Type 4s, the Type 3 fan is larger than the opening in the fan housing, thus the need to remove the housing in pieces.

After fan belt and three bolts are removed, fan will come off its hub with minimal effort. Puller isn't needed.

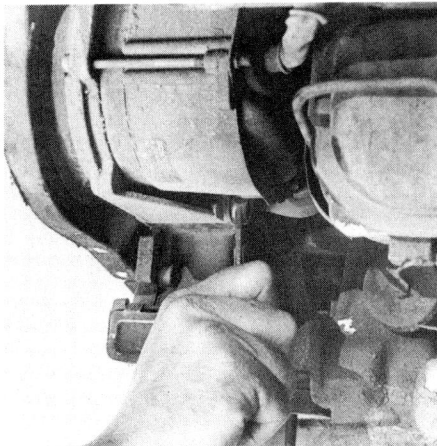

Lower alternator pivot bolt doubles as sheet-metal attachment (at left). Remove nut from bracket end of bolt; then extract bolt and remove sheet metal cover (at right).

Feed alternator wiring through sheet metal. Push grommet out of hole first.

Cooling air from fan keeps alternator from cooking so close to hot cylinder head. Remove cooling-duct clamp at alternator end only.

Finally, remove upper adjusting bolt so alternator will come free.

Use a screwdriver to pry off the domed plastic cap over the crankshaft-pulley bolt. Then use a socket and extension to remove the pulley bolt. Hold the crankshaft at the flywheel end to keep the crank from turning or use an air-powered impact wrench—this bolt is torqued to 94—108 ft-lb (13—15 mkg). With the bolt removed, pull the pulley off the engine.

The fastest and easiest way of removing this bolt is with an air wrench. If you don't have one, and didn't loosen this bolt before flywheel removal, temporarily mount the flywheel and lock it with a long bar or flywheel lock. Then remove the crankshaft-pulley bolt.

Now the seven fan-housing bolts can be removed and the rear fan-housing half taken off. This exposes the large fan. Use a puller to pop it off the crankshaft. You can cheat a little if you don't have a puller. Look at the fan's center and locate the two threaded 8mm holes. Rotate the fan until the holes are straight up and down: at 12 and 6 o'clock. Get two long 8mm bolts and thread them into the holes. Continue threading the bolts until they meet the engine case behind the fan and push the fan off the crankshaft. Don't try this trick unless the bolt holes are oriented as described. Otherwise, damage to the fan housing will result.

Under the fan are four nuts that hold the front fan-housing half to the case. Remove these nuts, and unhook the spring and air-control-flap linkage at the right flap. Lift the fan housing off the engine and set it aside.

Sometimes it's tempting to knock the fan housing with a hammer or apply extra force to it when it hangs up—don't. The housing is a thin aluminum casting and easily damaged or distorted by rough handling. Take it easy now and avoid fan housing-to-fan clearance problems later.

Type 4 Fan Housing—Pull or pry the plastic cover off the access hole for the alternator-adjusting bolt. Loosen the alternator-adjusting bolt, then below the alternator, loosen the pivot bolt. Swing the alternator toward the fan and remove the alternator belt. Unbolt the fan, pull it off its hub and set it aside.

Return to the alternator and feed any alternator wiring through the grommet hole in the side sheet-metal panel. Roll the engine on its side so you can see the alternator-cooling duct that runs from the fan housing to the alternator. Unclamp the duct at the alternator. Remove the pivot and adjusting bolts and take out the alternator.

Disconnect the air-flap-control cable from the air-flap linkage. Unscrew the upper cylinder shrouding and remove it. Here's your first

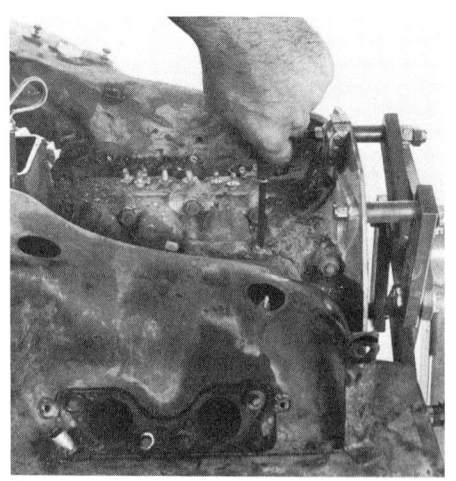

Upper cylinder air shrouds attach to case with small screws. Remove screws and lift off sheet metal.

You never know what you'll find under VW cylinder covers. Parking under tree filled this 914 engine with leaves, limited cooling.

Type 4 fan housing comes off in one piece. Remove four center nuts.

view of the upper side of the cylinders and cylinder head. If you've been parking your 914 under trees for a long time, you may find a bunch of dried leaves packed around the cylinders like I did. I use a car cover now, which seems like a strange way to make sure your engine cools.

The fan housing is held to the case with four bolts in the fan recess and a fifth on the oil-cooler end. This fifth bolt runs parallel to the cylinders. Remove the five bolts and lift the fan housing off the engine. Don't split the fan housing unless you really think you must polish the insides. A solvent wash and check for obstructions are all that are necessary.

Other Accessories—There are still some accessories left before you get to the basic engine. Next, remove the oil cooler. Undo the three fasteners on the flat Type 3 cooler and lift the cooler off the case. There's no need to remove the oil-pressure sending unit from Type 3 coolers. Depending on the Type 3 engine year, there may be a vacuum switch mounted on the case parting line, near the distributor. With dual carburetors, there is the throttle-linkage bellcrank to remove, and possibly a balance pipe that connects the carbs.

On Type 4s, start with the sheet-metal piece on top of the cooler. It just lifts off. The cooler is attached with three bolts at the inboard end. Also unthread the oil-pressure sending unit next to the oil cooler. Remove the crossmember and engine mounts. Then take off the oil filter and oil-filter mounting pad. The filter merely unscrews, the mounting pad is nutted to studs extending from the case.

Thermostat—Roll the engine over for access to the thermostat mounted under the left cylinders. It's the accordion-like tube with the wire or rod coming out of it. The first step in removing the thermostat is scribing the *exact* location

Single bolt at oil cooler end of fan housing must come out also. Then housing can be pulled off engine.

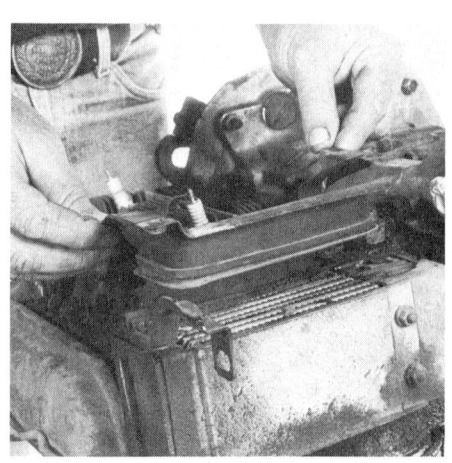

of the thermostat bracket on the case. Scribe along one vertical and the bottom edge. This will speed installation.

On Type 3s, unthread the thermostat from the rod connecting it to the linkage above the cylinders. This will free the thermostat from its bracket, so you can set it aside. Then unbolt and remove the bracket, as well as the bellcrank and linkage from the left cylinder head.

Type 4s are a little easier. After scribing the bracket position, unbolt the thermostat and bracket as a unit from the case. You'll need to unthread the cable from the sheet metal on some engines. Now, unbolt the cable pulley at the fan end of the case.

BASIC ENGINE

Teardown procedures are very similar for all

Type 4 sheet-metal air ducting behind oil cooler merely lifts out.

engines once the basic engine stage is reached, so I've grouped all engines together from this point out.

Air Ducts—While the heads are still bolted to the engine, remove the sparkplugs. Then look under the cylinders of Type 1—3 engines for a sheet-metal shroud. VW calls these large sheet-metal pieces with the spooned-up end *air ducts*. They attach to the case at two points and one end of the cylinder head. Unthread the small screws and remove both ducts.

Valve Train—Pop off the valve cover bales with a screwdriver. Completely remove the bales from the heads by prying their ends out of the cylinder heads. Then use the screwdriver to free the valve cover from the head.

The rocker arms and shafts are under the valve cover. Type 1—3 engines use a single

To remove Type 4 oil cooler, take off three nuts at crankcase end.

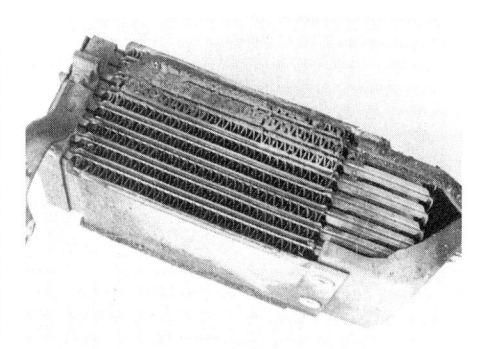

Dirt and oil in fan housing accumulated on cooler until about one-fourth of its capacity was lost or reduced.

Type 4 oil-pressure sending unit is no more invulnerable to dropped parts and swinging wrenches than others. Remove sender before it gets smashed.

Engine mounts come off now.

Oil filter mounting pad comes off case after removing two nuts. Lots of dark and wet dirt shows filter or cooler was leaking, but not bad enough to wash area clean.

Scribe thermostat's position to case before removing it. Front nut comes off with wrench, but rear one will require socket and extension.

rocker shaft per cylinder head. Type 4 engines have two separate shafts, one for each cylinder. Remove the rocker-shaft nuts and pull the rocker-shaft assemblies off their studs. On Type 4 engines there is a metal bent wire between the rocker assembly and the bottom of the cylinder head. Pull this wire out *before* removing the rocker-arm assembly.

Slip the pushrods out of the cylinder heads and set them in the valve covers. If reusing the original lifters, punch the pushrods through a piece of cardboard. Number the cardboard to keep the pushrods in order. Otherwise, it doesn't matter if the pushrods are kept in order.

On Type 4s, grasp the pushrod tubes with your fingers where they are exposed between the case and cylinder head. Twist and pull the tubes away from the case until they pop free.

Thermostat cable pulley comes off too. Remove center bolt and pulley comes off, but watch for bushing between pulley and bolt.

Air ducts under Type 1 engines secure to case with two screws. Remove screws and ducts to expose bottom of cylinders.

Pop rocker cover bales out of holes with screwdriver. Avoid levering screwdriver against rocker-cover gasket sealing surface. Gouges there can cause oil leaks.

Unscrewing two nuts will free rocker-arm assemblies, then slide them off their studs.

Push rods are next. Just slide them out of head. A slight head-low engine angle helps fish these slippery parts out, but beware of running oil.

After extracting wire retainer in rocker arm area, twist Type 4 pushrod tubes loose. Use no tools or thin tubes will crush.

Once loose, Type 4 pushrod tubes will slide through head.

Extract the tubes through the cylinder head and set them aside.

Use only fingers on the pushrod tubes, not pliers. You'll crush the thin tubes with pliers. Use a screwdriver to remove the sheet-metal piece screwed to the bottom of the cylinder head and case sides. Removing the pushrod tubes first makes it easier to remove this piece.

Lifters in Type 4 engines can be removed now if you wish. Usually an oil varnish buildup keeps the lifters from falling out of their bores after the pushrods are removed. This same varnish can also make removing them troublesome. If reusing the lifters, keep them in order, and the time spent extracting them from the case now is worth it.

If you leave them in, there's the chance they may haphazardly fall out when the engine is tipped or knocked around when the cases are being separated. If you're replacing the lifters, this doesn't matter. But this does make it difficult, if not impossible, to keep the lifters in order if you plan on reusing them.

To remove Type 4 lifters, use a small hooked instrument to get a toe-hold on the lifter's ID where the pushrod seats. Hydraulic lifters are easy because they have a wire clip in the lifter center. Just grab onto the wire and pull out the lifter. A very strong magnet will get them started if a hook isn't available. Rotate the engine on the stand so gravity is helping with the extraction. An egg carton makes a great container because it keeps the lifters separated and the price is right.

Cylinder Heads—Remove all the cylinder head nuts and washers. A magnet is helpful in freeing the washers from their recesses. Then grab the cylinder head with both hands and pull it straight off the cylinders and studs. The cylinders may begin to come off the engine with the head. If so, a light pry with a screwdriver should separate them.

Type 1—3 pushrod tubes will either pull off with the head or stick to the case. It doesn't matter. Once the head is off the engine, pull the pushrod tubes out of the head or case by hand and set them aside. Also look under the cylinders for a metal shroud. It isn't screwed onto the engine, but a screwdriver helps when popping it out of the cooling fins and cylinder-head studs.

Pull the cylinders off by hand. It might take a slight tug to break the *stiction* between cylinders and rings, but once the cylinders are moving, they will slide off easily.

Use retaining-ring pliers to extract the circlips retaining the piston pins. This job is most easily done with retaining-ring pliers that are bent 90°. The bend allows the pliers to fit between the cylinder-head studs. With the circlips removed, push the pin out of the piston until the pin clears the connecting rod. Then take the piston off the rod.

Rotate the crankshaft to expose each piston's circlips. Watch that you don't snag a piston skirt on the side of the case while rotating the crank. This will lock the crankshaft, at least, and may break the piston skirt. Have a helper support the pistons during crankshaft rotation.

Oil Pump—At the lower crankshaft-pulley end of the case is the oil pump. Only its cover and four attaching nuts are visible when it is

68

Like sparkplugs, CHT sensors are best removed before heads are taken off engine.

After removing the eight nuts and washers, heads will pull off studs. Four of nuts are in rocker box; other four are along top of head.

On Type 4s, remove lower cylinder air baffle. Two screws attach it to case, and one screw holds it to head.

On Type 1—3 engines, pull off pushrod tubes that didn't come off with heads.

Type 1—3s have this small but important baffle underneath cylinders. Pry it loose with screwdriver.

installed, so you may have overlooked it up to now. Remove the four nuts and pull the pump out of the case. Because the pump fits tightly in the case, this is easier said than done, but be patient. A series of side-to-side pulls will eventually wiggle the pump out of the case.

On Type 1—3 engines, remove the four nuts to take the pump cover off. This exposes the pump gears, which you should pull out. Now pull the pump body out of the case. Special pullers are made for this job and are handy to have around. But if you don't have a puller, start the pump moving by alternately prying each side of the pump body away from the case. Don't pry too hard or too long on one side or you'll cock the pump in the case. Just get it to wiggle. As soon as you can, use two screwdrivers for prying, one on each side. Beware of nicking the oil pump body with the screwdrivers. Gouging up the body will lead to oil leaks later, so take it easy. Keep wiggling the pump until you can do the job by hand. Be patient and remember the straighter the pull, the more effective it is.

An alternative method is to simply remove the pump cover and leave the pump in the case. Then when you split the case, the pump will come loose immediately and can be lifted out.

Engines mated to Auto-Stick transmissions use a two-stage oil pump. It is essentially two oil pumps sharing a common housing. The first pump is the normal engine oil pump. The second is a smaller one used to pump fluid into the torque converter. Removal is the same as the single-stage pump, except there is an extra set of gears and a plate to remove before pulling the pump body.

Type 4 pumps differ as they have two ears cast into the pump body. Use two screwdrivers to lever out the pump. It will come out as a unit, without any of the pump internals exposed.

Oil Screen—An easy job is removing the oil screen. The screen is in the bottom of the engine behind the round cover. All you need to do is remove the six nuts around the edge of the Type 1—3 cover, or the single nut in the center of the Type 4 cover. With the fasteners removed, pull the screen out of the case. It will be oily, so be ready with a drain pan.

Fan Hub—On Type 4 engines the fan hub should be removed now. Remove the center bolt and hook up a puller. Sometimes it will come off the tapered end of the crankshaft by hand, but most of the time a puller is needed. A common three-point harmonic damper puller like the one in the photos works well. Not much of a pull is needed, but it must be installed straight.

Splitting Cases—Now comes the exciting part of a VW engine teardown—*splitting the case*. Before getting into the job, a few words about the case's design should be kept in mind. Although you may think of VWs as cheap transportation, plenty of engineering skill went into the design and manufacture of the engine. The case—magnesium or aluminum—is an excellent example of the precision fit common to VWs.

The fit I'm referring to is between the two

69

With baffle gone, cylinders will slip off with light pull.

VW pistons are retained with circlips. Use needle-nose pliers to remove one clip from each piston.

Now piston pins can be slid out, away from remaining circlip. A pin sometimes needs a little help to start moving. This doesn't necessarily mean pin is too tight. Varnish and carbon are usual reasons.

case halves. There is no gasket at this junction, just a fine fit from precision machining and a thin layer of gasket cement. If you insert any object between the case halves, *YOU WILL DAMAGE THE PRECISION FIT AND OIL SEAL OF THE CASE.* It is very tempting to pry the case apart with a screwdriver, but don't do it. If you do, the chances are great you'll create future oil leaks.

Well, if prying is out, how do you separate them? It sounds ridiculous, but start by removing all *case parting-line fasteners.* Type 1—3 cases have 13; of these, three are bolts. The longest bolt is at the case's top; the others fit on the bottom. The remaining 10 fasteners are studs and nuts.

Type 4 cases have 20 parting-line fasteners. Don't overlook three particular fasteners. Take out the one at the bottom-front, in the bellhousing area. Extract another centered in the deep recess near the lifter bores in the left case half. And remove the fastener below the oil cooler, toward the fan, in the left case half. See page 125 for more info on case hardware.

I've seen cases stuck so full of screwdrivers, punches and chisels they looked like porcupines, but they still wouldn't come apart. Only after that last, overlooked nut hidden in the primordial goop on the engine was removed did the case halves fall apart.

Those fasteners that hold the case together fall into two broad categories. First there is the smaller hardware around the perimeter of the case. These clamp the edges of the case together, sealing in the oil. They run along the top of the case, and are on two sides of all openings, like the crank-pulley hole, oil pump bore, and oil screen opening. Follow the seam in the case and remove all nuts and bolts. You'll find bolts facing in both directions, plus studs

After removing its cover, oil pump will pull out of case. It's difficult to get started, but careful, concentrated wiggling and prying will do job. Pry only where pump overhangs case—*never* between sealing surfaces.

and nuts.

Once all parting-line fasteners are removed, move on to the second group of fasteners. These are the larger, heavier bolts and studs that clamp the main bearings together. Besides helping seal the case oil tight, these fasteners clamp the case tightly against crankshaft whip,

VW cast two ears onto Type 4 pumps, after recognizing oil pump removal wasn't easy. Use them to lightly pry pump out with alternating pulls. Prying too hard on one side and then other will only cock pump in its bore. Tap it back in and start over.

combustion pressure and other substantial forces.

These larger fasteners are near the cylinders. On Type 1—3 engines there are six studs anchored in the left case (looking from the pulley end) and nutted on the outside of the right case. On Type 4s, six bolts pass through both cases and are nutted on the right case side. Once the nuts are off a Type 4's case bolts, don't try to remove the bolts from the case. Plastic cylinders are wrapped around these bolts between the case halves where you can't reach them. If you try to remove the bolts, they'll hang up in the case, and tightly jam the plastic sleeves. A

70

Type 4 fan hub's tapered attachment always needs a puller to break loose. When aligned straightly, though, not much pull is needed.

Type 1—3 oil screen has six acorn nuts retaining it. Remove them and oil screen will come out. Lots of trapped oil usually does too, so watch out.

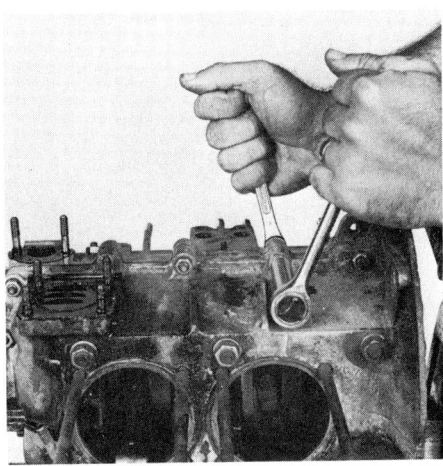

Work along case parting seam removing all hardware. Fasteners are a mix of studs and bolts.

Type 4 cases have two small nuts at top of oil breather area.

Overlooked hardware on Type 4 cases includes this bolt underneath oil breather area, and another bolt on bellhousing flange. Look on area covered by flywheel.

lengthy hammer, punch and bluestreaking session soon follows, which still won't get the cases apart. Just remove the nuts.

On all engines, remove the two fasteners in the lower, forward corner of the right case half. You may have spotted these while following the seam around the case, but chances are you overlooked them. Extract them before trying to separate the case.

To get the cases started apart, use two soft hammers—rawhide, lead, copper or brass mallets work fine. Find two case ears (overhanging flanges) on the crankshaft-pully end of the case where you can strike with the hammers. Look for a spot where one case half overhangs its partner.

Strike the two spots simultaneously in opposite directions. Using two hammers doesn't allow any wasted motion, like knocking the case around on the stand, or knocking the stand across the floor. Instead, all force is transmitted into the case halves. Therefore, you don't have to hit the case hard at all. A steady sequence of very light taps will get the case halves moving.

The case halves are dowel-pinned together. So, break the seal between the halves by pulling them straight apart.

Monitor the case seam to gage your progress. Once the seam is open 1/8 in. or more, stop hammering. Wrap your hands around the right-hand cylinder head studs and pull the right case half off the left. If the left case half is not securely bolted to the engine stand, there will be a lot of wasted motion. Recheck for fasteners or get a helper to hold the left half steady while you pull and wiggle on the right. If you are working on a bench, an assistant is a must.

If you're not making any progress—*STOP*. Look at the case for remaining nuts or bolts. Have your helper check for errant fasteners too. While splitting the case is not always easy, a total lack of movement means a bolt or nut is still installed.

Crankcase Internals—After the case is split, the task is easy—just lift out the parts. First out is the *crankshaft assembly*. Lift it by two connecting rods straight up out of the case. Next, lift the *camshaft* out. Don't bang the lobes on anything. Under the camshaft the *lifters* can be pulled free. If reusing the cam and lifters, keep the lifters in order so they can go back into the same holes. Old egg cartons are good for this job, as long as you mark which end is front.

Also remove the *camshaft bearings, center main bearing and main-bearing dowels*. Use a small container for storing the dowels, or they'll get lost. A container with a tight cover is best. I like the plastic cans 35mm film comes in. Slide the O-rings off Type 1—3 main bearing studs and discard them. Slide the *distributor driveshaft* out, being careful not to lose the two washers (Type 1—3) or washer (Type 4) under it. Lift out the camshaft plug.

The removal procedures are the same for

71

Two overlooked nuts on all cases are located at right, front-lower corner. Socket is on upper of the two in this photo. Six large, Type 1—3 main-bearing nuts also show here.

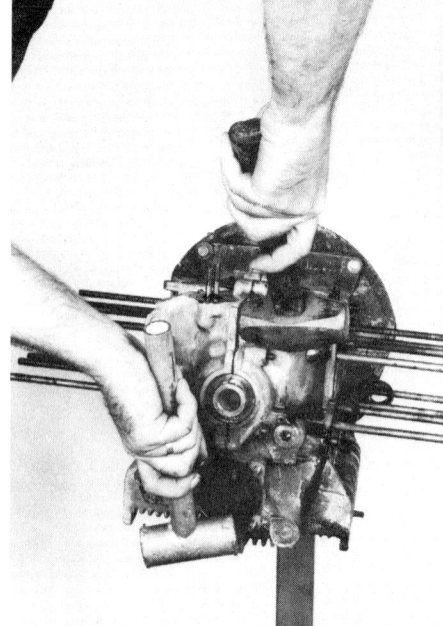

Starting cases apart is best done with two soft hammers. Case should begin opening quickly. If not, look for a forgotten fastener. Tap very lightly, and only enough to barely open parting line. Overhanging flange and oil-pump stud are used here.

Tugging on cylinder and lower bellhousing studs should be plenty to pull case apart. Slop in engine stand mounting will reduce efficiency and require extra effort.

After case is open, pull crankshaft assembly straight up using connecting rods.

Camshaft lifts straight out also. Don't bang lobes or gear against anything.

Type 4 engines except the *oil pump pickup* must be unbolted from the windage tray. Do this first so you can get the windage tray and pump pickup out of the way. The pump pickup will pull out of the case after unbolting from the windage tray because it is retained only by an O-ring at the case. Also, the lifters can be easily pushed into the case if they don't want to slide out of it. Again, keep them in order if planning to reuse the cam and lifters.

Don't forget to remove and store the lifters, bearings and dowels in the right case half.

Crankshaft Teardown—Put the crankshaft assembly on the bench with the rods out to the sides. Before removing the connecting rods, take feeler gages and measure *connecting-rod side clearance*. Do this by finding the feeler gage that just fits between the connecting rod and the crankshaft when you have pushed the rod all the way against the opposite end of its journal. This is the side clearance of that rod. Measure and record the side clearance of all four rods. Side clearance should be 0.004—0.016 in. (0.10—0.40mm), with a 0.027 in. (0.70mm) *wear limit*.

The wear limit is the point where the part can no longer be safely used. The first set of numbers is the desired operating range, and do whatever is necessary to keep the parts within that given range. Wear limits are the absolute end of the useful life for the part. So, building an engine with 0.025-in. side clearance means you'll soon have to tear it down to replace the rods because the side clearance will only get larger and exceed the wear limit.

Once the rod side clearance is checked and the results written down, use a socket and extension to remove the *rod nuts*. Remove the nuts from one rod, then pull off its *cap*. Remove the rod from the crank, then rejoin rod and cap and run the nuts on several threads. Do this to all four rods.

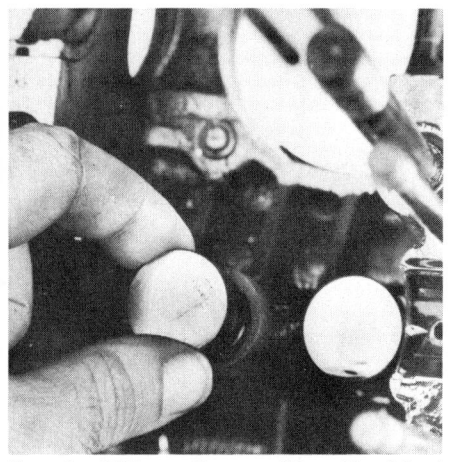

Type 1—3 lifters will now pull free. Varnished Type 4 lifters push out with little effort at this point as well.

Don't overlook cam plug. It pulls right out of groove.

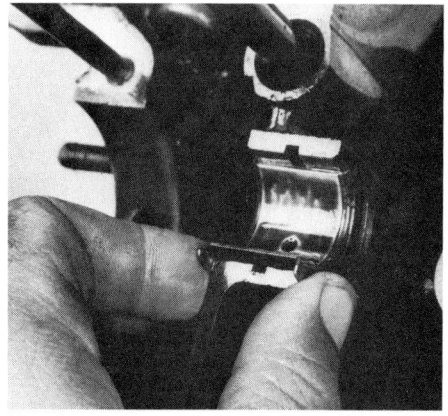

Bearings pull out of case by hand. Don't use prying tools or bearing bores will scratch. This cam bearing controls end play with its overhanging flanges.

Main-bearing dowels can be pulled out with small fingers, or needle-nose pliers. Store these tiny parts in a sealed container.

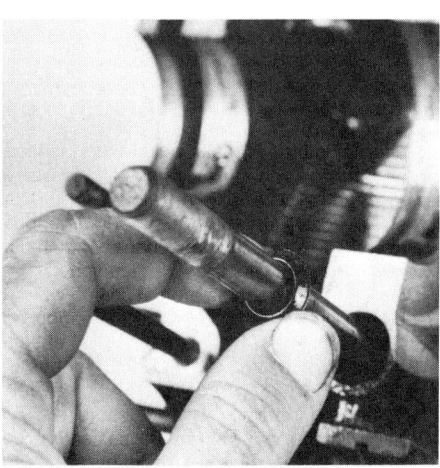

Type 1—3 main-bearing studs have rubber O-rings that should be removed now. They hide in recess at bottom of stud and are easy to overlook.

Push out distributor driveshaft and its washers. Type 1—3 engines have two washers, Type 4 only one.

It takes a fair amount of wiggling and a tight grip to start the rod and cap apart, so don't expect them to fall apart. It may seem like you need the grip of a gorilla, but don't use prying instruments between the rod and cap or you'll damage the precision mating area. To loosen a stubborn cap, lightly tap it with a soft mallet (rubber, rawhide), then use your hands to lift off the cap.

If there are *end-play shims* on the flywheel flange, remove them to a safe spot. These shims should be kept in a small box where they won't get lost or crushed by heavier parts. Next, remove the *front main bearing*, the one right behind the flywheel shims. The *middle main bearing* is a split-shell type, so you can remove its two halves now, too. At the pully end, remove the *oil slinger* and *Woodruff key*.

The oil slinger is the sheet-metal saucer and the Woodruff key is the halfmoon piece that locates the pulley. In front of these parts you'll find the small #4 main bearing. Remove it. Unless you have a press, leave the distributor gear, spacer, camshaft gear and remaining #3 main bearing alone. The gears are a press fit, so take the crankshaft to a machine shop to get these parts removed. This and other machine shop jobs are covered in the next chapter.

73

Bolt at 9:30 (left photo, arrow) must be removed from Type 4 oil pickup, so windage tray can lift out (right photo).

Check rod side clearance (0.004—0.016 in.) before removing connecting rods. Wear limit is 0.027 in. Excessive clearance means new rods, crankshaft or both.

Removing #4 main bearing is easy. It's small one at pulley-end of engine and simply slides off. Timing gear and #3 main bearing removal require a press, and are covered in Chapter 5.

CHAPTER 5

Crankcase & Cylinder Reconditioning

Volume rebuilders handling lot sizes like these have impressive knowledge learned by experience. It's often faster and cheaper to exchange major assemblies or buy kits.

Up to now, most of the rebuild work has been simply following procedures. But beginning now, you start making decisions about reconditioning parts. In this chapter I'll show how to recondition the crankcase and cylinders, which are the foundation on which every engine is built. The fundamental points to remember are simple. Patience and attention to detail will pay off; rushing the work and skipping steps will cause problems. They will then be difficult and expensive to fix when the engine is back in service.

Although you may not have the knowledge and tools to perform all procedures discussed in this chapter, carefully read each description anyway. Knowing what work is required will help you select a competent machinist. As you can see from the length of this and the next chapter, machining an engine is not a simple or easy task. Keep that in mind when evaluating a machine shop. The lowest price is often no bargain in the long run.

CLEAN & INSPECT CRANKCASE PARTS

Gaskets—Go over the engine and remove all gaskets and O-rings. Use lacquer thinner, a gasket scraper (the preferred tool) or a single-edge razor blade to remove sealant from mating surfaces. Be *extremely careful* not to gouge the case mating surfaces. Don't use a air- or electric-powered wire wheel on the case to remove sealant, especially around the part line.

Oil Valves—If you're reconditioning a very clean, lower mileage engine, there isn't a compelling reason to open the oil-control and pressure-relief valves. On all others, though, opening these valves is important. It allows better cleaning of the oil gallery and checking of valve operation.

Using a large screwdriver, remove the slotted plug or plugs on the bottom of the left case half. These oil-control and relief-valve plugs can be difficult to remove, especially without the right screwdriver. Because the correct screwdriver is a behemoth, there's usually nothing big enough in the average toolbox that works. Try a length of flat bar stock, a 1/2-in. drive socket-style screwdriver tip, or some other stout piece of metal that will fit in the slot. A heavy enough piece of metal with a foot of leverage works well. Once you apply the right tool, the plugs aren't so troublesome after all.

After removing the plugs, you'll get a spring

75

and valve out of each hole. Well, the spring shouldn't be a problem; you'll have to fish for the valve. Turning the case upright and tapping it on a wooden bench helps. Keep the springs and valves separate! Each is different, and must go back in its original hole.

Each plug has a sealing washer. If working on a dual-relief case with two plugs, you'll find later that most engine gasket sets have only one sealing washer for these plugs. So save one of the existing washers for reuse.

Initial Cleaning—To get parts clean enough to handle and for a quick visual inspection, they must be cleaned in solvent. If you do any amount of automotive work at all, a solvent tank is a major, but worthwhile investment. A pan, fresh solvent supply and cleaning brush will also work. Wash all parts free of oil and loose dirt. Don't worry about stains and baked-on deposits right now, just get the filth off.

Some parts you won't have to clean, unless planning on reusing them, are the pistons and cylinders. That also means there is no need to remove the rings from the pistons or scrape carbon from the piston crowns.

Unless you're installing case savers, don't remove the cylinder studs from the case. Doing so has no purpose, but does wear the potentially troublesome case threads, takes time, and stresses the studs.

Later, more comprehensive cleaning methods will be needed. The magnesium Type 1—3 case should be cleaned in a hot water and phosphate soap solution. Nothing more than soap and water are needed to clean magnesium. Don't use carburetor cleaner, and *don't hot-tank* the case. Those heated, caustic chemicals attack magnesium and weaken the case. In fact, hot-tanking will dissolve the case if it's left in the solution long enough.

Aluminum parts on VWs include the cylinder heads, oil-pump housing, flat-type fan housings and Type 4 cases. They can be cleaned in carburetor cleaner or a *cold tank*. This method uses unheated, less caustic chemicals. Aluminum parts and the hot tank don't mix; they dissolve, too, if hot tanked.

Just what is a hot tank? It's a tank filled with heated, caustic chemicals. It's just the ticket for cleaning iron or steel parts like the crankshaft and connecting rods. Consequently, many shops catering to all those cast-iron V8s still on the road use them. Make sure the cleaning method is compatible with the parts being cleaned. I want you to avoid disasters where the machinist's helper unknowingly dunks your VW alloy parts to their death.

Clean, Again & Again—Clean each part several times. There is the first solvent cleaning so parts can be handled, measured and machined. Then following machining, parts are cold- or hot-tanked. This removes machining grit and soaked-in oil. Then, before assembly, the parts need soap and water scrubbing to remove all traces of dirt, baked-on carbon and spray-on protective coating. While this last cleaning is often skipped, it is one of the biggest reasons most rebuilds don't last as long as the original engine. There's no reason why a rebuild can't easily match a new engine's longevity, but to do so, the parts must be perfectly clean.

Case cracks (noted by dark paint) are facts of Type 1—3 life. Top and bottom of case are oil pans; they don't add to case strength. Cracks there can be welded, but success isn't guaranteed.

Number 3 cylinder's vertical crack visible in bellhousing usually starts (and shows) inside case in main-bearing saddles. Semi-circle crack on left is common, too.

Case Cracks—Inspect the case halves for cracks. Use the photos to find the usual cracking spots, then closely examine these areas. Type 1—3 cases are magnesium, and sooner or later they will crack. It is just a fact of life with these engines. Magnesium hardens in normal engine use until it gets so hard and brittle it cracks.

Type 4 cases are aluminum (except for European versions that are very rare in the US) and aluminum doesn't harden with use like magnesium. So, Type 4 cases don't crack as a rule, and there isn't a photo of typical case cracks for this engine.

In fact, the general rule about Type 4 cases and everything inside of them is that they are bulletproof. But because no engine is perfect, perform the following tests and examinations on your Type 4 case. Nevertheless, don't expect to see cracks, pounded *bearing saddles,* or *fretting.* Fretting is the wearing away or reshaping of metal caused by hammering loads. It appears as a dark discoloration, accompanied by a change in the metal's texture; it is the result of movement between the case halves.

So, with the knowledge that sooner or later a Type 1—3 case is going to crack, carefully examine it. A big help in finding these cracks was shown to me at Southwest Import Rebuilders (SIR). To rapidly find cracks in the thousands of core engines they deal with, SIR uses a quick pass with a propane torch to open the crack.

Set a medium flame on a propane or oxyacetylene torch, then make a one second pass over the suspected crack area. If there is a crack, it will open up, and trapped oil and solvent will bubble from it. Don't overdo the heating or the oil will evaporate. Don't expect the crack to open up and yawn at you for five minutes, either. It will just part slightly for a moment, then close up again. You still must look for it, but the heat will definitely expose the crack.

Take some time when inspecting for cracks. You won't find them by turning the case over a couple of times while your eyes race all over it. Carefully examine every square inch of the case. Experience will help too, so slow down, especially if this is your first VW rebuild. And don't be afraid to admit you can't tell if the case is cracked. Have a machinist confirm areas you suspect are cracked.

Although the usual cracks are marked at their full length in the photos, it's likely you'll find a shorter crack. This means the crack is still growing, and hasn't reached its termination. If the crack, no matter how long, is clearly seen,

Cracks leading to oil passages (left photo) are serious because they cause oil pressure loss if plug or sender falls from loose, cracked case. Cracks around cylinder holes greatly weaken case. Inspect inside case, too.

then bite the mental bullet and admit the case is cracked.

Sometimes a cracked case doesn't mean replacement. Cracks running across the top of a case can be welded if they haven't spread much over an inch. In fact, a rod can be thrown through the top of a case and it can be made perfectly reliable by welding. The top of the case is nothing more than an oil pan, and carries no heavy bearing or case-clamping loads. A 2-in. crack confined to the top of the case should be repairable. On the other hand, cracks in the bellhousing area, especially behind cylinder 3 are real trouble.

This crack actually starts inside the case, at cylinder 3. It runs inside the case until it shows at the rear of the case, behind the flywheel. At first, the crack will appear at the top of the area marked in the photos. The crack will continue to grow longer in both directions, however, until the lower part reaches the oil gallery near the bottom of the case. When the crack reaches the oil gallery, disaster is the result. The gallery plug loosens as the crack opens the casting around it, and oil pressure knocks the plug free. Oil pressure immediately drops, of course, and major engine damage is done. So, if the case is cracked in the bellhousing area, replace it.

A crack on the left side of the bellhousing doesn't go to an oil gallery, but it does greatly weaken the case. Replacing the case is the cure for this one, too.

The crack shown on the left case half starts at the upper-rear cylinder-head stud and runs to the oil-pressure sender hole. Again, an oil leak results when the crack reaches the oil-pressure sender. Case replacement is the cure.

A corresponding crack at the upper, front stud hole is common in the right case half. It begins at the stud hole, runs forward and up to the top of the case, then "bends" around the corner by going diagonally rearward across the case top.

Aluminum Type 4 cases can be welded, welded again, and welded some more. Good thing too.

Welding Cases—This repair works most of the time on Type 1—3 magnesium cases, but it can be troublesome. Extreme heat from welding hardens the case, and many times the crack reappears—longer than before—five minutes after welding. Sometimes the weld just won't save the case; other times a seemingly hopeless case sawed apart by "grenading" internals can be saved. Your choice depends on new case availability and cost, subsequent machining costs and how good the welder is. Some people will spend a lot of money to save a German-manufactured case so they won't have to use a modern, durable replacement from Brazil.

Type 4 cases are aluminum and weld very nicely. How much damage you wish to repair on these depends only on price. Compare the price of welding and any necessary machining against the price of a replacement.

Main-Bearing Bores—Next, determine if the main-bearing bores are round and in alignment with each other. Look at the bearing bores in each case half. They should be the normal case color and have no dark discoloration or galled metal. You may see numbers or letters from the back of the bearing shell imprinted on the case. This means the bore has been *hammered* larger

CRANKCASE SPECIFICATIONS [-in. (mm)]

1200 (40 HP)

	New	Wear Limit
Main-Bearing Bore		
#1—#3	2.5590—2.5598 (65.00—65.02)	2.5601 (65.03)
#4	1.9685—1.9696 (50.00—50.03)	1.9700 (50.04)
Flywheel Oil-Seal Bore	3.5433—3.5453 (90.00—90.05)	
Camshaft-Bearing Bore	1.082—1.083 (27.50—27.52)	
Lifter Bore	0.7480—0.7485 (19.00—19.02)	0.750 (19.05)
Oil-Pump Bore	2.7559—2.7570 (70.00—70.03)	

1300/1500/1600

Main-Bearing Bore		
#1—#3	2.5590—2.5598 (65.00—65.02)	2.5601 (65.03)
#4	1.9685—1.9696 (50.00—50.03)	1.9700 (50.04)
Flywheel Oil-Seal Bore	3.5433—3.5453 (90.00—90.05)	
Camshaft-Bearing Bore	1.082—1.083 (27.50—27.52)	
Lifter Bore	0.7480—0.7485 (19.00—19.02)	0.750 (19.05)
Oil-Pump Bore	2.7559—2.7570 (70.00—70.03)	

TYPE 4

Main-Bearing Bore		
#1—#3	2.7559—2.7567 (70.00—70.02)	2.7570 (70.03)
#4	1.9685—1.9696 (50.00—50.03)	1.9700 (50.04)
Flywheel Oil-Seal Bore	3.7401—3.7421 (95.00—95.05)	
Pulley-End Oil-Seal Bore	2.4409—2.4429 (62.00—62.05)	
Camshaft-Bearing Bore	1.0826—1.0834 (27.50—27.52)	
Lifter Bore	0.9448—0.9456 (24.00—24.02)	0.9468 (24.05)
Oil-Pump Bore	2.7559—2.7570 (70.00—70.03)	

Main-bearing bores always show coloration from minor bearing movement, trapped oil and so on. Streaking here is normal and doesn't indicate align-boring.

than it should be and is now oval. Once a bearing bore is worn egg-shaped, it can't clamp the bearing halves as tightly as it should.

Pay special attention to the #2 main-bearing bore. Wear will show there first. The wear you see results from the uncounterweighted VW crankshaft. Below 5000 rpm, a stock crankshaft works well. Above that rpm, *crank flex* becomes a problem. Flex, or *whip,* is greatest in the center of the crank, right at the #2 main bearing. It is mid-span on the crank and is surrounded by two cylinders with their crank throws on the same side of the engine. This lopsided load pounds the case, displaces material throughout the bearing saddles and causes oval bearing bores. Wear is pronounced on hard-driven engines, but can appear in carefully operated engines after years of service.

A case that has been pounded hard, or for a long time, will be fretted along the mating flange. If you see fretting (displaced material) inspect the case very carefully. Chances are there are so many cracks, pounded bearing saddles, hardened metal and worn mating flanges, that replacing the case is the best course of action.

Excess wear can be detected by laying a precision straightedge in the main-bearing bores and trying to fit a 0.003-in. (0.76mm) or thinner feeler gage between the straightedge and bearing bores. Make sure there are no bearing inserts, dowels or grit in the way. This tests for alignment among the bores. It will also uncover an out-of-round bore, if you happen to stick the feeler gage in the same spot the bore is worn.

Otherwise, out-of-round is detected by torquing the case halves together and using a dial bore gage to measure the diameter of each bore. See Chapter 7, page 125, for torque values and procedures. Telescoping (snap) gages and inside micrometers also have the accuracy of a dial bore gage. If you suspect out-of-round, have the machine shop check for it with their measuring tools.

Actually, chances are your Type 1—3 case has round bores, but they aren't in line. This condition can get bad enough to see with the naked eye. Volume rebuilders don't even bother checking. They just pass the align-boring cutter through the case set at 0.020-in. If any uncut metal remains, they cut again at 0.040-in. or larger until the bores clean up.

Although an extremely rare condition, the bearing may have *spun.* This means the bearing inserts have come loose and rotated with the crankshaft. The bearing bore will be scored and burned very dark, the dowel pin will be rolled over into the bearing bore, and the crankshaft and related parts will be destroyed from oil starvation. It's time to start over with another engine.

Another align-bore troublemaker is extreme

overheating. Engines damaged by detonation, or run without the cooling fan are our concern here. Sometimes, such excessive heat practically remolds the case so the bores are drastically out of line. To save such a case, torque it together and bake at 425F (218C) for two hours. Then turn the oven off and let the case cool to room temperature. *Don't move the case while it's cooling.* Once cool, *repeat* this heat/cooling cycle. This will actually reshape the case to where align-boring the main- and cam-bearing bores is feasible.

Align-Boring—Short of a spun bearing, case damage from pounded or out-of-alignment bearings can be corrected by *align-boring*. In this machine shop job, the case halves are bolted together at the specified torque and a special cutter is passed through the bearing bores. Material is removed from the bearing bores until they are perfectly round and in line. Thicker main bearings are installed so the standard crankshaft can be used. The machine shop will tell you what size bearings you need, and most of the time can supply them.

For Type 1—3 engines, bearings are available in 0.020-, 0.040-, 0.060-, 0.080- and 0.100-in. oversizes. Type 4s have only a 0.020-in. oversize available, but usually don't even need that. It's not that the Type 4 crank is so much better than the Type 1—3 version, it's that the aluminum case absorbs pounding from the crank much better than the magnesium case.

Align-boring VW cases is fairly specialized work. Look for a machine shop that specializes in VW work for this job. Most shops use equipment that bores the holes round, but more or less on the original bore's centerline. The original centerline may not be parallel to the case centerline, but for stock engine rebuilding, this is accepted. Only for the most radical, all-out racing engine does the crank centerline have to be perfectly centered in the case.

End Thrust—Excessive end thrust problems can be cured, within limits, by cutting the case inside and outside the front main bearing. This machining step trues the bearing mounting area. A bearing with a thicker flange is then necessary to bring the overall bearing width to original specifications. On Type 1—3, #1 main bearings are available with flanges 1.0mm, 1.5mm and 2.0mm larger than standard. Type 4 oversize bearings come in standard measurements: 0.010-, 0.020- and 0.040-in.

This brings up an interesting point. Bearings for VWs come in standard and metric specifications, depending on where you buy them. Most standard and metric specifications are very close to each other, and you don't have to worry about mixing and matching machine work and bearings, but it isn't always true.

For example, a "standard" bearing of 0.010-in. is 0.0002-in. smaller than a 0.25mm metric bearing. This isn't enough difference to bother

Align-boring tool used by shops locates off oil-seal and crank-pulley bores. Micrometer adjustment at end of tool sets depth of cut.

Air drill powers align-bore cutter. Actual cut takes about three minutes, but setting up tool and case takes longer. This style tool cuts new bore concentric with center line of old bores.

with, so machinists ignore it. Therefore, if the case is machined with a 0.010-in. cut on the front main bearing, and fitted with a 0.25mm oversize flange bearing, all is well.

Now let's say the case is cut 0.080-in. A 2.00mm oversize metric bearing is listed in the parts book as equivalent. However, 2.00mm and 0.080-in. differ by 0.0013-in. This discrepancy is significant. To compensate for excessive end thrust, that means installing a different flywheel shim to take up the difference. This is one more reason you must check the fit of all bearings, no matter what the engine or who machined it.

End thrust is one of those problems in which there is little wrong, or there is plenty gone

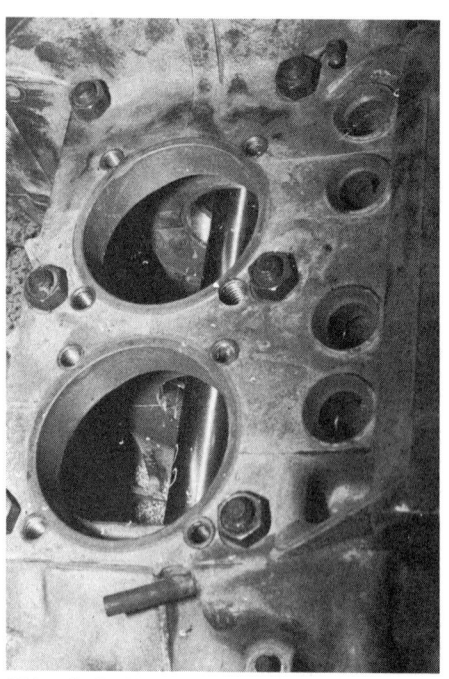

Chips fly inside case while cutter is rotating. Bearing at top shows wide shiny area already cut and thin dark band of material yet to go.

End thrust is cut with another massive tool. Both inside and outside of case are cut, dividing the amount necessary for clean-up between them. Depth of cut is set at other end of tool during setup.

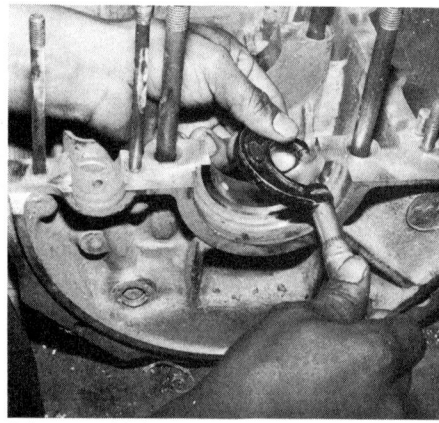

After machining, job is checked for total thickness. Metal removed is compensated for with thicker #1 main-bearing flanges.

wrong. The most common cause of end thrust problems is setting end thrust incorrectly in the first place. This comes from misinformation about how end thrust is set during engine assembly, or even what end thrust is. Invariably, under such conditions, the end thrust is set too loose or too tight, and that hammers or bears down on the case, wearing the end-thrust surfaces rapidly. The other common cause of excessive end thrust is very high mileage (typically over 150,000 miles), where the case just plain wears out from clutch pressure-plate force.

Precisely checking the case end-thrust surface is a machine shop job, using a dial indicator and fixture. So, if you have even a hint of end-thrust wear in the case, take it to the machine shop and have them check it.

Inspecting the bearings and case can tell you if substantial end thrust wear can be corrected with a simple bearing replacement. Hand-fit the #1 main bearing to the case and see if it is loose in front-to-rear motion. The bearing should fit snugly on the case. Any motion at all signals the need for closer inspection. If the crankshaft slid back and forth 1/2 in. during disassembly and the case looks beaten, then it's time to cut the end thrust. The machine shop can tell you how much cutting it will take to correct the condition.

Look for shiny wear on the case. If the metal has been moving, it will be bright. Sometimes you can catch a fingernail just outboard of the #1 main bearing. That means the case has been worked, too. Let the machine shop measure it. Also run a finger front-to-back in the bearing bore. Do this at 3 and 9 o'clock. If you feel any roughness or ridges at all, the bearing has been moving and end-thrust correction, plus align-boring are needed.

If clearances seemed OK during disassembly and the case, bearing flange and shims show no appreciable wear, then end thrust should be within tolerance. If you want to make absolutely certain, have a machinist check your case.

Camshaft-Bearing Bores—Except for some 40-HP models, all VW engines covered in this book use insert cam bearings. Thrust is taken at the third cam bearing, the one closest to the camshaft drive gear. These bearing inserts are cheap and easy to replace, so cam bearing wear problems are usually easily cured with new bearings.

Worn cam bearings cause low oil pressure. A large clearance between the cam and its bearings is like drilling more holes in the oil gallery; it lets out oil, reducing pressure. Consequently, to maintain correct oil pressure, change the camshaft-bearings. And measure the camshaft-bearing oil clearance. In regular practice, if the engine was merely tired, with no sign of oil starvation, changing the bearings is all that's necessary.

If oil starvation was a problem, as shown by worn main and camshaft bearings, check cam-bearing oil clearance. Obtain this oil clearance by measuring the camshaft with a micrometer. And compare it to measurements of the camshaft-bearing bore diameters, with the bearing inserts installed, of course. Use a snap gage and outside mike to measure cam-bearing IDs. Or, use Plastigage to get the bearing clearance. This means torquing the case together with the camshaft installed, then splitting the case.

Cam-bearing oil clearance should be between 0.0008—0.002 in. (0.02—0.05mm). The wear limit is 0.0047 in. (0.12mm), but keep the clearance less than that: 0.002 in. is preferred.

Adding 1200 Cam Bearings—If your early 40-HP engine doesn't have inserted cam bearings, they can be added by many VW machine shops. The original cam bores are bored oversize and a bearing tang cutout is machined so later style replaceable cam bearing inserts can be installed. This is worthwhile if the stock cam bores are worn out. After all, it beats the alternative, buying a new case, if the rest of the case is serviceable.

On the other hand, if there is nothing wrong with the cam-bearing bores in the original engine, there's no need to cut the case and install replaceable inserts. Many overhead cam engines, like the water-cooled one in the VW Rabbit/Jetta/Scirocco, run the cam directly in the aluminum head. Only when the cam bores are worn out do they need cutting. Unfortunately, there really aren't many 40-HP engines without worn cam bores anymore. By now, for practical purposes, all of them need insert cam bearings.

Case Savers—When a VW engine warms up, it grows. When it overheats, it grows some more. When it expands too much, something has to give. That something is often the threads in the case the cylinder-head studs are anchored in. The force of the expanding cylinders and cylinder heads pulls the studs away from the case. The weak link is the magnesium threads in pre-'73 Type 1—3 cases. They distort, and eventually fail; the studs pull out of the case. Then the cylinders are free to bang back and forth between head and case. This makes noise and ruins the case, heads and cylinders.

Case threads also wear when the engine is cold. During engine warm up, cylinder-stud torque is low because VW engines rely on thermal expansion to produce the correct torque at operating temperatures. But less torque at cold-engine temperatures allows the studs to vibrate. The vibration work-hardens the magnesium case threads until they get so brittle they break. Case savers work as *isolation dampers* between studs and case, damping harmful vibrations to the magnesium. This benefit can be lost if the studs are overtightened. Always install cylinder-head studs *snug,* and use Permatex 3H on their threads.

Don't expect the studs to fall out and the cylinders to pop off the engine. The studs will typically loosen enough to just relieve the tension on them, but it's enough to get the cylinders banging.

The cure is to install steel thread inserts in the case. These strong steel threads replace the weak magnesium originals and loose studs are then cured. With case savers installed, the cylinder heads warp when the engine overheats, but that's another story, see page 102.

The steel threads go by several trade names, but the one that stuck, and is used as the accepted term is *case saver.* No engine rebuilder of good repute will let a rebuild out the door without them. Any VW machine shop can easily install these threads in your case. Don't install case savers in a Type 4 case because they're not required. They aren't needed in '73 or later Type 1—3 cases either, because they come installed stock from the factory.

Make sure the case savers are threaded along their entire length, both inside and out. If they are not completely threaded in their ID, don't tighten the cylinder studs in them. This will only make the case saver part of the stud, and wear the magnesium case threads in the normal manner.

Also, there is confusion about stud diameter and the need to add case savers. VW used 10mm studs until 1973, then they added case savers and reduced stud diameter to 8mm. Nevertheless, all magnesium cases need case savers—period.

First step in installing case savers is to drill stud holes oversize. Properly set-up table, fixture and drill press ensure holes will be exactly perpendicular to case.

Next, holes are tapped to accept threads on case saver OD. High-volume rebuilders use a drill press to run tap, but twisting by hand works just as well.

Deep-Stud—A specialized case saver installation is performed when you ask the machinist to *deep-stud* a Type 1—3 case. This refers to drilling #3 cylinder's upper-rear cylinder-head stud hole deep into the case. A case saver is then installed in the hole, deeper than its usual location, much closer to the main-bearing saddle. A longer cylinder-head stud is necessary.

This procedure is supposed to help with the vertical case crack that occurs in the bellhousing area. It depends on whom you talk to, though, as to how effective it is. The factory obviously approves of it, as all stock Type 1—3 cases since '73 have been deep-studded. Deep-studding is not necessary on Type 4 cases.

Oversize Cylinders—If increasing the engine displacement by adding larger cylinders, it may be necessary to have the machine shop cut the cylinder openings in the case oversize. See Chapter 3 and your parts department for specific cylinder requirements.

Shuffle Pin—Another machine shop operation you may have heard of is *shuffle-pinning.* This is when a *shuffle pin* is installed around the two center main-bearing studs in Type 1—3 cases. A shuffle pin is a steel cylinder designed to fit over the studs in the right case half and slip into a counterbore in the left case half. Because the pin fits tightly in both case halves and is made of very hard steel, it then becomes very difficult for the case halves to move relative to each other.

Such motion comes from high cylinder pres-

Finally, thread inserts are threaded into case. Again, air-powered tools look fancy and work quickly, but hand-power works too.

Deep-studded cases are easy to spot. Note recessed case saver in #3's upper front stud hole. VW made deep-studding stock in '73, but it can be retrofitted to any Type 1—3 case.

Many big-bore kits require opening cylinder holes in case. This is a job for a large flycutter and perfectly flat table.

All these cranks are awaiting regrinding. Most crank problems are caused by dirty oil; someone isn't changing it frequently!

sures, high rpm, extreme heat and detonation. You can tell if the case has been moving by inspecting the #2 main-bearing saddles where they butt together at the case parting line. If you see fretting—dark, streaked, galled metal—the case halves have been vibrating against each other. Shuffle-pinning will definitely stop the vibration and save you subsequent align-bore jobs, but it's smart to find out what caused the vibration in the first place. If you don't, the shuffle pin may break out the entire #2 main-bearing saddle.

Shuffle-pinning is not necessary for a regular street rebuild. If you plan on running high compression or driving under high-heat conditions, like flat-out from Phoenix to Death Valley in summer, then shuffle-pinning will be useful.

After adding shuffle pins, you'll find they make getting the case together and apart again a real chore. In fact, when the case halves are together, it's more like a press fit than anything else.

Sand Seal—Another option is adding a proper oil seal to the pulley end of Type 1—3 crankcases. This isn't needed for street operation, but if driving off-road, it helps keep grit out of the crankcase. And if you are overly concerned about oil leaks, adding a neoprene seal may make you a little more secure.

Two types of seals are available. With one, the case is cut to accept the seal and the crank-pulley shaft is machined down to fit in the now restricted hole. With the second style, the seal bolts to the case, and the pulley again needs cutting. In the bolt-on seal installation, the *oil return thread* on the pulley is removed.

Full-Flow Oil—Because Type 1—3 engines have only an oil screen, a popular modification is to add a full-flow oil filter. With such a filter, more than just boulders get strained from the oil, and engine longevity increases. Have the machine shop drill and tap the main oil gallery at the pulley end of the engine so you can install a full-flow kit.

Full-flow oiling is one of the *best* modifications you can do to a Type 1—3 engine. It's a better use of your money than a big oil pump or a remote cooler.

Cleaning—Finally, have the case cleaned in the machine shop's wash tank. I'm not referring to dipping in a solvent tank or having the shop helper spray parts out back with a garden hose—you can do that at home. No, send the case through a hot water/phosphate jet tank. This is absolutely mandatory to remove machining grit, and money well spent even if no machine work is done to the case.

Later, right before engine assembly, hand-scrub the case using liquid dishwashing soap, even if it has been professionally washed. If you have the shop wash the case now, it will save you some effort during final cleaning.

CRANKSHAFT

Crankshaft Inspection—Under normal conditions, crankshaft reconditioning work is restricted to cleaning, polishing, and pressing the timing gears on and off. If the crankshaft has problems, they can be cured by grinding or replacement.

Most crank problems are caused by dirty oil. Although dirty oil is harder on the bearings than the crank, there is a limit to what a crank journal can take. Exceptionally dirty oil will scratch the journals. Deep scoring and grooving will result if large fragments are circulated through the oiling system.

Besides being scored, scratched or grooved from dirty oil, crank journals could be worn undersize, out-of-round or tapered. Additionally, the thrust face could be worn, especially if crankshaft end play was excessive. Or, there might be cracks hiding anywhere on the crank.

Press Off Gears—A press is needed to remove the distributor drive gear, spacer and crankshaft timing gear. Until these parts are removed, the #3 main bearing is held captive. Besides having to press the gears off for bearing R&R, you need to make sure the gears fit tightly to the crankshaft. Therefore, sooner or later, the timing gears must be pressed off.

It's your choice when to do so. I'm assuming you don't have a press at home, and this job will be handled by the machine shop. To save a trip, leave the gears on until after inspecting the crank for wear. You won't be able to check the #3 main bearing or journal. But if there were no bearing noises when the engine was running, and the other journals measure OK, I doubt there will be any surprises when the bearing is finally removed. If you suspect bearing problems, however, it's best to have the gears

Carefully inspect cranks. Rust and impact damage can ruin otherwise good cores. Brass distributor gear is vulnerable, note nicks.

Remove retaining ring before pressing off timing gears. Use large, sturdy expanding-ring pliers to open and remove ring.

Large press makes easy work of pressing off gears. Don't forget to hold crankshaft.

pressed off right away so crankshaft condition can be determined.

Get a hefty pair of external retaining ring pliers to remove the retaining ring in front of the distributor gear. This job is impossible without expanding pliers, so either get a pair or leave this chore to the machine shop. Remove the retaining ring, then set the crankshaft in a press. Back the crankshaft timing gear with a mandrel or the press bed, then press on the crankshaft snout. It won't take much to free the crank, so be ready for it and don't let it fall to the floor! Collect the Woodruff keys: two on Type 1—3 and one on Type 4 engines. Store them and the gear in a can so they won't get lost. Now you can slide the third main bearing off.

Check Timing Gear Fit—Before proceeding, check timing gear fit, particularly if removal was easy. The test is simple: try to slip the gears onto the crankshaft. Because these gears have an interference fit, they should *not* slide into position. If they do, try another crankshaft or gears. Repeated gear removal and installation will eventually wear the crank and gears undersize until the 0.0004-in. (0.011mm) interference fit is lost. Chroming or metal spraying (expensive machine shop procedures) the crank back to original size is also acceptable, but is more expensive than replacement.

Also, check the timing gear mounting surfaces on the crankshaft for *seizure marks*. These longitudinal score lines are where crankshaft metal was pulled by the timing gears when they were pressed off. The lines run parallel to the crankshaft centerline. If you find any, *carefully* file off the *ridges* of the score lines. Don't polish this area with emery cloth to try to remove the marks. You'll remove the seizure marks, but you'll also remove enough crankshaft material to reduce the interference fit.

Crankshaft Inspection—Start crankshaft inspection by checking the texture of the journals. Drag a fingernail the length of each journal. If there is any roughness, your nail will catch on it. Check the bearing inserts, too. If they are badly worn, the corresponding crankshaft journal should show some wear.

Don't assume you're doing something wrong if you can't feel anything. Usually, there's no roughness to feel. This is true if the engine was well cared for with regular and frequent oil changes. Assuming journal sizes are OK and the crank is straight, all you need to do is polish the journals to recondition the crankshaft.

If your fingernail catches on any deep scratches or scoring, those journals needs grinding. A special machine is used to precisely grind metal from the journal. How much is ground off depends on the depth of the scratches or scores. Also, the taper and out-of-round checks discussed later will show how much it will take to clean up the journals.

Even if only one journal is worn, all the main- or rod-bearing journals must be ground. Example: The crank is fine except for #2 rod journal; it's deeply gouged. All rod journals will be ground the amount it takes to clean up #2. This is done because bearing inserts are sold in sets; all inserts in the set are the same size. Therefore, it doesn't make sense to grind only one journal.

Machining just one journal doesn't save money, either. The initial cost of setting up the

83

Run fingernail across journal to feel for scratches. If you feel any snags or ridges, crank needs grinding. Note how small, dark dowel hole (arrow) is offset toward flywheel on #3 main bearing.

Specialized grinder with huge wheel is needed to grind crank. Grinding cures undersize, out-of-round, taper, and runout in one operation.

crank in the grinder represents the majority of the expense. And, it's the same for grinding one or four journals. You'd have trouble getting a machinist to grind only one journal anyway. Nevertheless, you can have only the main journals ground and not the rod journals, or vice versa. This is standard practice. Or, the rod journals may be ground 0.020-in. under and the mains 0.010-in. under, and so on.

Taper is usually the least concern when it comes to crankshafts. By the time a VW crank has tapered enough to cause trouble, the entire journal is probably undersize and out-of-round anyway. Therefore, you don't have to check for taper unless all other crankshaft measurements are good. The cure is the same for all, grinding, and grinding to the next undersize for out-of-round or undersize will automatically cure taper.

Journal Diameter—Before putting away the micrometer, establish the main- and rod-journal diameters. If you have snap gages or an inside mike, you can measure the main bearings while they're torqued in the case, and compare them to journal diameter. The difference will give you the exact *oil clearance*—distance between the journal and its bearing. You will need the journal diameters anyway to purchase the correct bearings.

Be aware that the crank could be undersize if the engine has been rebuilt. This is the case if the main or rod journals are grossly undersize by the same amount and in increments of 0.010 in. (0.25mm).

Also, after polishing the crank, remeasure the journal diameters. Depending on the duration and intensity of the polishing, a slight amount of material may be removed, perhaps 0.001 in. Even slight metal removal could result in excess oil clearance if journal diameter was worn to the wear limit. In that case, it would be better to grind the crank to the next undersize.

Once journal diameters are known, subtract that figure from the main- or rod-bearing ID. The difference is oil clearance. This space is filled with oil pressurized by the oil pump. If the clearance is too tight, not enough oil can reach the bearings. The crankshaft will touch its bearings, causing immense friction. Oil temperature will skyrocket and both bearing and journal will be damaged. If the clearance is too loose, the oil pump won't be able to supply enough oil, the oil pressure will be low, and the crank will touch its bearings.

If your measurements show excessive oil clearance, grind the crankshaft to the next undersize, or replace it.

Out-Of-Round—Bearing journals don't necessarily wear evenly, especially rod journals. They tend to wear into an elliptical, or oval shape. It's easy to see why when you consider the load applied to a rod-bearing journal. Load changes occur as the piston goes through its four strokes, or cycles.

The rod and piston push against the journal during the compression and exhaust strokes and particularly during the power stroke when near TDC (top dead center). They also pull against the journal during the intake stroke. Inertia loads change with the angle of the crankpin, or rod journal, and the relative movement of the piston in its bore. In short, varying loads cause uneven journal wear.

To detect rod journal wear (even or uneven) use a 2—3-in. outside micrometer, except when checking journal #4. For that one use a 1—3-in. micrometer. Measure the journal at several positions around its circumference, but in the same plane. If your measurements vary, the journal is *out-of-round*. Ideally, you should be able to make a measurement, then rotate the mike around the journal without changing its setting.

To determine how much a journal is out-of-round, subtract the *minor dimension* (smallest measurement) from the *major dimension* (largest measurement). The resulting figure must not exceed 0.0012 in. (0.03mm). If it does, the crankshaft must be ground.

Don't forget to measure the main journals. Although they are not as prone to out-of-round wear, check them anyway. The wear measurement is also 0.0012 in. (0.03mm)

Taper—When a bearing journal wears front-to-rear, or along its length, it is *tapered*. Measure the journal diameter at several points along its length. If these measurements vary, the journal is tapered.

Excess taper must be corrected. If not, the bearing insert will be loaded more at the big end of the journal than at its smaller, worn end. In effect, the load that should be distributed evenly over the length of the bearing will be concentrated in a smaller area. Bearing design load will be exceeded, resulting in rapid wear.

Taper is expressed in *thousandths-of-an-inch per inch*. So, make your first measurements at one end of the journal, then move to the other end and take another. Subtract the smaller from the larger to find taper. VW does not specify a maximum allowable taper, but 0.001 in. (0.025mm) or more definitely needs attention. Correct excessive taper by grinding.

Runout—Any permanent bend in a crankshaft is called *runout*. Check runout by laying the crankshaft in one of the case halves using only the first and fourth bearing inserts. Oil the bearings before you install the crank. A lathe can be used to support the crank, too. Set a dial indicator to read off the center main-bearing journal and slowly rotate the crank. Turn the crank until the *lowest* reading appears on the dial indicator. Zero the dial, then rotate the crank until the *largest* reading is obtained. Read runout directly on the dial indicator. Runout should be less than 0.001 in. (0.025mm).

Because the VW crankshaft is so short, it is resistant to bending, so runout is not a common problem. Many shops don't check runout, and chances are, you need not worry about it. If the engine has pounded main-bearing saddles, seen severe service, been hit in a rear-end accident or the crankshaft is new to you, have the machine shop check runout for you.

Also, check for runout if the engine has been severely overheated in the bottom-end area. This is especially true of Type 1—3 engines. Oil temperature is a good clue to bottom-end temperature. About 220F (104C) is the limit for dependable case longevity. At 220F the magnesium case is 2% elastic. At 240F (116C), elasticity reaches 6% and the case easily distorts from crank whip, inertia loads and combustion pressure.

As the bearing saddles of the case distort, they no longer support the bearings. The crank is then free to whip or bend. This lack of support can permanently bend the crank, which you read as runout. Don't confuse runout with *crank whip*. Crank whip is a momentary bending or flexing of the crank, after which the crank returns to being straight. Runout is permanent bending.

Unfortunately, VWs don't come with an oil temperature gage. Adding one is a smart idea while you have the engine disassembled; aftermarket suppliers offer different kits.

I assume you don't know exactly what the engine's oil temperature has been. In that case, go by smell and carbon deposits. Regular oil begins to lose its lubricating properties at about 250F (121C). When this happens the oil burns and leaves a rank smell. If you've passed an oil refinery, you know the smell I'm referring to.

Excessive heat will also bake carbon onto the case walls, leaving tell-tale crusty deposits in all the nooks and crannies. Examine the crankshaft for baked-on carbon. You won't find large flaky deposits on the crank because they would be flung off. Instead, look for a general overall blackness with very little gloss. Nor-

---

CRANKSHAFT SPECIFICATIONS
[in. (mm)]

TYPE 1—3
Main-Bearing Journal Diameter
#1—#3   2.1640—2.1648 (54.97—54.99)
#4   1.5739—1.5748 (39.98—40.00)

Rod-Bearing Journal Diameter
2.1644—2.1653 (54.98—55.00)

Maximum Out-of-Round, Rod- and Main-Journals
0.0011 (0.03)

Flywheel Maximum Lateral Runout
0.011 (0.30)

Flywheel Oil-Seal Shoulder Diameter
New   **Wear Limit**
2.7519—2.7598 (69.9—70.1)   2.7322 (69.4)

TYPE 4
Main-Bearing Journal Diameter (1700/1800/2000)
#1—#3   2.3609—2.3617 (59.971—59.990)
#4   1.5739—1.5748 (39.984—40.000)

Rod-Bearing Journal Diameter (1700/1800)
2.1646—2.1654 (54.98—55.00)

Rod-Bearing Journal Diameter (2000)
1.9677—1.9681 (49.97—49.98)

Flywheel Maximum Lateral Runout
at 8.27 (210) diameter: 0.0197 (0.5)

---

mally the crank is a deep brown. But when oil bakes on it, it turns black from the carbon. If the heat was extreme, like from a spun rod bearing, the crank will be burned blue at the hottest spots, fading to flat black in the relatively cooler areas.

If the crank fails this inspection, replace it. Crankshafts can be straightened by hammering, but the low cost of a VW crankshaft doesn't make this option that convincing. Your machinist can help you decide if straightening is worthwhile for the crank. Your decision will also depends on replacement cost and availability of the straightening service. Never straighten a crank in a hydraulic press. It will be straight, but probably crack in the process.

When buying a replacement crank, check runout at the counter. Take a case half and dial indicator to the store and do the measurement on the spot. Then, if the crank has too much runout, ask for an exchange. Besides getting a straight crankshaft, you'll probably be remembered by the parts people.

If you're buying a reground crankshaft, or yours is ground, you won't have a problem with runout. This is because the crank-grinding machine grinds the journals to a true circle relative to the crank's actual centerline.

Dowels—Ideally, the dowels on Type 1—3 cranks should be tight in the crank, where they'll stay during the entire rebuild. There is no need to remove the dowels, so don't try. Check them by wiggling each by hand to see if it's loose. Any movement is unacceptable. Their purpose is to position the flywheel to the crank flange and to resist torque loads between the flywheel and crank. The flywheel bolts simply clamp the flywheel to the crank.

If a dowel is only slightly loose, a new one may fit tighter. Try that first. If that doesn't fix the condition, talk to the machinist. Chances are, an exchange crank is the best way to get tight dowels. The alternative is to drill the crank and flywheel for larger dowels. An 11/32-in. dowel usually fits and works well.

Never run a crank with missing or loose dowels. You're begging for big trouble when the flywheel comes loose.

On Type 4s, the only similar part is the roll pin used for flywheel location. Again, there's nothing to do with this piece, so just leave it in place. It's purpose is flywheel registration (placement), but it isn't so critical that it be tight in the crank like Type 1—3 dowels.

Cracks—While most air-cooled VW cranks are not considered prone to cracking, there's

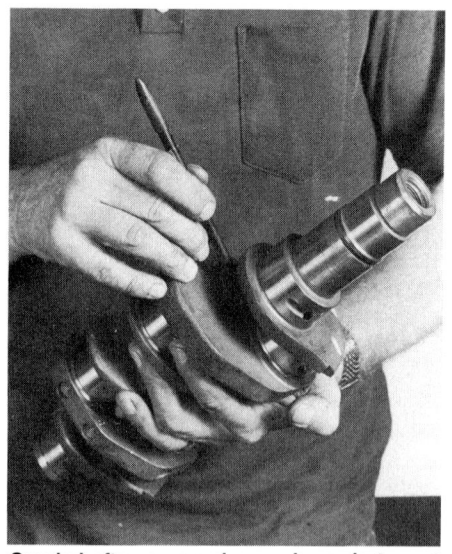

Crankshafts can crack anywhere, but most common spot is this hidden area where #4 rod journal meets its forward flange. Crank most likely to crack is 1600 cross-drilled unit.

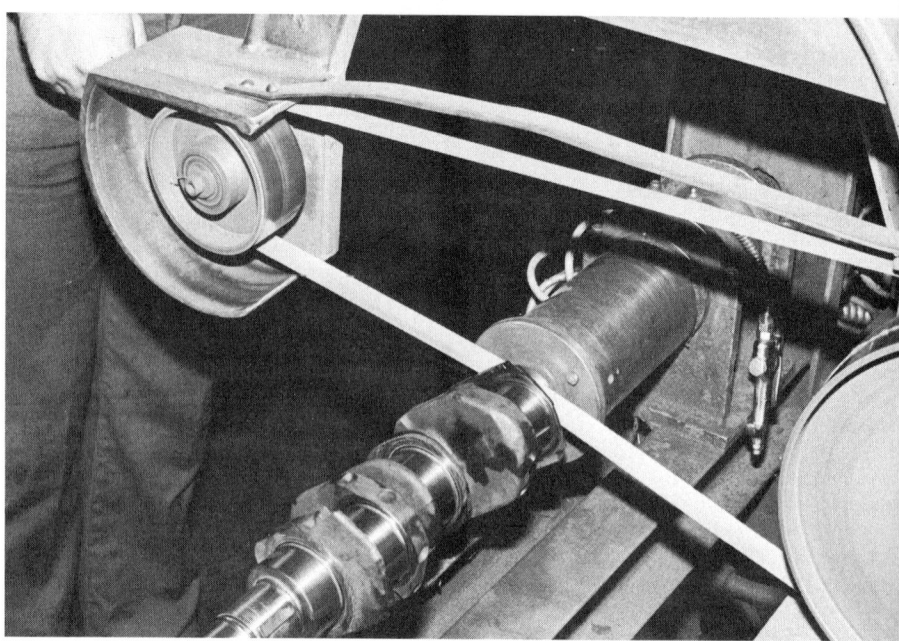

All crankshafts need polishing at rebuild time, ground or not. Machine shop uses power polisher and lathe to put a tooth on journals. You can do same with strip of emery cloth.

always the chance a crack is hiding somewhere on the crank. Cracks typically occur where a journal meets the flange. This area—the *fillet*—is radiused (has a curved corner) to distribute stress instead of concentrating it in a sharp corner. Just the same, it is the most highly stressed portion of the crank, and is therefore the most likely spot for cracks to appear.

This is especially true of the crossdrilled 1600 crankshaft. Always check the #2 rod journal; the one closest to the timing gears. This throw is particularly prone to cracks, especially down between the flanges where it is hard to see.

Because cracks are difficult to find by sight alone, they are best found by *Magnafluxing* or *Spotchecking*. Both are machine shop procedures, and should be used on all the crossdrilled 1600 cranks and any other crankshaft that has seen hard or long service. If a crack is found, replace the crank.

Purchase Bearings—You can buy bearings at an auto parts store, or many times, at the machine shop. Because of the machining possibilities with a VW case and crankshaft, you must have all crank grinding, align-boring and end-thrust cutting done before selecting bearings. Then you'll know which undersize bearing to use for mating with the ground crankshaft, and the right oversize bearing to fit in the bored case. This is an important point for those used to the standard undersize bearings available for other engines. Because it is common VW practice to grind both the crankshaft and remove material from the main-bearing bores, and to cut the end-thrust area, there are quite a few bearing size combinations available for VWs. Make sure you have all machine work done to the case and crankshaft before buying bearings.

When buying bearings, open the box at the counter. Inspect the bearing backs for a size marking. Make sure the box marking is correct. More than once, the wrong bearings have been put in the right box or vice versa.

Post-Regrind Checks—After the crank is back from the machine shop, make some more checks. Begin with the oil holes. The edge of each oil hole must be *chamfered* so it blends smoothly into the journal surface. A sharp edge will cause trouble. It could cut a groove in the bearing. If the oil holes are not chamfered, return the crank, complain, and get the job done right.

Another area where improper regrinding can cause trouble is at the journal fillets. If a fillet radius is too small, a crack could start in the sharp corner. The crack eventually causes a break. In this respect, the bigger the fillet radius, the better. However, a fillet radius that's too big will *edge-ride* the bearing. The flat surface of the journal is too narrow to accommodate the width of the bearing; consequently, the edge of the bearing would ride up on the fillet.

You may have seen old bearing inserts that had a line of babbitt around the edges. This was caused by edge-riding, but not necessarily due to an overly large fillet. On high-mileage engines, it is usually caused by excessive crankshaft *end play*—front-to-rear crank movement. If crankshaft end play measured at, or over 0.006 in. (0.15mm) during disassembly, expect to see edge-riding at the flywheel end of the bearing inserts.

The industry standard for fillet radii is 3/32 in. (0.0938 in. or 2.38mm). If the crank was ground, have your machinist check its radii with a radius gage. An alternate, but cruder method, is for you to fit a half of a new bearing insert onto the appropriate journal and check for interference between the journal and fillet. If there is interference, the fillet is too large. Return the crank to the machine shop for more measurements and corrective action.

Clean Threads—On Type 1—3 crankshafts, thoroughly clean the gland nut threads. An old toothbrush is a good tool, along with plenty of solvent. Aerosol carburetor cleaner cuts through and forces out all but the most stubborn dirt from blind holes like this, but beware of ricocheting material. Wear eye protection when spraying this strong cleaner. Take your time and remove all dirt, machining grit and leftover Loctite powder. These threads must be perfectly clean to support the huge torque load imposed by the gland nut and to provide a chemically clean surface for Loctite.

Clean Type 4 crankshaft flywheel bolt holes with brush and solvent. File off any raised metal around the bolt holes first.

On all crankshafts, don't forget to clean the

pulley or fan hub threads at the other end.

Final Polishing—If the crank doesn't need to be ground, reconditioning it is simple. Polish the journals and smooth any small scores or dings. There shouldn't be any dings unless the crank was abused while the engine was apart. But if there are, a needle-file works well to remove any projections caused by nicks or dings. With the file in hand, check the oil holes. Smooth any sharp edges where the oil passages open onto the journals.

If the crank was ground, it should have been polished by the machine shop. If not, you can do it yourself. Use narrow strips of 400-grit emery cloth as shown. Emery cloth is available at hardware or auto parts stores. It's packaged in long, narrow strips; 2-ft long by 1-in. wide is about right. If it's wider, tear it lengthwise.

Position the crank on the bench so it won't roll around, then loop the strip over a journal. Start at one point on a journal and move around its circumference, ending up where you started. Use light, even pressure and take care not to polish one spot too long. You must not flat-spot the journals. As you polish, the journal will take on a bright shine. Once the journal begins to shine, move on. The task is not to remove metal, but to clean and texture the journals. Polish the oil-seal area as well, but don't bring it up to a bright shine. As with the cylinder walls, the seal area should have some *tooth* to carry oil and lubricate the seal. This increases seal life and improves sealing.

Final Cleaning—Now that crankshaft reconditioning is complete, make sure it's clean. Even if you had the crank hot-tanked, brush out the all-important oil passages. A .22-cal. gun-bore brush works perfectly. Dip the brush in solvent, then push the brush completely through the oil passages before reversing direction. Otherwise, you'll ruin the brush. Continue this process until all crankshaft oil passages are clean.

Use a stiff nylon or brass brush on the counterweights if they're still dirty. After scrubbing, rinse the crank in clean solvent. Pay special attention to flooding the oil passages. Dry with compressed air if possible, or use paper towels. Avoid drying with linty rags.

When the crank is clean, a paper towel wiped over it will come off clean. Then you are ready to spray the crank with water-dispersant oil like WD-40.

Install Timing Gears—Now you're ready to install #3 main bearing, the crankshaft and distributor gears, spacer and retaining ring. Most machine shops press on the gears. That's fine for them because it's fast and easy, but then, they have a press. If convenience is your concern, have them press on the gears.

Another way of installing the gears is with heat. Fill an old pan with half a quart of vegetable oil. Put the timing gears in the pan, then heat it until wispy smoke starts to curl off the surface. Unless the smell of heated vegetable oil excites you and the ones you live with, do this job outside on a camp stove.

Have a helper hold the crank, or support it in a vise. Lightly coat the #3 bearing with moly lube, then slide it into position on the crank. Note that the dowel hole in the bearing is offset *toward the flywheel*. If you've got the bearing installed with the dowel hole closer to the pulley end of the crank than the flywheel, take it off and turn it around. Install the Woodruff keys. Don't oil the keys or keyways, just tap them into the crankshaft. Make sure their flat surface is parallel with the crankshaft, not slanted up or down.

Get a thick leather work glove on one hand, then dip into the hot oil with a screwdriver blade and extract the crankshaft timing gear. Working quickly now, let the excess oil drain off, orient the gear so the *chamfer* on the inner diameter *faces* the #3 main-bearing journal, and slide the gear onto the crankshaft with your gloved hand. The crankshaft gear has the aforementioned chamfer for reference, plus there is a *timing dot* on the opposite side. This timing dot faces to the *outside* of the case so you can line it up with the camshaft timing gear dot later.

Once the hot gear contacts the cold crankshaft it will cool quickly. You have adequate time to get the gear on the shaft—once. Make sure you press it all the way into place. If you

Inspect flywheel-mating flange for raised areas around dowel holes. File off any proud metal after removing pins. Four tight-fitting dowel pins are minimum requirement, so don't try to do without. If pins are loose and holes egged out, drill and install oversize dowels, or exchange crankshaft.

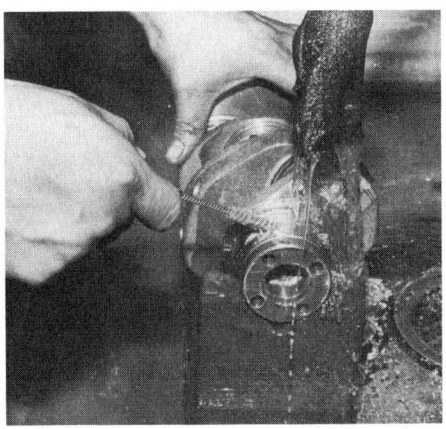

Once all grinding, polishing, filing, chamfering and so on is complete, wash crank. Use perfectly clean solvent and nylon brushes. Swab all oil passages, dowel pin and gland nut holes until perfectly clean.

Orient gear so timing dots face pulley-end of crank. Gear ID is also chamfered to clear snout-to-journal radius. Work quickly or gear will cool and shrink.

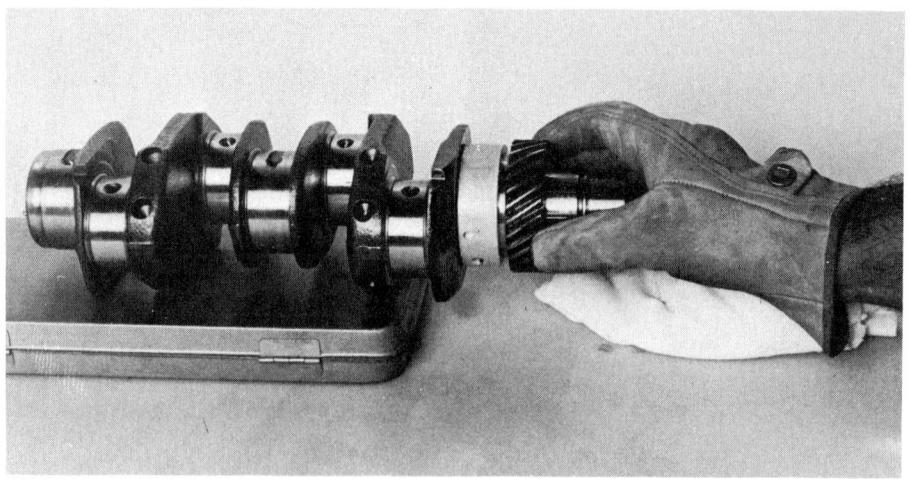

Quickly push hot gear into position. If hot enough, gear will slide right on, but cool to immobility in a second or two.

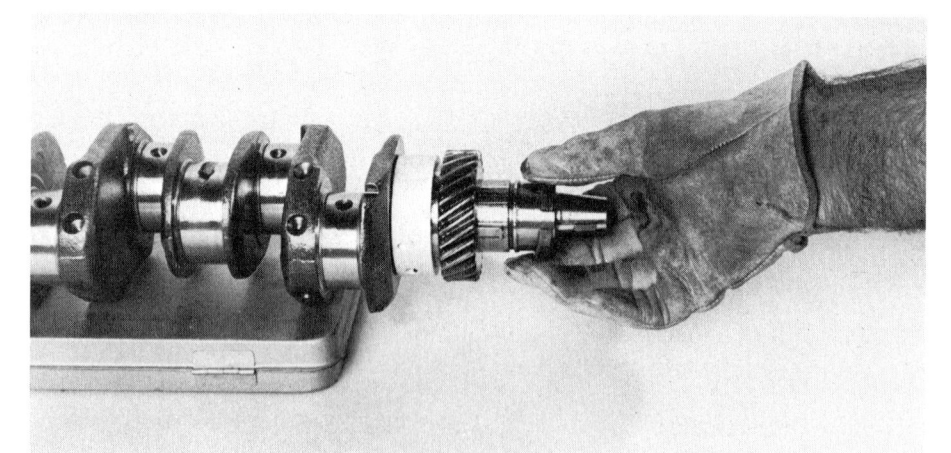

Spacer is easy to forget if not set out with gears ahead of time. Open section fits around Woodruff key.

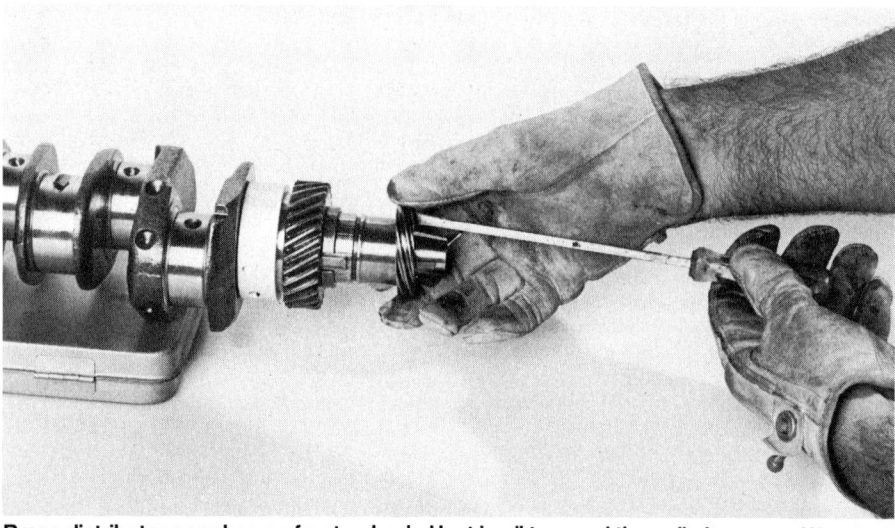

Brass distributor gear has no front or back. Heat in oil too, and then slip it on over Woodruff key and press against spacer. Rapid cooling will shrink it onto crankshaft in a hurry.

fail to do this, take the crank to a shop and have the gear pressed until it seats.

Slide on the spacer, then the distributor drive gear. There is no front or back to the distributor gear. Finish the installation with the retaining ring. Take care not to scratch the #4 main-bearing journal while installing the retaining ring.

Once the parts cool, respray the crank with oil, then store it in a plastic bag until ready for engine assembly.

Gland Nut—Unless the gland nut came off with stripped or galled threads, clean and reuse it. Remove the end ring and extract the felt ring and pilot bearing. Wash all parts in solvent, scrubbing the fine gland nut threads with a stiff nylon bristle brush. They should be perfectly clean so no dirt can mangle the threads.

Coat the gland nut with water dispersant oil: WD-40, CRC or the like. Put some motor oil on the felt ring, squeezing out the excess. Lubricate the pilot bearing with multi-purpose grease until all rollers are coated. Install the bearing, felt and end rings in the gland nut, then store the assembly in a plastic bag. The crankshaft should be bagged now, so just thread the gland nut into the crank for storage.

If the gland nut is stripped or galled, don't try to clean up the threads with a file and reuse it. The torque on this nut is too high for less than perfect threads. Besides, the threads are quite small and difficult to file. *Always replace a damaged gland nut.*

PISTONS & CONNECTING RODS

Unless reusing the cylinders, there is no need to inspect and clean the pistons. Just throw away the cylinders and pistons and buy new ones. If honing the cylinders, then you're probably planning to reuse the pistons. You must make sure they are reusable.

Don't make a snap decision about reusing the cylinders solely on their condition. To reuse the cylinders, both cylinders and pistons must serviceable. If the pistons are in poor or even borderline condition, go ahead and get a new cylinder set. In almost all cases this is the most economical choice. For a quite low price you will purchase new cylinders, pistons, rings and piston pins. The major benefit is as-new performance and longevity from the cylinders. You'll also get your VW back on the road more quickly because you won't be cleaning all those parts, honing cylinders, checking ring side clearance, rounding up new retaining rings, scraping carbon and so on.

Honing the cylinders and installing new rings is about half the cost of a new cylinder set. But at best you'll get half the life of new parts. The point to remember is that used bores and pistons are just that—used. Some of their working life has already been spent. You'll get the longest possible use from a rebuild with fresh bores and pistons.

Install snap ring to complete gear installation. Use proper tool stout enough to pry ring open. Above all, don't try screwdrivers and the like or you'll only scratch #4 main-bearing journal.

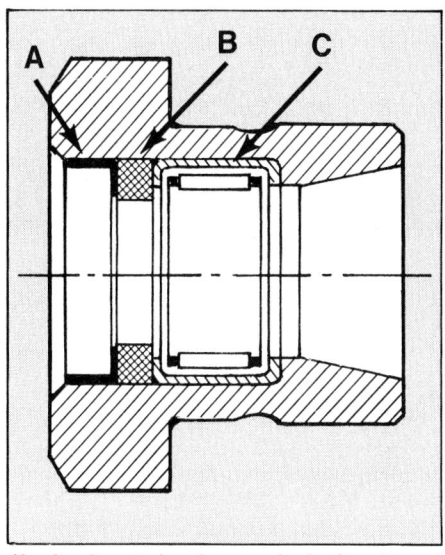

If gland nut is damaged don't attempt repair—replace it. Gland nut comes with end ring (A), felt ring (B), and needle bearing (C).

I'm not saying to throw the pistons and cylinders away on absolutely every VW being rebuilt. There are times when an engine must come apart for some oddball reason, and the cylinders still have plenty of life in them. In that case, hone the cylinders to break the glaze and use new rings. Or if the engine is still fresh, don't even take the pistons out of the bores during a teardown.

You can do this by pulling the cylinder off its piston only far enough to expose the piston pin. By removing the pin, the cylinder, piston and rings will come off as a unit, leaving their relationship undisturbed.

INSPECT PISTONS

If reusing the cylinders, inspect the pistons. The first step is to remove the rings. There are two ways to do this: by hand or with a ring expander. The ring expander is a plier-like tool that grips the rings by their ends and spreads them, allowing you to lift them off the piston. The advantage of using a ring expander is it lessens the chance of scratching the pistons. However, VW rings are supple, so spreading them with your thumbs works fine.

Ring removal requires holding the piston steady. The easiest way is to have a helper hold the piston while you remove the rings. You could also clamp them lightly in a vise. Clamp only on the pin-bore side, use wood, brass, aluminum or lead soft jaws, and clamp them very, very lightly.

Piston Damage—With the rings off, check for the following: general damage to the dome, skirt or ring lands, or excessive ring-groove, skirt or pin-bore wear. If any of these exist, the piston must be replaced. Also check for scuffing, scoring and collapsed skirts.

Gland nut houses parts that form transaxle input shaft pilot bearing. If gland nut threads are good, swab a very light coat of grease onto needle bearing rollers. New gland nuts are inexpensive, and have clean, unstressed threads.

Scuffing refers to abrasive damage to the piston skirt. *Scoring* is like scuffing, but is limited to deep grooves. Both types of damage are caused by insufficient lubrication, a bent rod, or very high operating temperatures, all of which cause skirt-to-bore contact.

The common denominator with these problems is excessive pressure and temperature at the piston and cylinder wall. The most common cause of scuffed pistons is that primary worry about VWs: overheating. Too heavy a right foot, missing sheet metal or engine-

Taper is being measured here at greatest wear point: at highest point rubbed by rings. End gap of new and used rings is also checked this way. Use upside down piston to square ring in bore.

compartment gasket, too hot a sparkplug, stuck cooling-air flaps, or a broken fan belt can all cause severe overheating.

In addition to visible scuff damage to a skirt, the shape and resulting piston-to-bore clearance may have been affected. The loads that caused the scuffing may also have bent or broken—*collapsed*—the skirt. Discard a piston with a collapsed skirt and start over. Additionally, a high-mileage piston will have lost some of its controlled-expansion qualities, even if it doesn't show bad scuffing. It's simply a matter of heating and cooling—*heat*

89

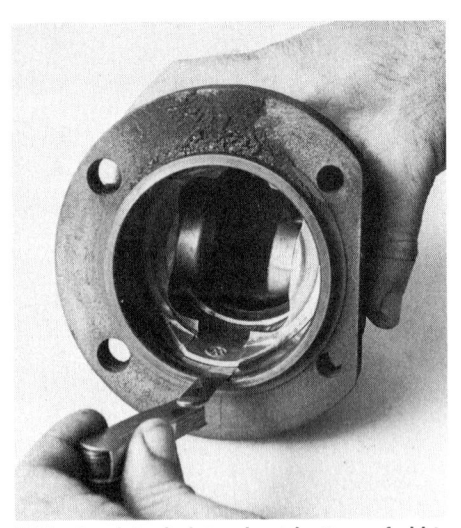

Widest point of piston is at bottom of skirt. Feeler gage is accurate enough to uncover collapsed skirts, but use micrometer to measure piston-to-bore clearance.

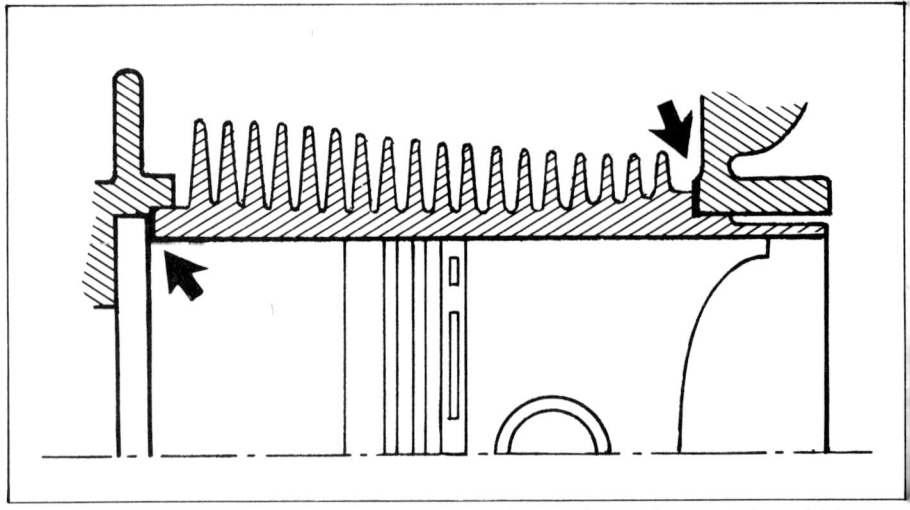

Another advantage to new cylinders is unworn sealing surfaces. Arrows point to high-wear areas on heads and case damaged by loose cylinders. Cylinders also sustain damage. Pulled studs and warped heads start wear cycle.

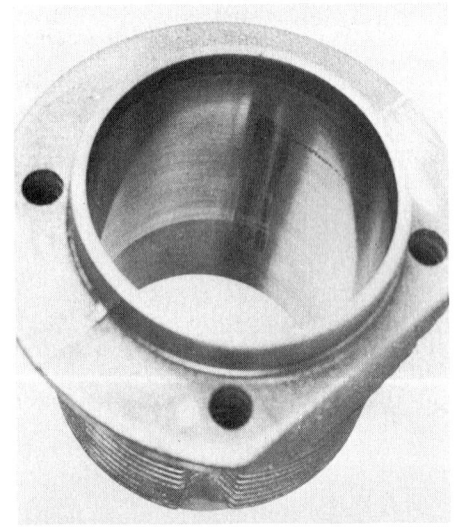

Scuffed cylinder walls mean scuffed pistons. Distance between piston and cylinder wall exceeded design limits. Don't reuse such parts when new ones are so cheap and plentiful.

cycling—thousands of times.

If a rod is bent or twisted, it will cause the piston skirt to wear or scuff unevenly from side to side. A little variation is normal, but when a piston skirt is scuffed left of center on one side and right of center on the other, you can be sure the rod is bent or twisted. Checking for bent rods is part of normal machine-shop procedure, but it never hurts to advise them of your findings. You can't check the rods accurately for straightness because you don't have the equipment—but you may have evidence that indicates they need checking.

Piston-Skirt Diameter—Measuring piston-skirt diameter is not so much to determine wear as to check for collapsed skirts. But, before measuring and judging the results, you should fully understand what piston-skirt collapse means.

Pistons are widest at the bottom of their skirts. This design allows a tighter piston-to-cylinder fit for better piston stability. Stability makes for quieter running and less hammering of the cylinders.

In normal service, piston and cylinder walls warm to operating temperatures and the piston skirts expand to a given size and shape. But, if the engine overheats, the piston skirts continue to expand until piston-to-bore clearance is gone. The piston squeezes out the lubricating oil, is forced against the cylinder wall, and begins scuffing. Scuffing increases piston heat even more, until the cylinder wall doesn't permit further expansion. The skirts are then overstressed, as they try to expand farther, and this causes the skirts to break or bend.

After the engine cools, the piston skirts contract smaller than before because they have been permanently deformed—they have collapsed. The result is too large a bore clearance, both during warmup and at normal operating temperature. The piston will slap during warmup, and possibly, after the engine is at operating temperature. The skirt may even break off.

Reusing pistons with collapsed skirts will cause a noisy engine. The oil-ring grooves may also be pinched, providing poor oil control.

Measuring Piston Skirt Diameter—Heavy scuffing is a common sign of collapsed piston skirts, but to be sure, measure the piston in two places and compare readings. First, measure across the skirt immediately below the oil-ring groove and 90° to the piston pin. Then, move to the bottom of the skirt and measure again. The skirts should be about 0.0005-in. (0.013mm) wider at the bottom. If the measurements are the same, or less at the bottom, one or both skirts has collapsed. The piston is junk. Don't even bother to measure the other pistons. Throw them away and buy a new cylinder set.

Inspect Domes—Because the melting point of aluminum is lower than that of cast iron, the piston dome is the first to suffer from heat caused by preignition or a lean fuel mixture. Such damage removes aluminum from the dome and deposits it on the combustion chamber, valves and sparkplug, or blows it out the exhaust. This leaves the dome with a spongy-looking surfaces that carbon adheres to.

Clean the piston domes so you can inspect them. If carbon buildup is heavy, remove it with a dull screwdriver or old compression ring. Make sure the screwdriver is old and worn. One with a sharp blade will scratch the piston easily. Don't use a sharp-edged chisel or gasket scraper on the pistons.

Follow screwdriver cleaning with a wire brush or wheel. Be extra careful to keep wire bristles off the piston sides, particularly the ring grooves. Light scratches on top of a piston won't hurt, but they will cause trouble in the grooves and on the skirts.

Piston-Pin Bore Wear—All air-cooled VW engines use *full-floating* piston pins. This means the piston pin is free to move in both the small end of the connecting rod and the piston-pin bore. The circlips at each end of the piston-pin bore retain the pin, keeping it from gouging

90

the cylinder wall.

The best way to determine piston-pin bore wear is to measure the bore with a snap gage and outside micrometer. It is unlikely you have these tools. An acceptable alternative is to fit the pin completely into its piston; now turn the piston on end and see if the pin will slide out of its bore. If the pin is a light push fit into the piston, but will not slide out, the clearance is fine. If the piston slides out with some resistance, the fit is still good. Only when the pin practically falls out of the piston is the piston bore worn out.

Don't be alarmed if the piston pins are a firm push fit. It may even be necessary to heat the pistons in 176F (80C) oil to cause enough expansion before the pins will slide in. This procedure is normal, and doesn't indicate too tight a fit. If you have a difficult time pushing the pins in the pistons, check for nicks or displaced material at the pin bore openings. One small raised spot can cause a very tight fit.

Circlip Grooves—After checking the piston-pin bores, inspect the circlip grooves. It is best to use a new circlip, but you probably don't have one handy. Therefore, use the least worn one you have. Fit it to each circlip groove in turn. See if the circlip will move nearly parallel to the piston pin. If it does, the groove is worn.

Piston replacement is the only realistic solution. Worn grooves are a sign of piston, pin, or connecting rod problems. Carefully check the piston wear patterns for a bent connecting rod and inspect closely for signs of detonation.

Ring-Groove Cleaning—There is no accurate way of measuring ring land wear if the grooves are dirty. You have a choice of two ring-groove cleaning methods. One is free, the other costs a little more. The free method uses the time-honored broken-ring method. The other uses a ring-groove cleaner.

Unless you're rebuilding an engine every now and then, use the broken ring. Break a compression ring in half and grind or file one end. Round off all ragged edges except the one cutting edge. This will keep you from removing metal from the ring lands. These surfaces must remain as tight and damage-free as possible for proper ring sealing.

Wrap tape around the ring so you don't cut your fingers on it. To make this job easier, regardless of the cleaning tool you're using, soak the pistons overnight. Put them upside down in a pan of water so the fluid covers the ring grooves. This softens the carbon deposits.

When using a broken ring to clean the groove, be *very careful* not to gouge or scratch the ring grooves. Use steady, relatively slow strokes. Keep the ring centered as much as possible at all times. This is important because piston rings rely on their grooves for sealing. Once a ring groove is damaged, the ring cannot seal well against the piston.

If you can afford some special tools, a ring-

VW pistons have a variety of markings. Arrow points to flywheel for correct thrust-face orientation, piston diameter in mm is stamped at lower right, and Std at left indicates piston is not an oversize. Two paint dots (at 2 and 10 o'clock) are for matching with cylinder paint dot and weight group identification when changing just one piston. Heavy weight group gets + visible at lower left. Normal weight is plain, light weight −.

groove cleaner is a useful one. It cleans the grooves much faster with less chance of damage.

Measuring Ring Grooves—The easy way of measuring ring-groove width or *ring side clearance* is with feeler gages and a new ring. However, don't spend the money for new rings just to discover that the old pistons are bad. Assuming you'd do the logical action and re-bore and install new, oversize pistons and rings, you'd be stuck with a set of useless rings. Therefore, use an old ring for measuring and compensate for its wear.

To measure ring-groove width, mike the thickness of an old compression ring. Subtract this from the *original* thickness measurement of that ring. Original thickness measurements are (mm):

40 HP	2.5
1300	2.0
1500	2.0 or 2.5
1600	2.0
1700	2.0
1800	2.0
2000	1.5 or 1.75

Add the difference to the specified ring side clearance. The result is the ring-groove *checking clearance*.

Now, insert the edge of the ring in the first compression ring groove and measure the distance between it and the groove with feeler gages. You have the checking clearance right away, but you must also check around the groove.

Slide the ring and feeler gage around the piston and you can easily tell whether the ring lands are bent. The feeler gage will tighten up in any constricted area and loosen where the ring lands are worn or bent. You've just killed two birds with one stone, ring side clearance and ring-land condition.

When you find the gage that fits snugly—not tight or loose—subtract this measurement from the checking clearance. The difference is *piston-ring side clearance*—the dimension used to determine if the piston-ring grooves are serviceable.

The side clearance for new Type 1—3 rings follows:

Top ring	0.0027—0.0039 in. (0.07—0.10mm)
Second ring	0.0020—0.0027 in. (0.05—0.07mm)
Oil ring	0.0012—0.0020 in. (0.03—0.05mm)

Type 4 side clearances are:

Top ring	0.0023—0.0035 in. (0.06—0.09mm)
Second ring	0.0016—0.0027 in. (0.04—0.07mm)
Oil ring	0.0008—0.0020 in. (0.02—0.05mm)

The wear limit for all engines is 0.0047 in. (0.12mm) for the top ring, 0.004 in. (0.10mm) for the second and oil rings.

Oil-Ring Grooves—Because oil rings and their grooves are so heavily oiled, compared with compression rings, they wear little, if any. Therefore, if the compression-ring grooves are OK, visually inspect the oil-ring grooves. If they look fine, they should be fine. I give the correct oil-ring groove widths, however, so you can measure their side clearance if you want. When measuring oil-ring side clearance, you'll have to install the oil ring into its groove. Otherwise, there is no way of obtaining the correct ring thickness.

INSPECT CONNECTING RODS
Obvious Rod Damage—When VW engines are overheated or run out of oil, the connecting rods are one of the first parts to suffer. This is because they are fed oil that passes through the main oil gallery, then the main bearing, and finally through the crankshaft oil passages to the rod bearings. If oil gets too hot and thin, or there isn't enough to go around, the rods are the first place the oil won't reach. Temperatures soar from metal-to-metal contact and a "thrown rod" results.

The bearing can be squeezed like toothpaste from between the rod and crank, and the crank journal beaten into a rough, dented oval. Both crank and rod will then turn black, blue and yellow from the heat. Carbon in the oil turns black, and where they're more badly burned,

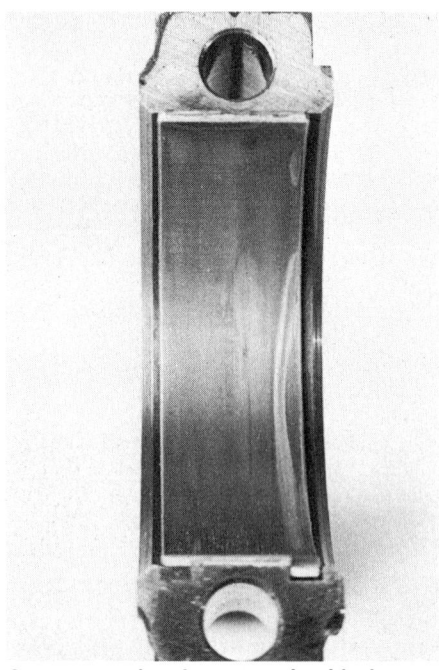

Any uneven bearing wear should trigger a close look at rod journal or rod itself. This wear is confined to top of bearing in relatively small arc and it's probably from power-stroke loads, not edge-riding or journal taper. Crank and rod checked out fine.

Dial indicator portion of rod machine shows wear instantly. A few seconds honing the big-end on the machine's precision hone and it will be perfectly round. This is called reconditioning, or *sizing* the rods.

you'll see blue. Once the parts have been heated to such extremes, there is no saving them. So, if you lost a rod and it is burned black and blue, don't consider reusing it. The metal in the rod has been weakened to the point where it's unusable. Throw that rod away and start with a new one.

If the engine is merely worn out, not a meltdown victim, you should be able to reuse the connecting rods. Typically, they will have the big ends reconditioned and new piston pin bushings installed. Type 1—3 rods are more likely to need big-end reconditioning than the massive Type 4 ones, but inspect every rod to be sure.

Check Big End—Think of a rod's big end like the main-bearing bores in the block: both can change shape. The conventional rebuilding fix for these problems is called reconditioning, or *sizing*. It is similar to align-boring or honing the main-bearing bores.

The rod-cap mating surface is ground, removing several thousandths of material. The cap is then installed on the rod and the nuts torqued to specification: 24 ft-lb (3.3 kg-m) for all engines. The bearing bore is now eggshaped. Next, the big end is honed round on a special precision hone. The machinist checks the progress with a dial-indicating fixture.

Reconditioning the rods ensures that the bearing bore is truly round, not tapered, and the correct diameter. Almost every Type 1—3 engine needs rod reconditioning at rebuild time, and shops perform this step as a matter of course. For Type 4 and low mileage Type 1—3s the best way of checking the big end is with a dial indicator. Take the rods to the machine shop and ask them to check them. They'll check the rods with a special dial indicator built into the rod reconditioning machine. Any out-of-round rods show immediately with this tester and a decision to recondition or not can easily be made.

An important part of VW connecting rod service is replacing the rod nuts. *Always use new rod nuts* for maximum strength and accurate torque readings.

Piston-Pin Bushing—All air-cooled VW connecting rods use a bronze bushing at the small end of the connecting rod. When the bushing is worn, it is pressed out and a new one pressed in. The new bushing is then reamed to exactly fit the new piston pin. Because almost every air-cooled rebuild results in new cylinders, pistons and pins, the bushings are not checked. They are changed as standard procedure, because it would only be a lucky coincidence if the new piston pins fit the old bushings. Therefore, checking for piston-pin bushing wear is considered a waste of time.

If reusing the pistons and cylinders, however, you will also reuse the pins. So, check the piston-pin bushings to see if they can be reused. Make sure the parts are at normal room tem-

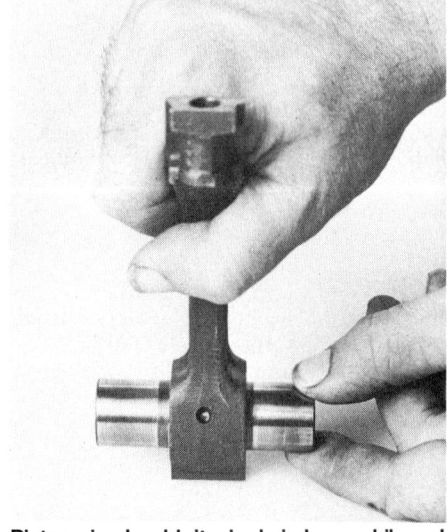

Piston pin shouldn't wiggle in bore while rod is held stationary. VW pins are a push-fit at room temperature. Any perceptible slop means new bushings. Volume rebuilders replace bushings in all rods, and chances are high you will too.

perature, then try sliding the pin through its pin bushing. It should be a light push-fit. This means you will have to push somewhat on the pin, but not force it. Also, if the pin slides in on

New bushings are honed to proper size on rod machine. Same process is used to resize big end, but smaller honing stones are used.

Bushing diameter is checked several times during honing. No matter how fancy the machine, it's still the machinist who makes decisions.

its own (too easily), the bushing is worn and needs replacing. Once the pin is centered in the bushing, try to rock it back and forth. There should be no motion; if there is, replace the bushing.

Bushings are replaced by pressing. It is best to install the new bushings with a hand-powered tool, like an *arbor press*. This allows the machinist to feel the amount of pressure needed to force the bushing into the small end of the connecting rod. The primary concern is that the bushing may be too loose in the rod. If the bushing is too loose, it may rotate in the connecting rod during engine operation. This will block the oil hole to the piston pin. More rapid bushing wear and an intermittent ticking noise will result.

After pressing in, the new bushings are reamed with a hone to fit the piston pins. Therefore, you must have the piston pins you will be using available for the machinist to measure.

Align Rods—Another machine shop job is checking *rod alignment*. Even in normal engine operation the connecting rods can twist and bend slightly. While most of these minor bends pose no significant threat to rod condition, it's smart to bend the rod back into perfect alignment so it won't bend any farther. As long as the rod is bent cold, without heating with a torch, no harm is done, although the procedure looks barbaric.

One of two fixtures is commonly used to

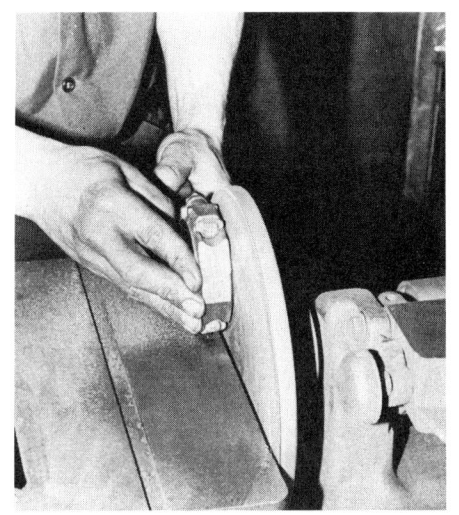

Touching rod to disc grinder removes nubs or edges left by sizing. VW specifies fairly wide rod side clearance, so too narrow a rod is rarely a problem. Little material is removed here anyway.

measure bend in the rod. One has a micrometer scale, the other, two flat plates the machinist can monitor for stray light. If any bend is found, the rod is securely clamped, then twisted or bent with long metal pipes or bars. Usually a couple of tugs will straighten a rod.

Numbers on connecting cap are useful. Number on cap and rod should match; if not, rod doesn't fit cap. Numbers match on same side of the rod, helping with cap-to-rod orientation.

VW rods don't seem to twist and bend as badly as rods from many other engines, so you don't always see air-cooled rods being aligned. This is especially true of the Type 4 rods.

Rod Replacement—When assembling an engine by cannibalizing the remains of others,

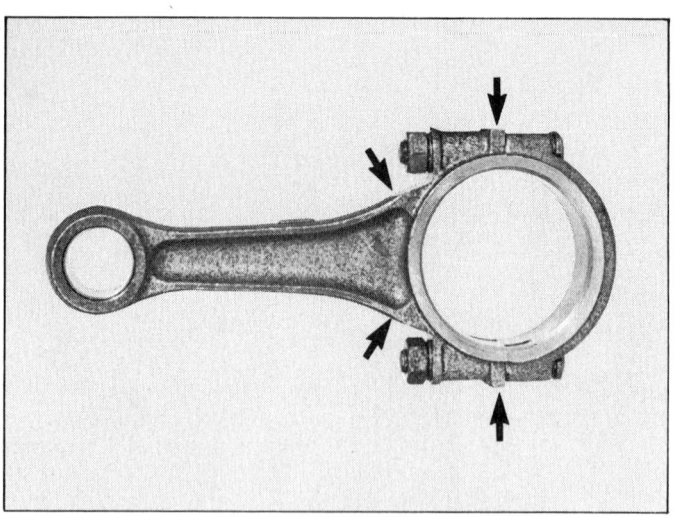

Arrows show where metal can be safely removed for balancing.

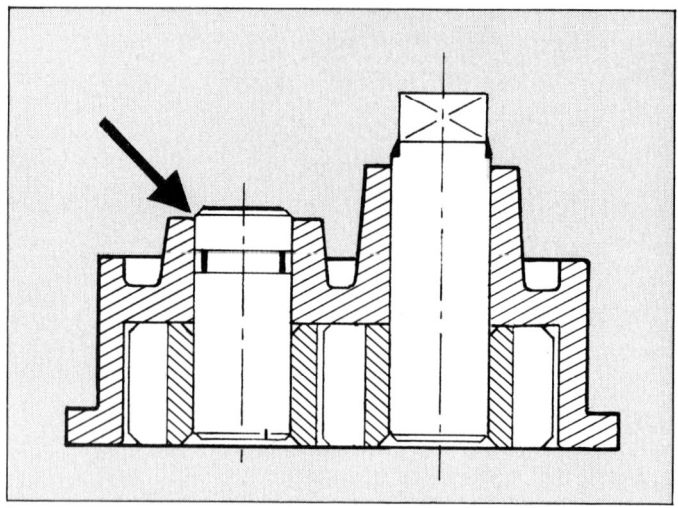

Driven-gear shaft should protrude from pump body 0.020—0.040 in. at arrow. If shaft is loose in body, peen body around shaft at arrow.

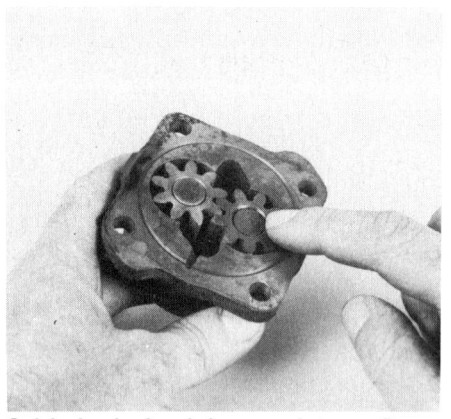

Quick check of end clearance is to run finger over gears and pump body. Very slight edge indicates some end clearance, which is OK, but there shouldn't be a large step here. Reduce step by lapping pump body.

you often end up looking for a replacement rod. This is fine, but remember to check it for total weight against the other rods.

Replacement rods from VW come in two weight ranges. They are identified by paint marks. On Type 1—3 rods, brown signifies a total weight of 580—588 grams (g), gray is 592—600g. Type 4s range from a white 746—752g, to a black 769—775g. No rod can weigh 10g more than any other rod in the engine, so note these paint marks.

On a used rod most paint marks have been washed down the hot tank drain and you'll have no idea how much the rod weighs. A good science grade scale, just like you used in chemistry class, can be used for determining total weight. Weigh the rods with nuts and without bearing inserts.

Alternately, you can use a beam scale to bring all rods to the same weight as the lightest of the bunch. You can't find the exact weight this way, of course. Reduce rod weight by judicious grinding. Carefully remove small amounts of metal from the side of the rod beam, right above the cap and along the concave portion of the shank above the big end. Check rod weight after each grinding effort and don't take too much metal off. Be careful not to erase any numbers or the bump on the shank.

OIL PUMP

Because oil pumps are so over-lubricated, they typically don't wear much. Thus, inspection and cleaning are all that are usually needed. But if the engine has been running dirty oil, the pump's utility is questionable because of the constant supply of grit that was moving through the pump with the oil. Gradually clearances between the pump gears and housing open up, and pumping efficiency is lost.

Disassemble & Inspect Type 1—3 Pump— The gears on this oil pump fit into the oil pump housing and are sealed with a gasket and cover plate. The cover plate was removed during engine disassembly, so slip the gears out of the housing. The gear with the shaft attached to it is the *driving gear*. The gear that fits over the shaft attached to the pump housing is the *driven gear*.

Discard the paper gaskets. Inspect the cover plate and housing for scratches, gouges and wear. When worn, these parts will have circles cut into them from the gears. If the housing is worn with deep gouges, or if you can feel a distinct ledge between worn and unworn areas, replace it. If the cover plate is worn, either replace it, or lap it smooth.

To lap the cover, get a piece of 220-grit Wet-n-Dry paper and set it on a perfectly flat surface. A very thick piece of glass, marble or granite will work. Flood the paper with water or solvent, then push the cover back and forth across the paper. Change directions often so you won't score the cover. Stop lapping after all signs of gear wear are gone.

Inspect the gears for wear. There will be a pattern polished onto the mating areas of the gear teeth, but no scoring or actual wear should be visible. Replacement is the only cure for worn gears.

Wiggle the driven gear's shaft, the one attached to the pump housing. It should be secure in the housing. If it wiggles, flip over the housing so you can see where the end of the shaft is surrounded by the housing. Using a small hammer and center punch, peen the housing around the shaft. This should stop the wiggling. If it doesn't, replace the housing.

Also check the fit of the drive-gear shaft where it passes through the pump housing. Slight movement is acceptable, but excessive wobbling is unacceptable.

With both gears in the pump, check for *backlash*. Unless you have wire gages, this turns out to be a subjective test. Just make sure there is not excessive clearance between the gear teeth. Maximum allowable backlash is 0.008 in. (0.20mm), and half that or less is preferable. To check this without gages, see how much one gear rocks back and forth while you hold the other stationary. Faint movement and no audible clashing is fine. If movement is very definite and you can get the gears to clink together, then backlash is starting to grow. Closely measure all other clearances, and start thinking about buying a new pump.

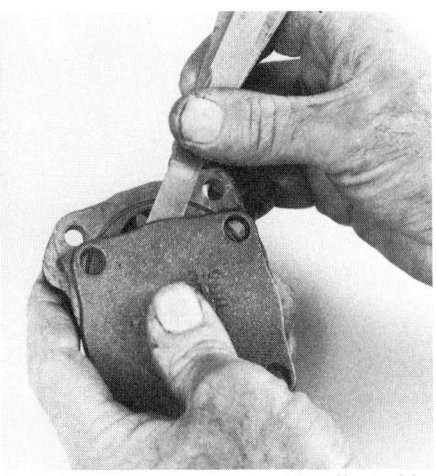

End clearance measurement is affected by dirt on pump body or cover, so have both clean. Maximum clearance is 0.004 in., but 0.000-in. clearance is even better.

Dual-stage oil pumps in Auto-Stick engines use intermediate plate with rubber seals. Remove Woodruff key (arrow) on driving-gear shaft before installing or removing intermediate plate. If key is left in place, it will tear seals, causing leak.

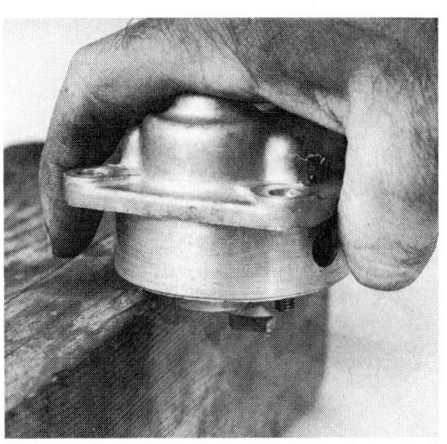

Rap small studs against wood block to start obstinate Type 4 pump apart. Once bearing plate is flush with studs, it will wiggle out of pump body by hand.

Now check *end clearance*. After washing any lapping grit or other dirt from the pump, reinstall the gears in the housing. Using no gasket, place the cover halfway over the housing so you can slip a feeler gage between the gears and cover plate. This measurement should be 0.004-in. (0.10mm) or less. If larger, end clearance is excessive. Reduce end clearance by lapping the pump housing like you did the cover plate. Lap gently and recheck often, as it doesn't take long to remove a thousandth or two.

It is very important to ensure there is no dirt in the pump when making these measurements. Just one piece of lapping grit can raise the cover and you'll measure a false, excessive end clearance. Then you would lap off too much material and end clearance would be insufficient. To cure that condition, you must lap the end of the gears.

End clearance can be reduced almost to zero with a disc sander or by lapping. Reduce end clearance with the gears in the pump body. There will immediately be light metal removal on the gear ends, but keep cutting the pump body until it's beyond a doubt the gear ends are being lapped or ground. Then you must deburr the gear ends with light filing and completely clean the pump with clean solvent.

Reducing end clearance all the way to 0.000 in. is not mandatory, but greatly improves the pump's efficiency. End clearance is not actually zero because the gasket provides the necessary clearance for the pump to work properly.

Double Oil Pump—The two-stage oil pump used in Auto-Stick cars gets the same service as the single-stage pump, but there are two gear sets to look after. Treat the first, or larger, stage just as you would a single-stage pump. The smaller gear set should be checked for gear tooth wear, plus plate and end cover scoring. Check end clearance both at the intermediate plate and end cover. Use the same measurements as for the single-stage pump. When removing and installing the intermediate plate, make sure the Woodruff key in the first-stage driven-gear shaft is removed. Otherwise, the seals will tear as they pass over it.

The end cover of a double pump is fitted with a pressure-relief valve. This valve controls pressure to the torque converter. Remove the valve plug and extract the spring and piston. Check for burrs, dirt, and wear. Make sure the piston is free to move in its bore, then reassemble the valve.

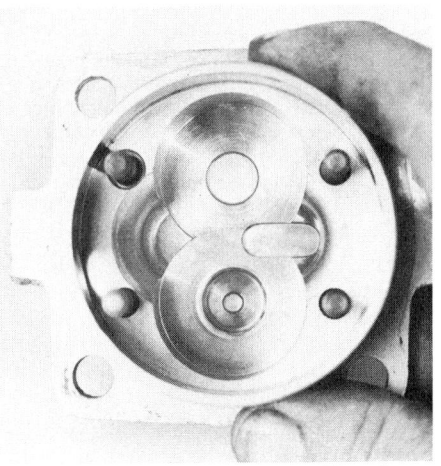

Inside will show some scoring. There's no way to lap here, so heavy scoring means pump replacement. These light marks are normal, and pump can be reused.

Type 4 bearing plate scores also, and like pump body is difficult to lap. Marks come from dirty oil (change it!). Pump has enough reserve capacity to overcome minor scoring like this.

Disassemble & Inspect Type 4 Pump—Servicing the Type 4 oil pump is similar to working with the single-stage Type 1—3 pump, just the parts are assembled differently.

Type 4 gears are carried on the *bearing plate*, which in turn fits into the pump housing. The trick is to remove the bearing plate from the housing. First, of course, remove the four nuts from the bearing plate backside. The bearing plate is now retained in the housing solely by an O-ring. Persistent wiggling and tugging on the drive-gear shaft will free the bearing plate. I like to very lightly tap the exposed stud ends against a wooden block. First tap one side, then the other. Eventually the bearing plate will

95

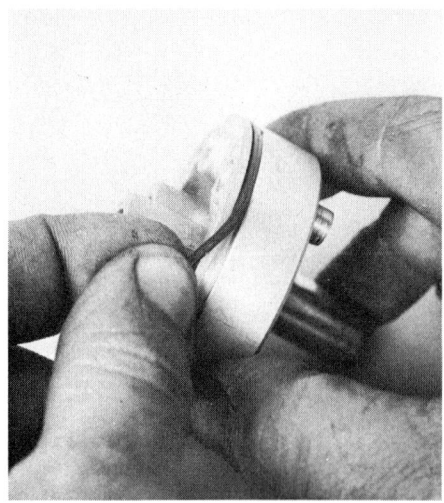

Peel old O-ring from Type 4 bearing plate and install new one from gasket set. Use no sealer. Note dowels around stud holes. If loose, store them in small box or coffee can to avoid misplacing them.

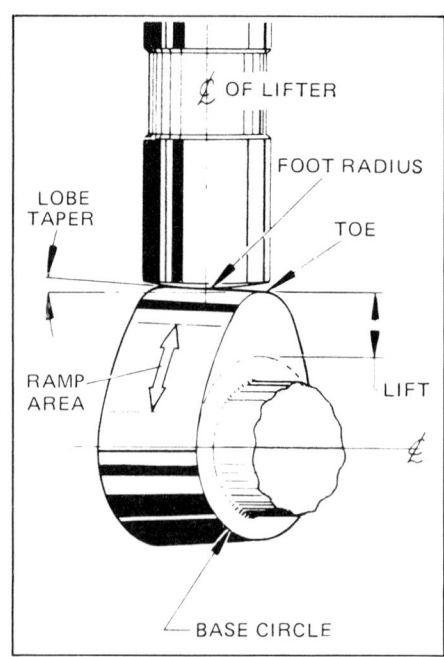

Drawing exaggerates cam lobe and lifter. Lobe and lifter are designed so lifter rotates in its bore. If cam-lobe toe is worn across its entire width and corresponding lifter foot is worn flat or concave, replace camshaft and all lifters. Drawing by Tom Monroe.

Typical Type 1—3 cam lobe wear is along toe of lobe and down one side. Pitting is sure sign lobe is worn, but lobe can be bad without it.

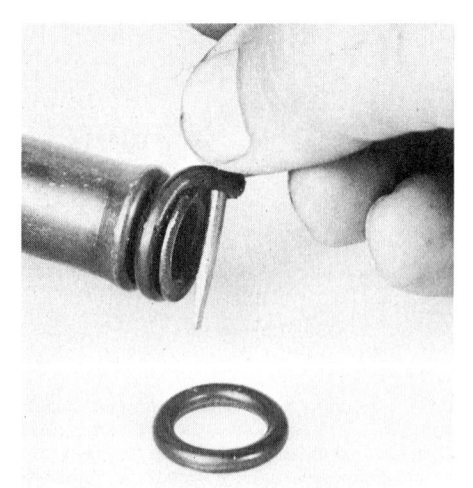

Install O-ring on case end of Type 4 oil-pump pickup tube. Use no sealer.

slide out far enough to uncover the O-ring. At that point the plate falls out, so watch for it.

Once the pump is apart, clean it and inspect for scoring and gear wear. You can't check the end clearance like a Type 1—3 pump, so go on the depth of scoring visible in the pump housing and breaker plate. You can't lap either one, so replacement is about the only cure for Type 4 pump problems. You can fit another bearing plate, if that is the only problem. This will cure most end clearance problems.

Parts or Assembly?—Most of the time oil pump service is nothing more than cleaning and checking. A light lapping with Type 1—3 pumps helps efficiency by reducing end clearance, but wholesale parts replacement or modification is not normal. If the pump needs plenty of parts, buy a new unit. This makes more sense than sticking a bunch of new parts in an old housing.

Oil Pump Blueprinting—Lots of good information on modifying stock Type 1—3 VW pumps is in HPBooks' *How to Hotrod Volkswagen Engines.* If you like to get the most out of a part, you should get the book. There just isn't enough room here to cover blueprinting completely.

Final Assembly—Pack the pump gears with white grease or petroleum jelly during final assembly. This ensures the pump will develop sufficient suction, or *prime,* to draw oil from the sump during initial engine start-up. Use only new gaskets and O-rings. If you bought a new pump, disassemble it and check it for grit and excessive clearances. Then pack it with petroleum jelly and set it aside until engine assembly. Store the pump in a plastic bag to protect it against dirt.

CAMSHAFT

A VW camshaft is serviced with its gear intact, so don't worry about removing it. If the gear or cam is worn, the new one will come with a gear installed.

Inspect Cam Gear—Stand the camshaft on end, with the gear flat against the bench. Hold the gear firmly in one hand and try moving the camshaft. If there is movement in any direction between gear and cam, the gear is loose on its rivets; replace the entire assembly.

Cam Wear—Cams lobes and lifters go for a long time with no measurable wear, then once wear gets started, they wear rapidly. This is because cams are *surface hardened.* Once the extreme operating pressures between cam lobe and lifter wear through this thin surface, the cam gets ground down. It doesn't make sense to reuse an old cam and lifters, especially if you're planning to use the engine for a long time. If you insist on reusing the old cam and lifters, and the cam does go bad, expect poor performance, noisy valves and complete engine disassembly to fix the problem.

If you had problems keeping the valves adjusted for longer than 500 miles, the cam and lifters are worn. You can confirm this by inspecting the wear patterns worn onto those parts. If valve adjustment wasn't a problem, use the same inspection to detect wear that is just getting started. This is especially true if you have an engine with hydraulic lifters. Then there are no easy signals the cam and lifters are wearing out.

Inspection—Look at the nearby drawings and photos for camshaft lobe nomenclature and

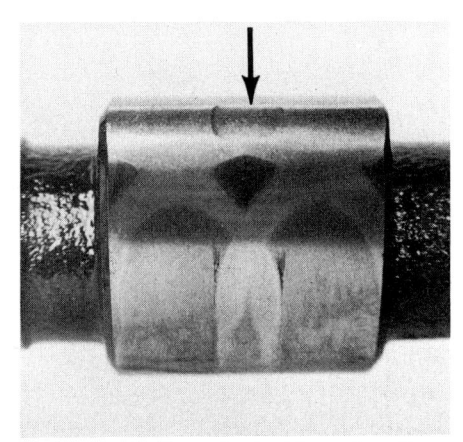

Type 4 cams wear in double-arch pattern. This one is worn out; note rectangular depression on lobe's toe (arrow). Depression is another typical Type 4 wear pattern, and means cam isn't acceptable core.

Using a micrometer or vernier caliper, measure across the lobe halfway between its heel and toe. This is the *minor* dimension. Then measure the *major* dimension, from the tip of the toe to the *heel*. Both measurements should be made at the points of maximum visible wear and the same distance for the same edge of the lobe. Now, subtract the minor dimension from the major dimension to get approximate lobe lift.

You can't depend on these measurements to accurately represent lobe lift, particularly with a high-performance cam. This is because you'll be measuring across the ends of the *opening* and *closing ramps*—transition sections of the lobe between the toe and the *base circle*. The reading you get will be greater than base-circle diameter; as a result, this measurement of lobe lift will be *less than actual*.

Measure all lobes and record the results. Compare the findings by placing all intake lift figures in one column, and the exhaust figures in another. All lobes should be within 0.005 in. (0.125mm) of others of the same type. If there is one bad lobe, the cam must be replaced.

By the way, the same cam profile has always been used on Type 1—3 engines and there is no performance gain to be had by swapping VW cams. In fact, you'll cause all sorts of headaches if you install anything but a stock cam in a fuel-injected engine. You can tune such a combination until exasperated, but there will still be flat spots and stumbles.

Runout—Checking runout requires supporting the cam in a lathe or V-blocks. Then use a dial indicator to measure runout at the center cam bearing while the cam is slowly rotated. New cams should measure no more than 0.0008 in. (0.02mm) and used ones less than 0.0016 in. (0.04mm). Runout is not a typical problem, and you shouldn't have any trouble with it.

End Play—To check end play, the cam bearings must be installed in the left case half. Of course this step doesn't apply if working on an early 40-HP engine because there are no cam bearings. Next, lay the camshaft in the left case half and set up either a dial indicator on the cam gear, or use feeler gages. With a dial indicator, tap the camshaft all the way in one direction, zero the instrument, then tap the cam the other way. Read end play directly.

If using feeler gages, tap the cam all the way in one direction. Then find the gage that just fits between the thrust bearing and the fence on the camshaft. The *thrust bearing* is the third cam bearing—the one with the lip overhanging the bearing saddle. The *fence* is that part of the camshaft standing out from the shaft right next to the thrust bearing. Next, tap the cam the other direction and measure between the cam gear and thrust bearing.

Besides being easier to use, the dial indicator is more accurate. End play should be 0.0016—0.0051 in. (0.04—0.13mm) with a wear limit

This is what 99% of Type 1—3 lifters look like when removed from engine. Rays streaking outward at edge are from previous regrind. Worn center and dark middle ring indicate this lifter is shot, and needs regrinding again.

of 0.0063 in. (0.16mm). Remember, installing new cam bearings will reduce these clearances somewhat, but if the measurement is well over the wear limit, it's time for a new cam.

For those 40-HP engines *sans* cam bearings, the only choices to correct excessive end play are to replace the cam, the case or have the case modified to take cam bearings. Having the case modified to accept cam bearings is probably the best course because it will restore the correct oil clearance and help restore the correct end play. Also, at any later engine rebuilds you can simply change bearings.

Cam Replacement—Because VW cams are rarely separated from their drive gears, cams are bought and sold with a drive gear riveted in place. Don't bother with removing the gear.

Minor variations exist between cam gears. A range of cam gear diameters is made, so they can be matched to specific engines. These variations compensate for differences between the centerlines of camshafts and crankshafts as they mount in their bearings. In other words, slight variations in case construction and align-boring make slightly different camshaft gears necessary.

All you need to remember is to get the same number gear as you already have, assuming camshaft gear backlash was correct in the old engine. Look on the cam gear for ink markings like −2, −1, 0, +1, +2, +3, +4, +5 and so on. Make sure the new cam has the same number inked on its gear.

For the technically minded, each numeral specifies the difference in pitch radius from nominal in 1/100mm. So a +2 gear has a 2/100mm larger pitch radius than a 0 (zero)

wear patterns. Start the inspection at the *toe* of each lobe. Look for rounding of the lobe when viewed from the side. Also note the wear or polish pattern on the lobe. On Type 1—3 engines the pattern will run across the toe and down one side. Type 4 lobes wear with two arcs, one passing over the other. These patterns are difficult to read all by themselves, so it is difficult to say just when a lobe is worn. The definite clue is when you see pitting on the toe or in the wear pattern.

Pitting is the first sign the hardened surface is being worn through. If any pitting is showing, the cam is shot and needs replacement. If pitting is spotty and not very deep, the cam is acceptable for regrinding. You'll be able to trade it in on a regrind. If pitting is deep and widely dispersed across the toe, however, the cam won't be a usable core.

Lobe Lift—Your preliminary inspection will usually tell you what you need to know, whether or not to replace the cam and lifters, almost every time. And almost every time the camshaft and lifters will need replacing. This isn't because VW builds lousy cams and lifters, just that these moving parts work under extreme pressure.

If you don't trust a visual inspection alone to be sure if a cam is worn, measure *lobe lift*—unless it's obvious they're all worn away, of course.

Measuring cam lobes is a comparison check, one lobe to another. The purpose is to compare relative *wear*, not *lift*. Because cam-lobe wear accelerates once the surface hardening is worn through, you should have little trouble finding a bad lobe.

Type 4 lifter wear is more subtle than Type 1—3. If there is any pitting, replace it. Inspect for polished, concave wear pattern in center of lifter. Reflecting light is best way to spot worn, non-pitted Type 4 lifters.

Be suspect of all factory Type 4 lifters and camshafts. SIR reports 30—40% of these cams are so worn they're not acceptable as cores.

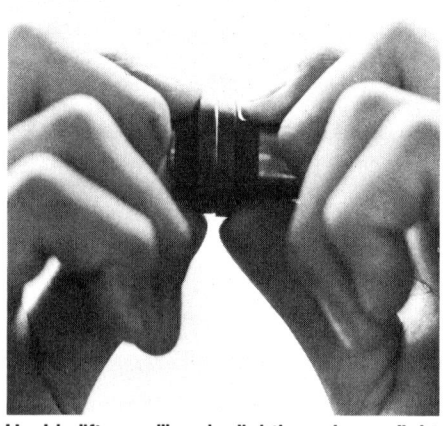

Usable lifters will rock slightly and pass light between them. This is a pair of reground lifters.

Rock test with worn Type 1—3 lifters results in light-tight fit between lifter bottoms. If no light passes, or no rock is felt, replace them.

gear. These are slight variations, and engines have been successfully run with gears different from those supplied at the factory.

You can check *camshaft gear backlash* by laying the cam and crank in their bearings in one case half. Hold the crank stationary and try and rock the cam gear back and forth. Backlash should be barely perceptible, if at all. More cam gear backlash information is found on page 122 of Chapter 7. If unsure whether the backlash was correct in the old engine, check it now before buying a replacement cam. Then if it doesn't seem OK, get a larger or smaller gear as required.

Lifter Inspection—Skip this step if buying a new cam, because then you must install new lifters. At the risk of boring you: Keep those lifters in order or you must buy new ones! It's all too easy to forget their correct order if you start inspecting them at random; use order and precision as you work. Do them one at a time.

Inspect the lifter bottom for damage. Look especially for pits or depressions. If you find any, the lifter is junk. Don't bother going any further, just replace all the lifters. Chances are the others are in a similar state. You can check them too, but the wiser (read: cheaper in the long run) action is to throw them out and buy a new set, or turn them in for a reground set.

If a lifter is worn enough to pit, you'll see pronounced wear marks. Look at its bottom from different angles and watch the reflections. Make this visual check even if it appears fine. If you see a circular pattern at the center which reflects differently from the rest of the bottom, that's an indication the lifter's bottom may not be convex.

VW lifters are commonly reground by rebuilders. These lifters have a grinding pattern that radiates from the bottom's center outward. As the lifter wears, the radiating lines are erased, until a well-worn lifter will show marks only at its edge. Replace such a lifter, and check for a corresponding amount of wear on the camshaft.

Double-check the lifters with the "rock-test." Butt one lifter bottom-to-bottom with another. If you can't rock them, they are worn beyond use. Also sight between the lifter bottoms. The slight convexity of good lifters will allow a little light to pass at either edge as you rock them. Worn lifters pass no light or a little in the middle. A lifter worn flat or concave like this also has a worn cam lobe. Rechecking the companion lobe toe should show some wear. If this is the case, both cam and lifters are junk. However, if you can rock the lifters, you're in luck. Keep on checking.

Clean Lifters—Almost all VWs have plain old mechanical (solid) lifters. They are a single piece, and need only a good scrubbing with solvent. A couple of hours soaking in carburetor cleaner helps get the varnish off.

Type 4 engines with hydraulic lifters are another story. These lifters are a collection of small, precision parts. The correct way to clean them is to disassemble, dip in carburetor cleaner, and assemble them. This takes time, but the tight, precision-ground fit of the plunger to the body keeps solvents from penetrating when the lifter is assembled.

Get a helper to assist when disassembling or reassembling hydraulic lifters. While holding the lifter body, depress the plunger with a pushrod, then remove the wire lockring. Invert the lifter over a clean rag, then lightly tap it to coax out the pushrod cap, plunger, ball check valve, check-valve spring, check-valve retainer and plunger spring. These exact parts may not be in the lifters because some engines have had aftermarket hydraulic lifters installed, but there will be corresponding parts inside.

The first step is to keep the parts in order until you've got a firm idea of how they all go back into the lifter. Spend some time examining how the various chambers fill and empty of oil to constantly keep the valve clearance at 0.00 in. (0mm).

Clean all parts in carburetor cleaner, rinse in solvent and dry with lint-free paper towels. Don't use rags, old tee shirts, diapers or other cloth products. These all leave lint behind, lint that will clog the tiny oil passages in the lifter. Paper towels leave behind a residue, but it dissolves in oil and is no threat.

Reassemble the lifters in a pan of clean 30W oil. This is a real mess, but is the best method of filling the lifter internals with oil. Get your hands clean before you start, or the oil will lift the dirt off them and put it inside the lifters. Also have a stack of towels ready when you finish, or you'll track oil around the shop looking for a rag.

Clean the lifters one at a time to keep the internal parts with their original lifter body. This will also help you keep the lifters in order with each other.

If the engine was kept clean with frequent oil and filter changes, and is disassembled now more for a valve job than major reconditioning, then you can *probably* just clean the lifters without disassembling them.

Fill a clean pan with fresh solvent. Submerge the lifters, then work the plungers up and down with a pushrod. This will pass the solvent through the interior chambers. After discolored solvent ceases to drift from a lifter, remove it and work the plunger out on the bench to get all the solvent out. Then submerge the lifter in fresh 30W oil and work it again with the pushrod. This will help prime the lifter.

Adding Hydraulic Lifters—If your Type 4 has mechanical lifters, consider changing to hydraulic ones. This isn't the cheapest swap because you have to buy the corresponding camshaft, lifters and different pushrods, but it has benefits.

First, there are no more valve adjustments. This alone may sway you. The valves are kept in perfect adjustment all the time, which helps extend their life. More power is available because the valves are properly adjusted and not waiting for you to get around to adjusting them.

Hydraulic lifters are considered low-performance parts by most people because many highly tuned engines use mechanical lifters. But in reality, the only performance limitation of hydraulic lifters is their lower rpm limit. This isn't a consideration with stock Type 4s, or even many with aftermarket carbs and other tweaks. Hydraulic lifters are stable to around 6000 rpm. That's sufficient for almost any need; if you're running above that, then you're on your way to the races and need HPBooks' *How to Hotrod Volkswagen Engines*.

In fact, you'll gain power with hydraulic lifters because they automatically adjust the valves to varying engine temperatures. Any mechanical-lifter valve adjustment is a compromise. The adjustment must be tight enough to take maximum advantage of the cam profile, but loose enough to provide clearance when the engine is hot and swollen like a balloon. That means the mechanical-lifter valve adjustment is at its optimum at only one engine temperature. The hydraulic lifters stay at optimum from the second they fill with oil at engine start-up to engine shut-off. They keep the valves perfectly adjusted in Arctic winter or Sahara summer.

Hydraulic lifters will make the engine considerably more quiet. It makes touring in your Bus and especially 914s and Type 4s a lot more enjoyable. Your neighbors might thank you, too.

If you want to install hydraulic lifters, see your parts supplier. You'll need a camshaft for hydraulic lifters, lifters and a kit. The kit contains shorter steel pushrods and some rings and washers to replace the springs in the rocker-arm assembly. All parts are straight replacements, with no grinding, welding, threading or the like required.

Pushrods—Cleaning and inspecting the pushrods is all that's usually necessary. Be sure to pass solvent through the hollow pushrod centers. Check the ends for wear. If the oil holes are broken open, replace the pushrods.

The most common pushrod problem is bending. Check for bent pushrods by rolling them on a flat surface. I use the corner of the bench or kitchen table. Then, just watch the end of the pushrod as it rolls. If the end wobbles up and down, the pushrod is bent. Replace it.

CHAPTER 6

Cylinder Head Reconditioning

If you rebuild this many heads a day, you know what works and what doesn't. About 15 of these weren't cracked when they arrived. Welding restores them into useful parts.

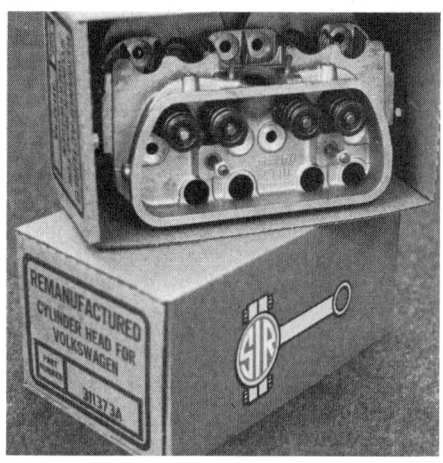

Exchanging heads for high quality rebuilt pair could be best course when rebuilding.

Ask experienced air-cooled VW engine rebuilders where most of their work is and they'll respond: the cylinder heads. There are several reasons for this, but the primary ones are *heat* and *moving parts*. The cylinder heads on any engine are one of the hottest running parts, and on the air-cooled VW, this places quite a strain on the aluminum castings. Thermal stress exercises the metal until cracks start to relieve the strain. Heat also erodes the valves. Add moving valves and rocker arms to the picture and you can see there is a lot of wear in the cylinder heads.

A lot of wear means a lot of worn parts, so don't be too surprised by the number of parts that must be discarded. It may seem like a lot, but this is necessary to fully restore your engine's power and longevity.

While reading this chapter you'll see there are several good reasons to farm out cylinder-head work. First, there's the equipment required: valve grinder, valve-seat grinder, valve-spring compressor and spring tester. With the possible exception of the valve-spring compressor, these tools are so specialized and expensive no one rents them. And you sure can't justify buying them for doing only one engine job, or even 20.

Second, even if you had access to a machine shop full of head-reconditioning tools, you'd still have to learn how to use them. It's easy to say "grind the valves" and something else to do it.

So, if you don't have the tools or skills, deliver the heads to a reputable machine shop and have them do the job. It won't take long, and if you consider the time saved, the job doesn't cost that much.

Another variation to consider is exchanging the heads for reconditioned ones. You simply deliver or ship the old heads plus a fee to an engine rebuilder in exchange for a completely rebuilt pair. This option offers greater speed because you don't have to wait around while the machine shop completes the head work. It's probably cheaper, too. The disadvantages are about the same as hiring any machine shop: you may not know the rebuilt head quality, or where the part came from in the first place. You can answer these questions by talking to the rebuilder, inquiring about the guarantee and calling consumer information services.

When shopping for a machine shop, remember price isn't everything. The VW parts and service business is very price-conscious, and competition among shops drives parts and service costs down. Read through the popular VW magazines to establish the going rates for engine work. Don't expect to beat these prices

You get what you pay for. Voids around sparkplug hole (left photo) are from poor weld, which should have been cut out and redone. Using these heads will result in combustion leaks around plug, until it's finally blown out of head. Shims in combustion chamber (right photo) are required for excessive flycutting.

anywhere, and don't automatically reject a shop because it charges $5 more for the same part or service. Read the fine print, or call up and ask them why their parts are more expensive. Chances are they have some good reasons. Also beware of super-low prices. There are also good reasons for too-low prices; the shop may skip a few steps or use inferior parts.

Just remember that price shouldn't be the main consideration. It usually pays to go to a somewhat more expensive, but faster, cleaner and more professional shop. The extra care and time spent on the heads from *your* car cost money; so don't worry about spending a little extra now. Better a little more now than a lot more later.

To make the first stages of cylinder-head work more pleasant, clean the head with solvent. Spray-on degreasers and the garden hose work well, but if you don't want to make a mess at home, use a car wash. Spray the head with degreaser and use the pressurized spray to clean the many awkward shapes in the head castings.

DISASSEMBLY

The first step in head reconditioning requires a *valve-spring compressor*. The most common type is the *C-type*, so called because it looks like a huge C-clamp. It is designed for use with the head off the cylinder bores, unlike a *fork-type*, which are best used with an installed head.

A C-type compressor fits around the head. One end butts against the valve head, the other end straddles the valve-spring retainer. A lever on the compressor moves its two jaws together

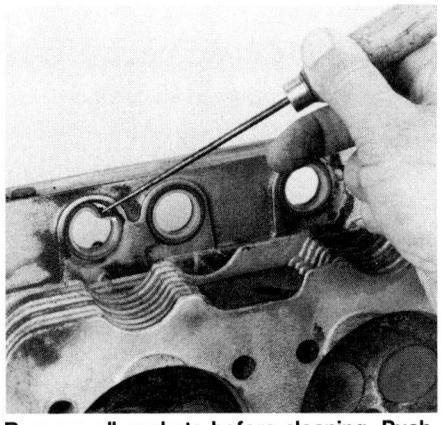

Remove all gaskets before cleaning. Push-rod tube seals can stick in head. Remove them with pick or screwdriver.

Compress valve spring just enough to extract keepers. Magnet helps remove these slippery little devils.

against the spring with an over-center effect. This compresses the spring, forcing the spring retainer down the valve stem to expose the two *keepers,* also known as collets, locks or keys. Their purpose is to lock the spring retainer to the valve stem with a wedging action, so the retainer holds the valve spring in compression.

A common problem when using a spring compressor is breaking the retainer loose from its keepers. If you try to force the compressor, you may bend its frame. To prevent this, lever down with the compressor and tap the retainer/compressor with a soft mallet. This will break the retainer loose, allowing the spring to be compressed.

Be careful when doing this; parts tend to fly out of sight, especially keepers. It's best to point valve springs at the wall behind the bench. Wear safety glasses and keep your head and body out of the way, but watch the keepers' trajectories. Find or replace any lost keepers before continuing.

There's no need to arrange the keepers and retainers in order. Just dump them into a container so they won't get lost. But you must keep the valves, springs, spring seats and any shims in order. You can make a holder with a 2x4. Bore eight shallow holes in it large enough for the eight valve stems. Space the holes at least 1-1/2 in. apart. Arrange the valve hardware in a

101

Pull spring and retainer off head. Keep any shims under spring with that spring.

After spring is off, pull valve out of head. Keep valves in order so guides and seats can be evaluated using their own valves.

quiet corner on the bench.

If you find any shims under the valve springs, someone besides VW put them there. Shims are used to correct low spring pressures or height, as I'll discuss later in this chapter. For now, just keep any shims with the springs you found them under.

Missing Cooling Fins—Before investing any time in the heads, look for missing fins. Two adjacent missing fins are acceptable, but three or more are cause for replacement. Bent fins are common, and don't adversely affect cooling, so you can use a head with bent fins.

Cleaning—Wash all parts in solvent, but don't worry about carbon in the combustion chambers. It is too baked-on for the solvent to do much good. Knock the largest carbon pieces off with a blunt screwdriver or chisel. Don't nick or gouge any part of the head, especially where the cylinders seal against the head, the valve seats or sparkplug hole. Remember, this is an aluminum head; treat it accordingly.

Bead-Blasting—In the machine shop, the head can be *bead-blasted*. Only the combustion chamber and sometimes the ports are bead-blasted. Doing so to the rest of the head is not necessary, takes lots of time, and fills voids with blasting residue. Bead-blasting quickly removes carbon deposits without danger of gouging critical surfaces.

If the head is bead-blasted, make sure it gets a thorough cleaning afterward. Blasting residue collects in all bolt holes and corners, and it takes some time to get it all out. Force it out with high-pressure water, steam or solvent. Chase every thread with a tap to ensure all grit is removed. If this junk is not washed away, you could strip some threads. Also, leftover blasting grit will get in the oil, score the bearings and journals, and ruin the piston rings within several thousand miles.

The machine shop should clean the head after bead-blasting, but sometimes this means only a few shots of compressed air. Always check this step yourself. In fact, cold-tanking followed by thread-chasing and tapping is what is usually required to get the head spotless.

Cracks—All VW cylinder heads are very prone to cracking. If the heads are overheated, or after thousands of heat cycles, they will relieve the stress running through them by cracking. Type 1—3 heads crack a little more easily than Type 4 heads, but you still have to closely inspect each type.

Cracks don't mean you have to throw the head away! In fact, some of the cracks don't even affect engine operation, other than to make an exhaust-leak noise. But most cracks do cause serious damage. Nevertheless, unlike water-cooled engines, the aluminum VW head is easily repaired by welding—if the correct procedure is followed. You've probably heard horror stories about welded heads and how they cracked the first time the engine was started. Those stories could be true, but then, those heads probably weren't welded properly.

The necessary, but often overlooked step when welding VW heads is to heat them first. If the heads are placed in an oven at 400—500F (205—260C) for a half hour before welding, then they're prepared for the white-hot thermal shock of the arc welder. The head doesn't expand explosively when the arc welder heat is applied. Instead, the aluminum expands more slowly, placing less stress on the part. With the stress reduced, cracks don't start.

All competent VW engine rebuilders weld cylinder heads. If they didn't, there wouldn't be many cylinder heads left. Southwest Import Rebuilders estimates 95% of all core Type 1—3 heads are cracked when they arrive there. Some of those heads are sent to scrap because they are practically cut in two from cracks, or they have other problems; but the majority are welded. These heads work just as well as new ones, so don't dismiss a shop's work because they weld heads. When correctly welded, the new material becomes part of the head, adding strength. Just make sure the shop heats the heads before welding.

Look at the photos to identify typical VW cylinder head cracks. Then examine your heads for the same cracks. Chances are good you'll find some, so don't gloss over this inspection.

Cracks usually run between the sparkplug hole and the valve seats, directly between the valve seats, and inside the exhaust port. Surprisingly, exhaust-port cracks don't affect head strength, and it can still be used without harm. Depending on how large the crack is, you may or may not hear an exhaust leak from this type of crack. For this reason, large-volume rebuilders weld these cracks even though the engine doesn't run any better for it. You might decide to let such a crack remain if it's the only one you find.

Cracks running between the sparkplug hole and the valve seats or between the valve seats are a lot more trouble, and can lead to loose valve seats and sparkplugs. These cracks weaken the head. Therefore, such cracks must be welded.

Before welding, the cracked area must be completely cut out. Only original, non-cracked material may remain, or the stressed, cracked area will continue cracking after welding. Cracks are removed with high-speed die grinders. As you can see in the photos, rebuilders aren't shy when using the grinder.

After grinding, the head is preheated, then welded. The welder merely fills the voids, adding plenty of material. The sparkplug hole, for example, is completely filled. After the head

Missing cooling fins will get welded up, like crack between valve seats. Three missing fins must be corrected; either weld replacements or buy another head.

Sparkplug hole-to-valve seat cracks usually follow shortest path (left photo). Valve seats and sparkplug threads have been removed. Repair requires completely removing cracked metal (right photo).

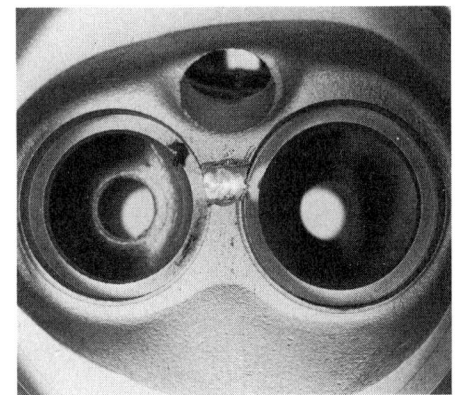

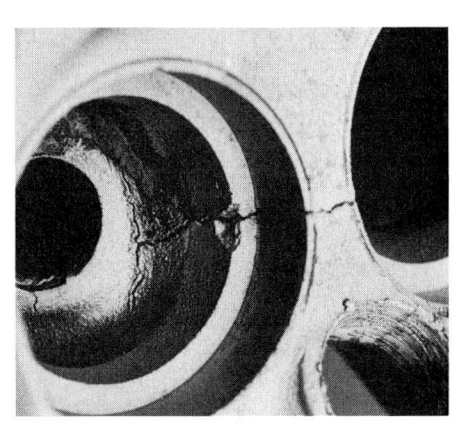

Another typical crack runs between valve seats (left photo). Again, metal must be cut out until all traces of crack are gone (right photo).

Type 4 heads can crack with same vengeance as Type 1s. Exhaust port has "T" crack. Vertical section between seats and into port is serious, but horizontal section inside port just causes noisy leak.

has cooled, it's flycut, drilled, tapped, and seats are installed and hand-ground back into specification. There's a lot of work in getting most heads back into shape, which should make you appreciate the low prices for rebuilt heads.

Flycutting—Sometimes it is necessary to machine the area where the cylinders seat in the heads; this is done to clean up damage done by loose cylinders. The cylinder hole can also be enlarged in diameter to accept oversize cylinders, or the cut can be made deeper to let the head set lower on the cylinders. This increases the compression ratio. The machining process is commonly called *flycutting* the heads, because a flycutter in used to remove the metal.

Check the cylinder-to-cylinder head mating area for damage. This area will normally be about the same color, texture and shape as the rest of the head. You'll be able to see where the cylinder mates against the head, but you should not see any shiny material. When the cylinder seating area is shiny, it means the cylinder has been rubbing against the head, and that means the cylinders are loose. When the cylinders have been pounding, the mating area will be depressed, brightly polished and look generally beaten. Pay careful attention to the case threads if you spot this damage.

Don't mistake the head gaskets between the cylinders and heads on Type 4 engines for damage. Sometimes they stick inside the head and you don't even notice them.

There is a limit as to how much material can be removed by flycutting without weakening the head. Cut too much and the head will split open at a head stud or between two cooling fins. Unfortunately, they don't split open until after you put the engine together and fire it up. But for repair purposes, there is plenty of material on most heads. Cuts up to 0.100-in. (2.540mm) deep are fine. You'll create compression problems (the compression ratio will be too high for pump gas) before you run out of metal to cut. Cutting wider for larger cylinders depends on the heads. Of the Type 1—3, any 1300, 1500 or 1600 head can be so cut, but not the 1200 40-HP head. All Type 4 heads can be enlarged to the next larger cylinder diameter.

Removing enough material to clean up the sealing surface should be no problem. But flycutting the heads changes a lot of basic engine parameters. Besides reducing combustion chamber volume and raising the compression

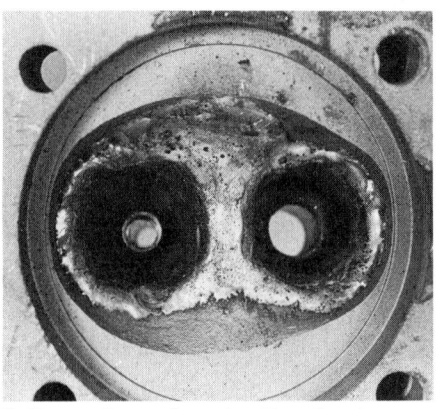

Entire sections of combustion chamber can be repaired by welding. Sparkplug hole is filled and bridge between valve seats is deep.

New valve seats and sparkplug threads are installed, but excess weld hasn't been removed. With proper penetration, weld has original material strength.

Hand removal of excess weld is an expensive part of head rebuilding. Original contours of ports and combustion chambers are carefully maintained.

Flycutting cylinder heads restores sealing surface and quickly shapes welded chambers, but alters head geometry if overdone.

ratio, flycutting moves the cylinder head closer to the crankshaft. This changes the valve train geometry; in effect, making the pushrods too long. This extra length forces the rocker arms up at the pushrod end, and down at the valve end. This alteration can be corrected by fitting spacers between the cylinder heads and bores. Another method is to shim the *rocker-arm stands* to compensate for the extra pushrod length.

If you were building a race engine, there would be several more steps than simply shimming the rocker stands or cylinder heads, but with the relatively small amount of material removed on a street rebuild, shimming will supply the necessary correction. Shim the same amount removed during the flycut.

Rocker-Arm Stud Breakage—All early 40-HP heads and '62—65 Type 2 1500 heads have a problem with breaking the rocker-arm studs. These studs don't thread into the rocker-arm-box floor, but pass through it and thread into the cylinder head proper, above the combustion chamber. Vibration and flexing rub the studs against the rocker-arm box where they pass through it. This wears away the rocker-arm box, causes oil leaks, and establishes a fulcrum to break the studs.

Various kits are available to fix the problem. With one style, the long studs are removed and inserts threaded into the rocker-arm-box floor. Then new, shorter studs are threaded into the inserts. The problem with this repair kit is the insert OD and its threads result in poor thread mating between the insert and the cylinder head. Valve spring pressure can then rock the insert in the head, which loosens the insert even more.

The other style of insert is sometimes called the *pull-in* type. No threads are tapped into the head with this kit. Instead, the shouldered insert is pressed into the bottom (combustion-chamber side) of the rocker box from the outside. The rocker studs are replaced by cap screws (bolts) that pass through the rocker-arm pedestals and thread into the inserts. Thus, the cap screw and insert sandwich the rocker-arm pedestals and cylinder head between them. All threaded connections with this kit are steel-to-steel, and don't work loose.

Sparkplug Thread Inserts—Another common problem is stripped sparkplug holes. The differential expansion of steel sparkplugs and aluminum cylinder heads sometimes galls the hole's threads enough that they are destroyed. Galling is *snowballing,* where a small amount of aluminum tears off the cylinder-head threads, then begins pushing larger and larger amounts of similar material in front of it.

The cure is to drill the sparkplug hole, tap it and install a steel thread insert. This is the same procedure discussed in the sidebar on page 105. The steel threads are much stronger than the original aluminum ones, and are found on some rebuilt heads.

Large-volume engine rebuilders may or may not install a thread insert in a head, depending on how any welding comes out. First, in those cases where the threads are fine and there are no

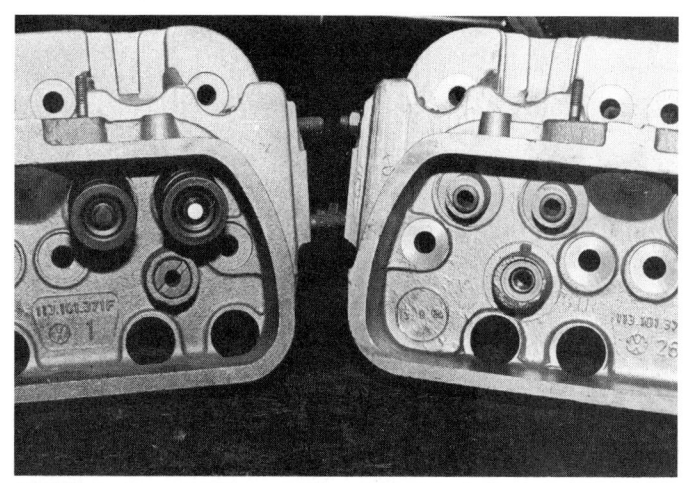

40-HP stud repair kits are either thread-in (left) or pull-in (right). Threaded insert suffers from poor insert-to-head fit, and often breaks head threads.

Pull-in inserts press into bottom of rocker box from outside head. There are no threads to pull out or vibrate loose. If using "long stud" 40-HP heads, this is repair kit to use.

CAUTION: ALUMINUM THREADS

When a steel bolt is threaded into aluminum, the softer aluminum threads are easily damaged. Common problems are crossthreading, overtightening and galling. Crossthreading is probably the number-one cause of destroyed aluminum threads. However, if the threads are clean, of the right size, and reasonable care is taken starting the fastener in its hole, the chance of crossthreading is small.

When a fastener is overtightened, something has to give. In the case of a steel bolt in an aluminum head, the softer aluminum threads usually give.

Sometimes, aluminum threads strip even without crossthreading or overtightening. This is a result of galling. It's common with sparkplug threads. The problem is heat-expansion rates. When an aluminum head is hot, it grows several thousandths of an inch in all directions. Thus, the pressure between the sparkplug threads and those in the sparkplug hole increases. This forces the aluminum into voids in the sparkplug threads, locking the plug in its hole. When the plug is subsequently removed, its threads tear or gall the aluminum threads, particularly in the case of a hot engine.

How do you restore stripped heads? It can be done with a steel thread insert. Thread inserts fall into three groups: steel-wire coil, threaded sleeve and locking threaded sleeve.

The procedures for installing inserts, sometimes called *thread savers,* differ. Basically, the old threads are drilled out, the hole is tapped oversize and the thread insert is installed. The new steel threads are actually stronger than the aluminum ones they replace. For this reason, race-car builders often install thread inserts in light-alloy components as a matter of course. It's a good preventative measure in any component that is removed and installed frequently.

The flip side is there is absolutely nothing wrong with steel thread inserts. The common complaint about them is they are a heat barrier between the sparkplug and head, causing plug overheating, which is baloney. Look at Porsche cylinder heads for proof; they come stock with inserts. Adding sparkplug thread inserts is a job you can do yourself. You'll have much stronger threads, and won't have to worry about galling again.

VALVE GUIDES & STEMS

Next, let's consider valve-guide wear and how to measure it. As a rocker-arm pushes its valve open, a side force is created that forces the valve stem against the guide. This action wears both guide and stem. Keep in mind that when someone says, "The guides are worn out," they actually mean the clearance between guides and stems is excessive—both stem and guide could be at fault.

A quick and fairly accurate check for wear uses a valve in the guide. Insert the valve in the guide so the stem is flush with the rocker arm end of the guide. While holding the valve head, rock the valve back and forth. Move it in different directions until you find the one that gives the most movement. This will be the direction the valve is pushed by the rocker arm. Measure the movement, or *play* at the valve head with a dial indicator. The depth-gage end of a vernier caliper can be used to take this measurement in a pinch. Another method is to divide this play measurement by 3.5 to get the *approximate* stem-to-guide clearance. This method doesn't tell you if the valve stem or guide is the culprit; it could be either.

Type 1—3 valves should rock 0.009—0.011 in. (0.23—0.27mm) when new, and no

cracks radiating from the sparkplug hole, the threads are left alone. But if there are cracks or bunged threads, the sparkplug hole is reamed out and completely filled with weld. The hole is then redrilled and tapped to accept a sparkplug. Thus the head has aluminum sparkplug threads, like stock. But sometimes there may be small voids in the weld, and the newly tapped threads come out incomplete. Then the rebuilder installs a steel thread insert, unless the welder was so incompetent the entire weld must be redone.

There isn't anything inherently bad about aluminum sparkplug threads, except they are a tad on the weak side and will gall. You can avoid galled sparkplug threads by not removing the sparkplugs when the engine is hot. This is the most common cause of destroyed plug threads. Letting the engine cool first allows the dissimilar metals to contract to their original sizes. This also lowers sparkplug torque so you don't have to use a breaker bar to loosen the plugs. And it demonstrates why you should *never overtorque* sparkplugs. Sparkplug torque increases with engine heat, so too tight when cool means much too tight at operating temperature. Use anti-seize compound on plug threads as insurance against galling.

105

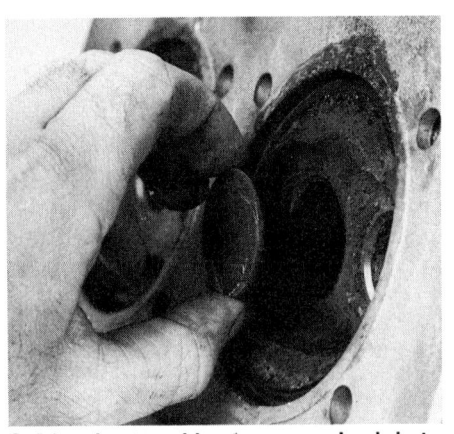

Quick valve-to-guide clearance check is to wiggle valve in its guide. Place valve another half inch closer to seat than shown to obtain most accurate reading. Method doesn't pinpoint wear on valve or guide, but usually guide is worn most.

Drilling guide before driving helps break grip between guide and cylinder head. With less wall thickness, guide stretches more easily, which means it gets thinner and breaks loose from head.

more than 0.031 in. (0.80mm) when worn. Type 4 valves rock 0.018 in. (0.45mm) new and 0.035 in. (0.90mm) worn.

Of course, if the valve is replaced for other reasons, there's no need to figure out how badly its stem is worn. The new valve will be at specification, leaving only the guide as the source of excessive clearance.

In rebuilding practice, exhaust guides are replaced and the intake guides are often reused. Unless the heads have very low or very high miles on them, this rule of thumb should hold. Exhaust guides wear for the same reasons the valves do, heat and lack of lubrication. The exhaust guide runs hotter than the intake because there is no cool intake charge of fuel and air to carry away heat, and little oil is available for lubrication because exhaust pressure in the port sneaks up the valve stem, blowing oil out of the guide.

Friction that wears the guides comes from the rocker arms pushing against the valve stems. If you study the VW rocker arm and valve stem geometry, you'll see how the rocker arm must slide across the valve-stem tip when opening the valve. Even though there is oil between the rocker and valve, a lot of this sliding motion is altered by friction into a push against the valve-stem tip. The valve is then forced against its guide at each opening, wearing both guide and stem. Because the guides are softer than the valve stems, they wear the most.

Another wear pattern from the rocker arms is a depression on the valve-stem tip. Many aftermarket rocker-arm adjusting screws are very hard. Too hard, actually, because they dig a hole in the valve tip. This makes it practically impossible to adjust the valves, because a feeler gage can't measure the actual gap that is down in the hole in the valve tip. Valve clearance will be excessive and you'll hear a lot of valve noise. If so, tighten the adjusting screws one-eighth of a turn. If this quiets the valves, you can bet the valve tips are *dished*.

A valve job is the best cure for this condition, along with a set of genuine VW rocker-arm adjusting screws, or substitute the swivel-foot type. Swivel-foot adjusters use a half-round ball where the adjuster meets the valve. The flat section of the ball rests against the valve tip, and the round section is contained by the adjuster. The adjuster swivels as the rocker arm changes angles with the valve during an operating cycle. Side loads are greatly reduced, along with valve noise.

Guide Replacement—Guides are replaced by driving them out with a hammer and stepped punch or drift. Always drive old guides out from the *combustion-chamber side*. New guides are driven in from the *rocker-arm side*. The drift must have a shoulder that butts against the guide end, but be small enough to pass through the guide bore in the cylinder head. Before driving out the guides, measure how much they project above the rocker side of the head, and reduce the valve guide ID by reaming or boring. This weakens the guide's wall and loosens its grip on the cylinder head. Bore or ream to within 1/4 in. of the combustion-chamber side of the guide. This leaves enough material to hammer against without distorting the guide OD. If the guide OD becomes mushroomed, grind or file off the distorted metal before continuing.

Another way to assist guide removal is to tap the rocker end of the guide ID with a 3/8-in. tap. Tap about 3/4 in. into the guide, then thread a 3/8-in. bolt into the guide. Select a drift that passes through the guide from the combustion-chamber end and butts against the bolt. Hammering via the drift onto the bolt will stretch the guide, decreasing its OD. This will help break its grip on the cylinder head.

If the guides won't budge, heat the head to 300F (149C) in an oven for about one-half hour and try again. Don't burn yourself! A note of caution: don't bake the head unless it is perfectly clean or you'll stink up the house for a week.

Another removal trick is to use a pneumatic hammer, but only if you are an experienced machinist. This tool takes a deft touch. Be sure to use a stepped drift and keep your weight against the gun so it won't jump around. A 5/16-in. Allen-head cap screw will work as a stepped drift if you don't have one. It's best to use one of the manual methods outlined earlier as it's just too easy to mushroom the guide end with a pneumatic hammer. Then when the mushroomed end passes through the cylinder head it *broaches*—widens by rolling the metal aside—the guide bore.

VW machine shops have a large variety of guides with larger than stock ODs. If the guide hole is enlarged during removal, or if the original guide was loose in the head, get oversize guides. The correct fit is a 0.002-in. (0.050mm) interference. The rule of thumb for how much force to apply is for each moderate hammer blow, a correctly fitting guide will sink 1/4-in. into the head.

Installing guides is easier if they are chilled and the head heated. Again, bake it for one-half hour at 300F (149C). This will expand the guide bores. Have the guides in the freezer at the same time. While it takes time to heat and chill the parts, this method makes driving in new guides easy, and lessens the chance of galled guide bores and guide distortion from excessive hammering.

Many machinists don't take the time to heat and cool the parts; they just hammer harder. I prefer the heating and cooling method. No matter which method you choose, apply some sort of lubrication on the guide OD and bore. A thin coating of white grease, moly lube or anti-seize compound works well.

Drive the new guide in until it projects from the rocker-arm area the same distance as the one it replaced. Most aftermarket guides have a step in the OD which butts against the head when fully installed; it eliminates any guess work about how deep to drive the guide. With non-stepped guides, you can save time by fitting a section of tubing over the guide during installation. Cut the tube to the guide's proper installed depth, then drive the guide until it's flush with the tube. Or, use a machinist scale to measure progress—just don't get carried away and hammer too much without measuring.

Reaming—Once installed, a new guide is reamed to fit the valve stem. As sold, its ID is smaller than the valve stem. This allows for reaming the guide for an exact fit with its valve. It is done with a long reamer chucked in a drill. Apply cutting oil for a cleaner cut and to carry away spent material.

Try fitting the valve in its guide. It probably won't go. Oil and ream the guide again for several seconds, then retry the valve. Ream as necessary until the valve barely wiggles in the guide. Intake guide-to-stem clearance should be 0.002—0.003 in. (0.050—0.076mm). Exhaust clearance should be greater to prevent valve sticking: 0.0025—0.0035-in. (0.064—0.089mm). This is the stem-to-guide clearance, not the amount the valve will rock. Valve rock should just be perceptible. Flush the guide with oil and a small brush, if possible. Check for burrs on the valve tip or guide opening. If the valve still won't start after all the reaming, look for a burr holding it up. The valves must remain in order so they'll end up in the guides that were reamed for them.

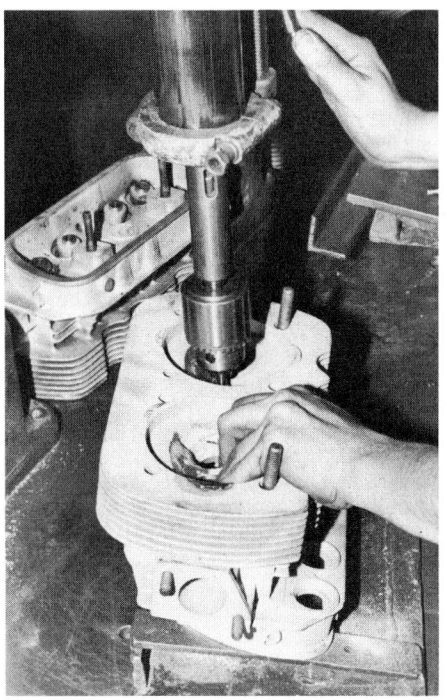

INSPECTING & RECONDITIONING VALVES

The first step in evaluating air-cooled VW valves is learning which valves to automatically replace. They are all Type 1—3 exhaust valves and all Type 4 non-sodium-filled exhaust valves. That's right, all but sodium-filled Type 4 exhaust valves are non-reusable. All intake valves can be reused if they pass inspection.

Expect 50,000 miles from typical Type 1 exhaust valves, as little as 30,000 miles from the exhaust valves in fuel-injected Type 3s.

The problems with exhaust valves are material, design and thermal load. These exhaust valves are prone to breaking where the head meets the stem; one of those failures that causes engine paralysis. The only cure is to replace the valves. Prices for these valves are reasonable, so there isn't any great financial hardship about replacing them. Besides, new valves have new stems and sealing surfaces. That means less measuring, cleaning and machining.

Luckily, VW exhaust valves stretch before they break, so if the valve clearance starts to close up (decrease), look out. The valve clearance will also close if the valve burns, so either way, a tight exhaust valve means real trouble that should be corrected quickly.

What about those sodium-filled Type 4 exhaust valves? These are special valves, with a special price, about three times the cost of conventional valves. The valve stem and part of the head is hollow to form a chamber. This chamber is partially filled with sodium, which travels from one end of the stem to the other during engine operation. Because sodium quickly absorbs and dissipates heat, it picks up heat at the valve head and conducts it away to the far end of the stem. This part of the stem is always in contact with the valve guide; in turn, it transfers the heat into the guide. The guide radiates the heat into the head, where ultimately it ends up in the cooling air. Therefore, a sodium-filled valve is able to shed heat faster than a conventional valve and it runs cooler. The cooler the valve runs, the longer it lasts. A sodium-filled exhaust valve has a useful life through one rebuild. If the heads are rebuilt a second time, even sodium-filled valves must be replaced.

There are no sodium-filled valves for Type 1—3 engines, and the Type 4 pieces won't fit. If rebuilding a Type 4 engine, pay the extra money and get the better valves. Both conventional and sodium-filled valves are available, but the longer life and greater cooling of sodium-filled valves pay for themselves in the more powerful Type 4. This is especially true for driving heavily loaded Buses and high-speed, long-distance 914s. You can identify sodium-filled valves by the small hole centered on the valve head. This is where the sodium is added during manufacture.

Check Valves—After discarding the exhaust valves, give the remaining intake and sodium-filled exhaust valves the once over. Reject any

New guides are reamed to fit. Production-line tooling uses drill press, but hand-held drill works fine, too. Reaming is checked by wiggling valve in guide.

Not adjusting valve clearance can have serious consequences. After burning, head can break off exhaust valve, ruining head, piston, cylinder wall, valve train parts and scattering metal particles throughout engine. Beware of perpetually tight exhaust valves. They are stretching while getting ready to break.

Head broke off Type 1 #3 exhaust valve and danced on this piston. Complete engine rebuild was required. Photo by Ron Sessions.

valves with obvious problems, like burning. Look at the tips for wear. If a tip was severely pounded by its rocker arm, it may be *mushroomed*—it will resemble the hammered-end of an old chisel. Or it may be dished. Replace the valve and check the rocker-arm adjusting screw. As mentioned earlier, if the

Classic burned VW exhaust valve is missing pie-shaped section. Valve began burning on head exposed to combustion chamber. Because face cools when on seat, it's coolest part and hasn't yet burned away completely.

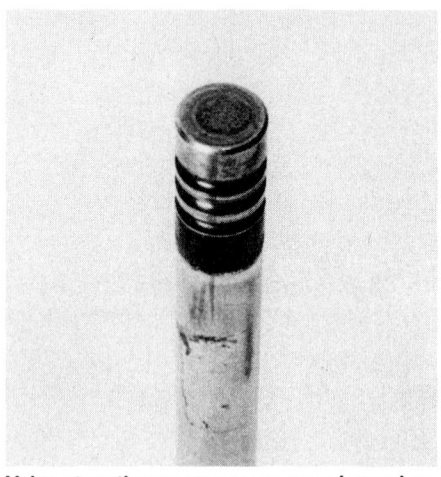

Valve stem tips are worn concave by rocker-arm adjusting screws. Extremely hard aftermarket adjusters cause rapid wear, make accurate adjustment impossible and ruin valve.

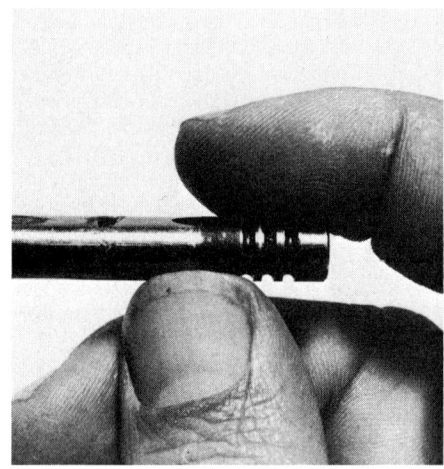

Sharp edge felt on bottom edge of keeper grooves is normal. Replace valve only if burrs have been raised by keepers.

engine has aftermarket adjusting screws and the valve tips are dished, the adjusters are too hard. Replace them with stock VW or swivel-foot adjusters.

The tips may also show some *pitting*. Very light pitting will clean up in a valve-grinding machine, but heavy pitting means the *hardface* is worn; replace the valve. You may be able to clean the pitting off, but I'll bet the overall valve length will be marginal and not to spec. And then once the valve is returned to service, the tip will wear quickly because the hardfacing is gone. Heavy pitting is a sign the valve is worn out.

Hardened caps for the valve stem are available from VW to restore the valve-stem tip. These caps slide over the valve tip after material is removed from the valve tip to compensate for the tip's thickness. These valve caps are really a temporary fix for service shops to repair worn tips. If mechanics find a destroyed tip, they can fit a cap without disassembling the engine. Because you already have the engine apart, you can easily replace the valve and don't have to use temporary measures. Another good reason not to use caps is after a while the valve tip may wear more. The cap will then come closer and closer to the keepers. Eventually the cap may rest against the keepers and release the valve to fall into the cylinder...more engine paralysis.

Keeper grooves sometimes get burrs after moving against the keepers. All keeper grooves will feel slightly sharp when running your fingers up the stem and over the grooves. This is normal. But watch for a large, visible burr on one of the keeper grooves. This signals a worn-out groove and time to replace the valve. The keepers themselves seem very resistant to wear, even if they have worn a burr on the groove. They are almost always reusable, but are very inexpensive to replace if worn beyond use.

Look at the valve's sealing surface. Heavy pitting, and wear so excessive the surface is concave when viewed edge-on, mean the valve is finished. A few pits here and there are OK, though. They'll disappear after grinding.

After weeding out the obviously shot valves, use vernier calipers to measure the dimensions shown in the drawing above. By now you may have only one or two valves that have passed muster. You've reached the point I always seem to end up at: tossing out all the old valves

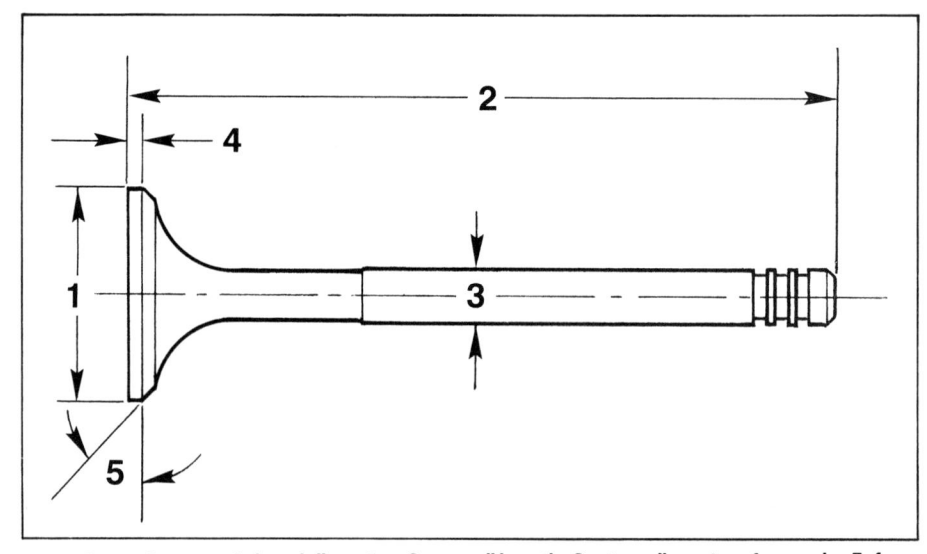

Valve dimensions are 1: head diameter, 2: overall length, 3: stem diameter, 4: margin, 5: face angle.

and installing new ones. My logic is that if six or seven of eight valves are no good, the remaining one(s) can't be all that useful. Then again, if you took good care of the engine, and it hasn't got excessive mileage, measure on.

Before measuring, clean the valves with a wire wheel. Remove all carbon, oil and crusty deposits. Avoid using the wheel too much on the keeper grooves.

Overall Length—Start by measuring the overall length. VW valves tend to stretch. Once they start stretching, you know they are preparing to break off at the head-to-stem intersection. Replace any long valves.

Stem—Next measure the stem diameters.

VALVE DIMENSIONS
[-in. (mm)]

TYPE 1—3

1. **Head Diameters**—see page 47
2. **Length**   4.409 (112.0)
3. **Stem Diameter**
   - All   Intake   0.3126—0.3130 (7.94—7.95)
   - '61—74   Exhaust   0.3114—0.3118 (7.91—7.92)
   - '75—79   Exhaust   0.3508—0.3512 (8.91—8.92)
4. **Margin**
   - Intake   0.031—0.059 (0.79—1.50)
   - Exhaust   0.040—0.067 (1.00—1.70)
5. **Face Angle**
   - Intake   44°
   - Exhaust   45°

TYPE 4

1. **Head Diameters**—see page 47
2. **Length**   4.6063 (117.0)
3. **Stem Diameter**
   - Intake   0.3126—0.3129 (7.94—7.95)
   - Exhaust   0.3508—0.3512 (8.91—8.92)
4. **Margin**   0.020 (0.50)
5. **Face Angle**
   - Intake   29.5°
   - Exhaust   45°

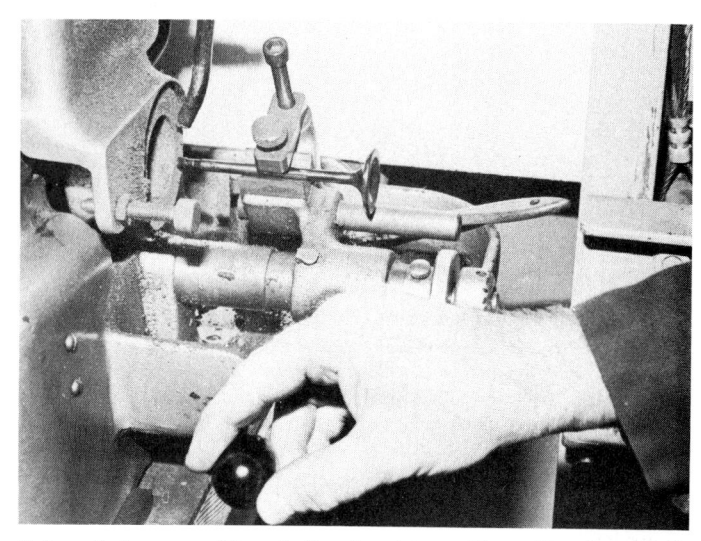

Valve grinder reconditions both valve stem and face. Stem is ground first (top photo), so it is sure to seat squarely in chuck used to clamp valve during face work. Grinding wheel and valve rotate when resurfacing valve face (bottom photo). If valve is bent, it shows as a wavy line on valve face.

Check the chart, directly above, for the original diameters. Or compare worn and unworn portions of the valve stem. Mike the unworn portion of the stem between the keeper groove and the *wear-line*—where the stem stops at the top of the guide when the valve is fully open. You might feel a small step here with a fingernail. Record you findings. Then measure immediately below the wear-line. Subtract this last measurement from the first to get *valve-stem wear*.

Because the greatest valve-stem wear occurs farthest from the valve head, you should always take the second micrometer reading right below the wear-line in the swept area. The farther away from the wear-line you take the measurement, the less accurate the reading will be.

How Much Wear?—Keeping in mind there is only 0.003 in. (0.076mm) *total* stem-to-guide clearance to work with, you can see there isn't much room for valve-stem wear. For instance, if you replace the valve guides, but install valves with stems worn 0.001 in. (0.025mm), then about half the allowable wear is already used. So even though you could use a stem worn as much as 0.003 in. (0.076mm), it wouldn't be long before the engine started burning oil. Use 0.001 in. (0.025mm) as the stem-to-guide clearance limit—less for optimum durability.

The important point to remember is the looser the stem-to-guide clearance, the greater the valve-stem and guide wear, and the sooner the engine will burn oil. Also, maintaining proper stem-to-guide clearance helps maximize horsepower and fuel mileage, particularly with the intake valves. This is because vacuum is lost through a loose guide during the intake stroke. Then less fuel/air mixture is drawn into the cylinder. It is replaced by blowby and oil vapor sucked in through the intake-valve guide.

Valve Grinding—Normal rebuilding practice is to grind valves to restore fresh sealing surfaces. When grinding limits are exceeded, the valve is replaced. With air-cooled VWs, this means the intakes and sodium-filled exhausts are usually ground.

After grinding, each valve must retain the minimum *margin*. This is the distance from the outer edge of the *valve face* to the valve-head surface: it's the thickness of the valve head at the OD of the valve face. The valve face is the angled portion of a valve head which closes against the valve seat in the cylinder head. Minimum margin is dimension 4 in the drawing. If the margin is below this minimum specification, replace the valve. Otherwise, the valve may burn.

Reconditioning a valve requires resurfacing

109

its tip and face. This is a job for a machinist with a valve grinder. The tip end of the valve stem needs attention first. The tip must be trued first because it pilots the valve, centering it in the grinder. An untrue tip will result in an untrue valve face.

As shown in the photos on page 109, the valve stem is clamped securely and the tip is passed squarely across the face of a spinning grinding wheel. A small amount of material 0.010 in. (0.25mm) or less—is ground away. This is only enough to remove small irregularities. Otherwise, the valve would be ground too short. Large irregularities, such as mushrooming, require discarding the valve. Finally, the tip is chamfered to remove the sharp edge or any burrs.

Valve faces are reconditioned by grinding away small amounts of material at the correct angle until all flaws are removed. See the chart on page 109 for Type 1—4 face angles. This work is also done in the valve grinder. Major components of a valve grinder include an electric motor to power the grinding wheel, coolant pump and a chuck to hold the valve. The valve is installed stem-first in the chuck.

The valve is turned slowly while the spinning grinding wheel passes back and forth across the valve face. While the valve is being ground, coolant is pumped over it and the grinding wheel for cooling and to wash away grinding chips. Coolant is water with water-soluble oil added.

It is important that no more than a minimum of material is removed from the valve face to retain a sufficient margin. If the margin is thinned excessively, the valve will overheat and burn or warp easily, particularly exhaust valves. A vernier caliper is best for measuring valve margin.

VALVE-SEAT RECONDITIONING

Valve seats are ground using special power tools and stones or cutters. If the valve seats have been ground before and there isn't enough seat material left to complete all necessary cuts, the old seats are removed and new ones installed. Both grinding and seat replacement are jobs for an expert with the proper tools.

VW uses hardened-steel valve-seat inserts that are a tight press-fit in the head. If a seat is damaged or worn out, VW recommends *replacing* the entire cylinder head. They contend it is cheaper to replace the head than to use the specialized equipment necessary to replace the seat. There is also the chance the replacement seat may fall out of the head if it isn't installed correctly, or if the machinist isn't experienced with VW seat replacement. Stock VW seats don't fall out because they are shrunk into place using sophisticated production equipment. Volume rebuilders press replacement seats into heads and have no failure problems. If they're organized into a production

Pilots use valve guides to center valve seat grinding stones. Thus, guide work is completed before seats are ground or seats won't be concentric with guides.

line, the cost to you is less then replacing the head.

A local machine shop, however, may find it easier, cheaper and wiser for them to simply replace the head. You will probably find it easiest to exchange the cylinder heads for a rebuilt pair if yours are full of cracks, worn guides and seats. With such labor-intensive heads, the volume rebuilder's production-line techniques become quite cost-effective.

Not only does valve-seat grinding restore the seat, it has another very important benefit. Because the grinder *pilots* off a shaft installed in the valve guide, it makes the valve seat concentric (has the same center) with the guide. However, if you didn't do any guide work and the valve seats are OK, leave the seats alone. Work on the guides can upset the guide-to-seat geometry; the seats must be ground after guide work.

When correctly done, valve seats are ground at several different angles. All Type 1—3 seats are ground at 15°, 45° and 75°. Type 4 intake seats are cut at 15°, 30° and 75°; the exhausts at 15°, 45° and 75°. The valve face seals against the 30° or 45° cut while the other two angles establish seat width and diameters. The 15° and 75° angles also improve gas flow into and out of the combustion chamber.

To restore valve seats, the machinist first makes the 45° or 30° cut with a *dressed stone*—shaped—at that angle. The top of the seat is then cut with a 15° stone to establish valve-seat OD. Seat OD is cut about 1/16 smaller than valve face OD. The 15° cut is called the *top cut*.

Next, the *bottom cut* is made with a 75° stone. This greater angle allows the stone to cut below the seat to establish valve-seat ID and width. Seat width on Type 1—3 intake seats measures 0.051—0.063 in. (1.30—1.60mm); the exhausts from 0.067—0.078 in. (1.70—2.00mm). Type 4 intakes are 0.070—0.086 in. (1.80—2.20mm); the exhausts 0.078—0.098 in. (2.00—2.50mm).

A machinist uses special equipment to quickly and accurately check valve-seat work. For instance, the valve-seat area is coated with Dykem or a similar-type metal dye (blue or red) so the seat stands out from the bright metal. This makes it much easier to make measurements because one cut is easier to distinguish from another. Width and diameter measurements are made with dividers and a 6-in. machinist's scale. Also, a special dial indicator, or *runout gage*, is used to measure valve-seat concentricity. With its mandrel centered in the guide and the indicator plunger positioned square against the valve seat, the dial indicator is rotated. If the indicator needle moves, the seat is not centered or concentric and must be reground.

Lapping Valves—Once a valve and seat are properly ground, there is no need to *lap* them together. But do lap a new valve being fitted in

Actual time spent grinding valve seats is short; usually a couple of seconds for each of the three stones is all that's necessary. If seats don't clean quickly, they could be worn out, non-concentric with guide or grinding stones need dressing.

Other steps are sometimes necessary to restore heads to usefulness. Valve spring seating area can get chewed up by spring and require surfacing, as shown. Broken studs or mangled threads need attention, plus rocker cover needs cleaning, especially where gasket seats.

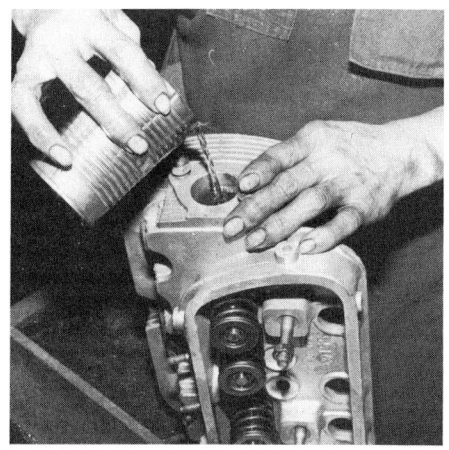

Check valve seat sealing with solvent after valves are installed. Fill port, then eyeball valve seat. Very slight leakage is cured by lapping. Running leaks require regrinding, new seats or valves.

When wear on rocker-arm adjusting screws (note tips) becomes more than noticeable, it's time for new screws. Use only original equipment or swivel-foot adjusters.

an old seat that hasn't been ground. Lapping is the process of manually grinding a valve face and its seat together with lapping compound. Lapping compound or paste is put on the valve face and then the valve is rotated back and forth in its guide while light pressure is put on the valve head.

If a valve job is done correctly, the angles will be correct and the valves will seal from the start. On the other hand, if the job is done incorrectly—the seat and guide may not be concentric—lapping won't cure the problem. It will, however, expose the problem, enabling you to fix it before the engine is assembled and compression is found to be way down. The seat-contact pattern will transfer to the valve seat and face—a real help when grinding. For instance, if this pattern is not consistent on the valve or seat, the guide and seat are not concentric; the seat or valve must be reground.

Racers lap valves regularly for the very last fraction of horsepower, plus they don't want to be continually grinding those carefully contoured valve seats. If lapping the valves is within your time budget, go ahead and do it. You certainly won't hurt anything and should pick up one horsepower or two by eliminating any grinding scores on the seat. A stone-ground seat is not quite as tight as a lapped one. You can prove this by assembling the valves, springs and keepers, and filling the ports with solvent. A stone ground seat will leak just a little, a lapped one will not. If nothing else, lapping is a good check of the machine shop's work.

Lapping a valve to its seat is simple enough. Merely apply lapping paste to the valve face, insert the valve in its guide and attach a rubber suction lapping tool to the valve head. Rotate the valve and tool back and forth while holding the valve lightly against its seat. Check your progress frequently and lift the valve off the seat every few seconds to let fresh lapping compound get between valve and seat. When the valve has a uniform gray band on its face, the lapping job is complete. Be sure to remove all traces of the abrasive lapping compound from the valve face and seat with solvent, or rapid valve and seat wear will result. Most parts store sell lapping paste and the suction tool. Get fine compound if you have a choice of fine and coarse.

ROCKER-ARM SERVICE

Rocker-arm service is limited to the adjuster, rocker-arm socket, rocker-arm ID and rocker-arm shaft.

After cleaning the rocker in solvent, turn the assembly upside down on the bench. Inspect the adjuster's working surface, which is the bottom of the adjuster that touches the valve stem. This area will be shiny, but should still be convex. If the adjuster is worn flat, scored or grooved, replace it. You may also need to replace it if the jam nut is rounded, or the threads are worn out. And replace valve adjusters you discovered to be too hard in previous inspections. If valve clearance changed a thousandth or more when the jam nut was tightened during valve adjustment, then the adjuster is stretching; replace the adjuster.

At the other end of the rocker arm is the socket where the pushrod is retained. The socket will show wear, but there should not be a ledge worn into the socket. If there is, replace the rocker. You'll probably need to replace the pushrod, too, if this wear exists.

Now disassemble the rocker-arm assembly. On Type 4 engines, you've probably had a difficult time keeping it together during the previous inspections because nothing holds the parts on the shaft once the mounting studs are removed. Just slide the parts off the shaft. Do

111

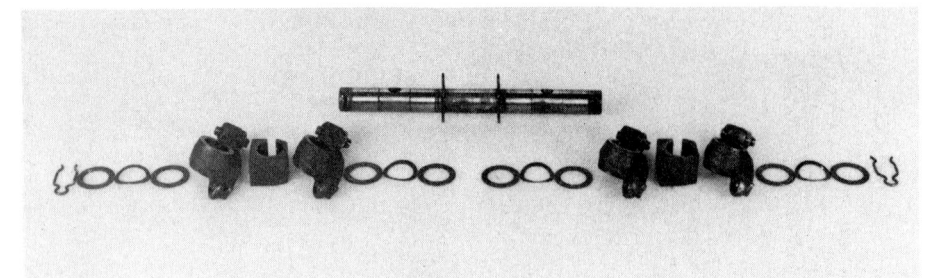

Type 1—3 rocker-arm shafts disassemble after removing clips at either end. Clean parts in solvent, then check for wear. Keep spacers, washers and clips in order to avoid confusion at reassembly.

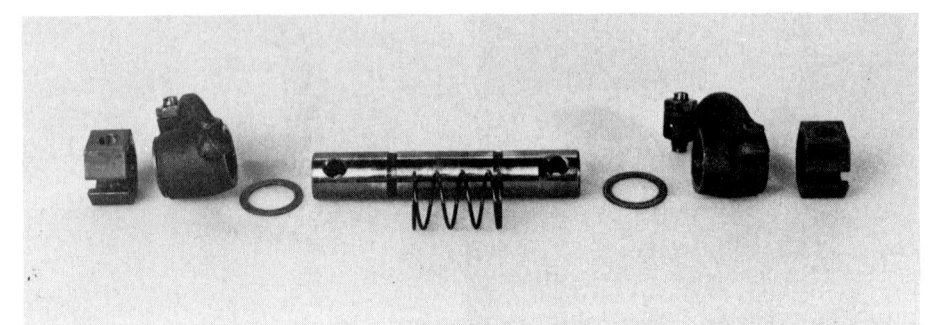

There are no clips or retainers on Type 4 rocker-arm shaft assemblies; the parts just slide off. Clean and inspect for wear. Unless oil was dirty or engine had extreme mileage, these parts should be OK.

Swivel-foot adjusters flank stock VW adjusting screw. Ball and socket of swivel foot reduces sideways push on valve guide.

this over a clean rag and keep all parts in order. On Type 1—3 engines, remove the circlips, then slide the parts off the rocker shaft.

When inspecting the parts, take the time to keep them in order and arranged in their relative positions. In other words, don't turn the shaft end for end. Look for worn rocker-arm IDs and shafts. An outside mike will indicate how much wear is on the shaft if you compare worn and unworn areas. Use a snap gage and an outside mike to measure the rocker-arm IDs.

There will usually be little wear between rockers and shaft. As long as there was a constant supply of fresh, clean oil to the rockers, the parts won't have any problems. But if the oil pressure dropped from a low oil level, bad pump, pressure-relief valve or blocked passage, then the shaft and rockers will be worn. If you don't see any wear, or can't feel any ridge on the rocker shaft, then there is nothing to worry about. But if you do see wear, or can feel a ridge, then the assembly is worn. Replace all worn parts.

Unless there was absolutely no oil in the rocker cover, the springs, washers and rocker stands will be fine.

If you find no wear, thoroughly clean all parts, oil and assemble them. NOTE: On Type 1—3 engines, the open slot on the rocker-arm stands faces *up,* and on Type 4 engines, the slot faces *down.* The adjuster end of the rocker arms is the up end, so on Type 1—3 assemblies the slot faces the adjusters, and on Type 4 ones the slots face the pushrod end.

One useful modification to make to the rockers is to substitute swivel-foot adjusters for the stock types. These adjusters have a ball-and-socket base that touches the valve stem. Part of the ball is ground flat, so it distributes the load across the valve stem. The advantage of the swivel-foot adjuster is it takes some of the side load off the valve. This increases valve-guide and valve-stem life, with attendant increases in engine efficiency over time. Swivel feet will also quiet the valve train, which is appealing, too.

Swivel adjusters are available from many VW parts stores and mail-order houses. It may be necessary to raise the rocker arm slightly off the head to accommodate the increased height of the swivel-foot adjuster. Determine this value from the parts source, or directly measure how much of the swivel-foot extends below the rocker arm compared with the standard adjuster. If the swivel foot is longer, compensate by shimming the rocker-arm pedestals. Ready-made shims are available for this, or you can make some.

VALVE-SPRING INSPECTION & INSTALLATION

Next up for inspection are the valve springs. The following steps must be performed to verify that the springs are usable. Those that aren't must be replaced. Good valve springs are critical for good engine performance.

There certainly is a lot to evaluate when checking valve springs, including:
- Spring rate.
- Squareness.
- Free height.
- Load at installed height.
- Load at open height.
- Load at solid height.

A *spring tester* must be used for making some checks. And, although it's a lot easier to use a spring tester, some checks can be made with a ruler and bathroom scale.

Or you can do as the volume rebuilders do and replace all valve springs. This eliminates practically all tests and saves time. The rebuilders certainly don't want any work returned, so they're willing to fit all new springs as insurance, if nothing else. VW valve springs aren't noted for being the industry's strongest, so replacement makes good sense. Also, VW doesn't publish many specifications for checking their springs, so not all the spring parameters listed above can be accurately tested. Most of the time, a fresh set of valve springs is the best course.

SPRING TERMS

If you choose to replace the valve springs, there isn't much to check. Inspect the new springs for obvious defects, rejecting any pitted springs. It's smart to check the springs for load with a spring tester, but modern production-line techniques rarely produce a bad spring, so few shops check new springs. But by checking *free height,* you can discover any problems with new springs.

When reusing springs, you need to understand what to look for when testing a spring. So let's define some spring terms.

Spring Rate—Basic to a spring's operation is its *rate.* Spring rate is how much force a spring applies for a given amount of deflection, usually expressed in pounds per inch (lb/in.) of compression for a coil-type valve spring. As a spring is compressed, its resisting force increases. For example, if the valve spring rate is 100 lb/in. then it exerts 100 lb when compressed 1.00 in.

Valve-assembly *inertia* must be controlled by the valve spring. Inertia is the assembly's resistance to change in movement. The more the valve weighs, the more inertia it has. Also, the faster the valve moves—resulting from engine speed and valve train geometry—the greater the inertia. So, above a certain rpm, valves will *float.* To raise rpm above that, higher-rate springs are needed.

Valve float occurs when the inertia of the valve train exceeds the ability of the valve spring to close the valve. The spring ceases to hold the valve against the rocker, and the rocker against the pushrod, lifter and cam lobe. When this occurs, compression and combustion pressures are lost out the intake or exhaust port, or worse, the piston touches the valve. In the first instance, power is lost; in the second, valve train or piston damage results.

Engineers try for the lowest-possible spring rate, even on high-rpm race engines. Only enough "spring" should be used to close the valves, not much more. It takes horsepower to turn the camshaft and raise the valves against their springs. Too-stiff springs mean wasted horsepower and unnecessary valve train wear.

Consequently, there's little reserve strength in a valve spring. After exposure to high heat and millions of openings and closings, springs *fatigue,* or lose their load-producing capability. More precisely, the springs will *sag*—much like a chassis spring. When this happens, the valves float at lower and lower rpm.

Free Height—This is simply the unloaded height, or *length,* of a spring. That is, how high or tall it is when sitting on the bench. The force or load exerted by a spring is directly proportional to how much it is compressed. Consequently, a shorter, fatigued spring has less free height—it exerts less force because it's compressed less.

Because you probably don't have a spring tester to determine spring force, or spring load at different heights, use free height to judge valve-spring condition. You can be sure a spring is defective if its free height is low.

Load At Installed Height—This is the force

Expensive industrial spring tester accurately measures valve spring output. If concerned about valve spring condition, have machine shop test springs.

an installed spring exerts when the valve is closed. *Installed height* is spring height measured from the spring seat to the underside of the spring retainer. It is measured with the valve closed.

Load At Open Height—This is the force a spring exerts when the valve is fully open—or when the valve spring is compressed more than its installed height by the amount of maximum valve opening.

Solid Height—If a valve spring is compressed until each of its adjacent coils touch and it can't be compressed any farther, the spring is at its *solid height*—it is *coil bound.*

In actual driving, a valve spring that *goes solid* or *coil binds,* creates great havoc with its valve train. Bent pushrods and broken rocker arms are the usual result. In air-cooled VWs, coil bind only happens when a cam with more lift than the original is used.

Squareness—How straight a valve spring stands on a flat surface is called *squareness.* If a spring leans too much, 1/16 in. or more measured from the top coil to a vertical surface, it should be replaced.

SPRING TESTS

Let's apply these basic valve-spring terms and inspection information to put the valve springs through two tests: squareness and open load.

But before testing, check the springs for the slightest sign of pitting, rust or corrosion. Such springs are worn and will break eventually, so don't even consider reusing them.

You probably don't have a spring tester. An industrial type costs $500 or so, and this puts it out of the occasional user's pocketbook. Therefore, take the springs to the machine shop for the installed load test.

There is a valve-spring tester, however, that is within the average do-it-yourselfer's price range. It's the bathroom scale. Although not as convenient to use, all you need in addition to the scale is a vise and a vernier caliper or ruler. Clamp the spring and scale in the vise, compress the spring to the specified height (measure with the vernier caliper) and read spring load directly.

There are also small scales designed specifically for testing valve springs in a vise. Ask a parts department, or look in tool catalogs and tool specialty stores for this style spring tester. If unavailable from these sources, try:

C-2 Sales & Service
Box 70
Selma, OR 97538

or

Goodson
4500 West 6th St.
Winona, MN 55987

Squareness—Spring squareness is the easiest check, so let's start there. The equipment needed is a carpenter's framing square or any right-angle measuring device. Stand the spring upright in the corner of the square. Turn the spring until you determine its maximum *lean*—distance between the top coil and the vertical leg of the square. Maximum allowable lean is 1/16 in. Replace any spring that leans more than this.

While the springs are lined up for the squareness check, sight across their tops and look for any short springs. This is the poor man's free-height check.

Measure Free Height—Springs shorter by more than 0.100 in. (2.54mm) of the others have fatigued. They should be replaced or at least checked in a spring tester for installed load.

Springs that fail the free-height test by .0625 in. (1.6mm) should also be replaced. I don't recommend shimming VW springs. By the time a spring needs shimming, it is worn, and VW springs don't have much reserve left. If they show wear, replace them.

If you choose to shim a spring to restore its installed and open loads, then check for coil bind. Because the spring will be compressed more than its specified open height, its reserve travel is reduced when the valve is open—the distance between coils will be reduced. This is applicable to high-performance engines fitted with high-lift camshafts. Stock VW cams don't lift the valves anywhere near coil bind, so it isn't a concern. Only when camshaft lift reaches .450 in. or higher should you check for coil bind. There must be 0.010 in.(0.254mm) between coils when the spring is compressed to its open height.

Load At Installed Height—Check this figure with a spring tester or the bathroom scale. Compress the spring to its installed height, then

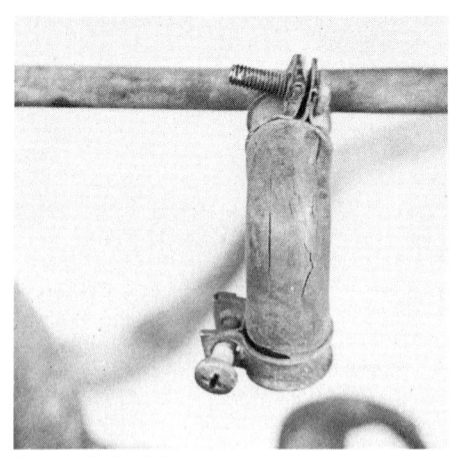

Cracks in fuel lines . . .

. . . lead to barbecued piles of junk. Replace every piece of fuel line on fuel-injected engines. Clamp every connection, even if there wasn't one there to start with.

read how many pounds it is exerting. The 40-HP engine valve springs exert 89—102 lb (40.8—46.8 kg) at a height of 1 5/16-in. (33.4mm). The remaining Type 1—3 valve springs should yield 117—135 lb (53.2—61.2 kg) at a height of 1 7/32-in. (31mm). Type 4s register 168—186 lb at 1 9/64-in. (29mm). Replace any spring that fails this test.

CYLINDER HEAD ASSEMBLY

You'll need these items to assemble the cylinder heads: a spring compressor, ruler and possibly some shims. Don't buy the shims until you determine that they are needed—chances are good they won't be.

Installed Height—It's time to check spring *installed height,* the height of an installed spring with its valve closed. Directly related to load at installed height, correct installed height ensures that the springs are compressed enough to exert correct closing force to prevent valve float.

In other words, the spring must be compressed enough when at rest, or it won't be compressed enough when the valve is open to prevent valve float.

Don't be confused about the possibility of having to shim the valve springs again. The shimming you may have done earlier was to compensate for sag or to restore installed and open loads. This checking and possible shimming now is to ensure that the distance between the spring seat and its retainer is at the specified installed spring height.

For instance, if a valve stem extends too far out of its guide—if the valve seat or valve face needed considerable grinding—the spring won't be compressed to its installed height. The spring must be shimmed to compensate for the difference in installed height. Therefore, you may need to add more shims to springs that already have them; one set to correct for installed and open load, and possibly one for installed height.

There is nothing wrong with having to shim for installed height. Buying new springs won't compensate because you aren't correcting a problem with the spring, but adjusting the head and spring relationship. It changes as the valve seats are ground or replaced. Consequently, even if you replaced the valve springs, check the installed height. There should be no problem, but you want to make sure all measurements are to spec.

Check Installed Height—Installed height is checked with the valve installed in its guide without the spring.

Insert the valve into its guide. Now, install the retainer with its keepers. With the valve against its seat, pull the retainer up snugly against the keepers. While lifting firmly on the retainer, measure the distance from the spring seat to the underside of the retainer. This should be 1.5 in. (26.67mm) on all engines. One of two methods can be used for measuring this distance.

If you have a snap gage, use it and a 1—2 in. micrometer for checking gage length. Or, make a gage from a piece of coat hanger, welding rod or other heavy wire. Snip out a straight section, then grind it to the specified installed height (1.5 in.). Stand the gage between the retainer and the spring seat. If there's a gap, use a feeler gage between the gage and spring retainer to determine the required shim thickness.

If using a telescoping gage and micrometer, subtract specified installed height from the actual measurement to find the shim size needed, if any. Ignore anything less than 0.015 in. (0.381mm). Compensate for differences over 0.015 in. with a shim of the same or less thickness. Don't over-shim a spring.

Record the thickness needed on a narrow piece of paper and slip it under the valve spring that needs the shim. Also, write it down on your shopping list. After you've checked all springs, go buy the shims.

Now, install them under each correct spring. Lightly lube the guide and valve stem with oil, then insert the valve. With the valve in its guide, the stem-seal ring can go on. You can leave the seal ring off Type 4 engines, even if it had one on it to begin with. Oil the valve stem and as gently as possible, push the ring over the stem and down onto the guide.

Retainer-To-Guide Clearance—The bottom of the valve-spring retainer should clear the top of the valve guide and oil seal by at least 0.060 in. (1.5mm). I mention this only for those with a high-lift, aftermarket camshaft. Stock engines don't come close on this dimension and don't need checking. If you have a high-lift cam, check by opening the valve—while holding the retainer and keepers in place—to its full-open position. Use a machinist's 6-in. rule against the retainer to check valve opening. Eyeball the retainer-to-guide clearance. If it appears to be remotely close, measure it with a feeler gage.

If there's insufficient retainer-to-guide clearance, the top of the guide must be machined shorter. Have a machine shop do this.

Cylinder Head Assembly—After purchasing

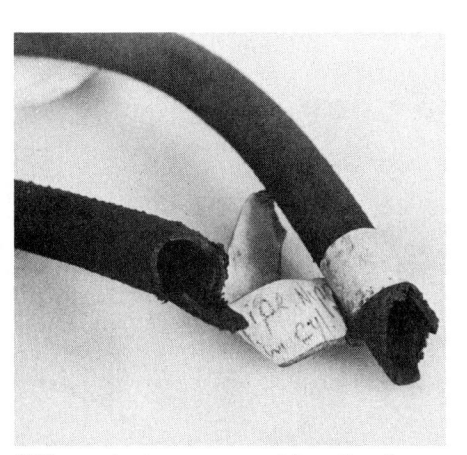

Stiff, cracked vacuum and breather hoses don't cause fires, but they affect driveability and emissions. Replacing all hoses because engine heat and ozone in smoggy areas ruins them.

the shims, you should have everything needed to assemble the head.

Set the heads with the valves, seal rings and any shims installed in front of you. The springs should be in order behind the heads. Remember, the springs, shims and valves must be installed exactly as they were machined or checked.

Use a valve-spring compressor to install the springs. Remember to compress a valve spring only enough to install the keepers, no more. Also, one end of the valve springs has its coils closer together than the other end. This is the bottom of the spring, and it's placed against the cylinder head. After the spring and retainer are compressed, install the keepers. A dab of grease on the keeper groove will hold the keepers in place. This allows you to use both hands to release the compressor.

Double-check that all keepers are in their grooves. When finished, lightly rap the retainer with a plastic, lead or brass hammer to guarantee the keepers are seated. Once the heads are assembled, spray a protective coating of oil on the valve springs and valve heads. Then store the heads out of the way, wrapped in protective plastic bags.

INTAKE & EXHAUST MANIFOLDS

Working on the intake and exhaust manifolds might not seem like part of the typical valve job, but it is a logical extension of cylinder head work, so I've included it here.

Because the manifolds are non-moving parts, they don't wear out. All you have to do to the intake manifolds is clean them. On single- and dual-port upright engines, give the intake manifold a good solvent bath and leave it at that. Replace the rubber gaskets where the dual-port runners meet the aluminum castings that bolt to the heads. If the gaskets are cracked or broken, replace them.

The real work is with fuel-injection manifolds on the flat engines. On these you must replace all fuel lines, and make sure each line is clamped to its fitting. VW doesn't clamp these lines, leading to the old saying, "There are those that have burned and those that will." Replacing fuel lines and adding clamps will help keep your VW engine from spraying fuel over itself and starting an engine fire.

Use high quality fuel line and worm-drive clamps. Get the intake manifold as complete as possible in front of you on the bench. Start at one corner and work your way around the assembly, changing each hose. Do yourself a favor and cut each hose a little long. You can always come back and cut off some hose if it gets in the way later.

You should be able to change the entire fuel ring, the short hoses leading to the injectors and even vapor hoses right there on the bench. This may not be such a huge advantage to Type 3 owners because the top of that engine is easily accessible in the chassis, but anyone who has spent the day upended in a 914 engine compartment or on their knees behind a Bus will appreciate getting the job done on the bench.

Don't worry about the fabric covered connectors between the sheet-metal intake plenum and individual intake runners unless they are old, dry and cracked. If they look fine and have some resistance to being slid around the runners and air distributor nipples, they are still sealing.

Exhaust Manifolds—There's even less to do with the exhaust manifold. Apply a wire brush to the flared pipe ends and clamps to make installation neater. New sealing rings and gaskets should be in the engine gasket kit; if not, get them. There is an exhaust hardware kit available from most aftermarket suppliers with new clamps, complete with bolts, nuts and so on. Use it during assembly because the old hardware will be worn out from heat cycling.

Consider replacing the muffler or tail pipes now if they were starting to deteriorate before the engine overhaul. It's easier to change them now, than after the engine is installed in the chassis.

CHAPTER 7
Engine Assembly

Now comes some excitement. You're going to put your engine together. All the parts are clean, and the running around to machine shops and part houses should be all over. Your shopping list should be completely checked off, and all lubricants and sealants at hand.

PREPARATIONS

Torque Wrench—There are some special tools needed for engine assembly. An accurate *torque wrench* is a must. Typical automotive torque wrenches are 1/2-in. drive with a 30—150-ft-lb range. Such a wrench is great for torquing 110 ft-lb Chevy main-bearing bolts, but is too large for accurate work at the low torque levels VWs are assembled with. A better tool is a torque wrench with a 5—50-ft-lb range. Your best bet is to buy a 3/8-in. drive torque wrench calibrated in in-lb. To convert ft-lb to in-lb, merely multiply by 12. Thus, 18 ft-lb equals 216 in-lb. To convert from in-lb to ft-lb, divide by 12.

A 60—600 in-lb wrench will be adequate for everything except flywheel or driveplate hardware. Because Type 4 engines use conventional flywheel/driveplate bolts, maximum torque is 80 ft-lb, or 960 in-lb. You might have trouble finding a wrench that handles the entire torque range of the engine, but a combination of in-lb and ft-lb wrenches will work fine.

At the other end of the scale is the 300 ft-lb needed to torque the gland nut. If money is no problem, or you do a lot of VW engine work, invest in one of the large torque wrenches sold through the popular VW magazines. Otherwise, get a breaker bar and pipe "cheater" extension so you have 2 ft of leverage on the nut. You'll also need a 36mm or 1-7/16-in. socket to fit the nut.

Flywheel/Driveplate Locking—Some method of locking the flywheel/driveplate is required when torquing it. This may not have been necessary if you used an impact wrench at disassembly. Because a torque wrench is required for assembly, the flywheel must be locked. VW flywheel locks are available in parts houses or by mail order. Alternately, you can fabricate a locking tool with a 5-ft metal bar with holes drilled in it for the pressure plate bolts. Mount it to the flywheel to hold it steady.

Flywheel locks will hold a flywheel fixed, but how do you keep the driveplate stationary on cars with automatic transaxles or an Auto-Stick? A driveplate doesn't have a ring gear (it's attached to the torque converter, which is with the transaxle) so a conventional flywheel lock is useless. So, fabricate a lock by using a metal bar with two holes in it. One hole fits over a bellhousing stud and the other to a driveplate hole. Put a bolt into the driveplate and tighten a nut to the bellhousing stud.

Ring Compressor—This tool is needed when installing pistons in the cylinders. Depending on how and when you do this job, you can use almost any ring compressor, or you may need the type that opens enough to get it past the connecting rod.

If you don't have any sort of ring compressor, buy the very simple and inexpensive band-type. This is just a thin band of spring steel with the ends turned up. Pliers are used to grasp the ends and compress the rings. If you are short of money or just like roughing it, a large hose clamp will work, but it's slow going and can scratch the piston.

Clutch Tool—A clutch *centering tool* is required, too. If you have some spare VW parts, saw off the input shaft from an old transaxle. Very inexpensive wooden-dowel centering tools are available at parts houses. My favorite is the inexpensive plastic version of the input shaft made by Kingsborne. It fits much better

All your preparation comes together here. Be patient, check all clearances, and pre-lube those parts! Your engine will return your diligence with long service.

and lasts longer without rounding off than a wooden one.

Assembly Lube—You also need some sort of engine *assembly lubrication*. There are all sorts of these and you may be partial to one particular type or brand. The point is, some sort of lube is needed. For VW engine assembly I use multi-purpose white grease. Because it is thick, it stays in place. This is particularly useful if the engine sits around the shop a day or two before installation. It is commonly used for camshaft and lifter lubrication, although an assembly lube with molybdenum disulfide, *moly* for short, is better.

The white grease offers convenience when bought in a small can with a replaceable plastic lid. You can cut a small hole in the lid and insert an *acid brush,* which is used for applying acid before soldering, through the hole. Then you don't have to bother with a cap or lid each time you need some grease, and dirt is denied wide-open access.

Loctite—Another essential is Loctite. A small tube of Loctite Stud & Bearing Mount (red) is needed for rod nuts and other low-torque

116

Bolting crankshaft to flywheel and standing the two on end makes a stable assembly stand.

Lay Plastigage across rod journal and make sure it extends the entire width. Any taper will show as an unevenly flattened stripe.

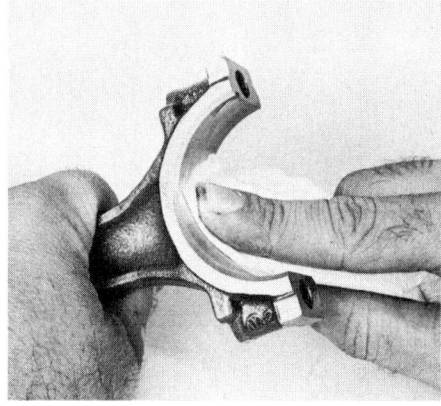

Wipe connecting rods clean and dry before installing bearings. Use paper towels to clean engine internals, not cloth rags. Cloth fibers can clog oil passages; paper residue will dissolve.

fasteners that absolutely must not come loose.

Gasket Sealer—Some form of gasket sealer is also needed. The entire engine can be assembled with Permatex 3H, or a variety of sealants can be used for specific jobs. Some people specify Gasgacinch for sealing the case halves, for example. I don't have any absolute recommendations, except 3M Weatherstrip Adhesive should be avoided. It's a fine sealer and some people swear by it, but it's tedious to remove if the engine is ever taken apart again.

More Parts—You probably still need to buy some parts, too. Get a muffler hardware kit for Types 1—3 and early Buses. This has the two-piece clamps that secure the muffler to the heat exchanger slip joints. The kit is cheap and beats working with old, bent, rusty hardware. Also buy new intake manifold gaskets for Type 3s. These are the gaskets that go between the phenolic blocks and heads. They are not included in gasket sets. And purchase a new fan belt, particularly for Type 3s.

More Patience—There is one other ingredient you must have for this job—patience. All the best assembly lubes, torque wrenches and expensive parts won't do any good if you slap them haphazardly together. Unlike a large domestic V8 with lots of reserve power and displacement, a VW engine relies on careful assembly and constant checking to produce the maximum available power from modest displacement. Take your time, and check all clearances during assembly. Don't be afraid to double- or triple-check clearances, or return parts that don't fit. After all, it is your engine, not the parts supplier's.

CRANKSHAFT ASSEMBLY BUILDUP

Begin engine assembly with the crankshaft. You need a stand to hold the crank while you work with it. The cheapest method is to fit the flywheel to the crank and attach the gland nut or flywheel bolts. Then the crank can be stood on the flywheel, which makes a nice wide, stable base.

More convenient is to weld an old gland nut to a pipe and fit the pipe to the engine stand. You can also clamp the pipe in a bench vise. These methods keep the crank horizontal, which makes it easier to work with. Some shops use a round mount so they just have to stick the gland nut threads over it and let the crank's weight cock the crank on the fixture to hold it.

Of course, you can always do without a stand and build up the crank while it sets on the bench. Have a helper ready to hold the crank when torquing the rods with this method.

If the timing gears and #3 main bearing are not installed yet, put them on now. See page 87 in Chapter 5.

Plastigage Rods—Before installing the connecting rods *Plastigage* the rod bearing oil clearance. This takes a minute, but quickly and accurately determines the oil clearance. Thus, it is an excellent double-check that the crankshaft was ground correctly, the rods were properly reconditioned, and the correct bearings have been fitted.

First get a strip of fresh Plastigage. This is a thread of precision wax, available at parts houses. Several sizes are available. You want the green variety for measuring 0.001—0.003-in. clearance.

Next install the rod bearings. Remove the rod cap from the rod and wipe the bearing bore dry. Install the rod-bearing inserts in both rod and cap, be sure to align the tang on the bearing with the slot in the rod. Press the bearing insert firmly home with your thumbs. Bearing inserts must be dry on their backsides where they mate with the connecting rod, so don't oil them. Just wipe their backsides off with a paper towel. Leave the working surface of the bearing alone for now.

Snip a length of Plastigage and lay it across a crankshaft connecting-rod journal. Without lubing the bearings, fit the connecting rod onto the journal, install the cap and nuts, and torque them. Torque all the rods in three steps, 15-, 20-, and 24 ft-lb, except for early 40-HP rods with capscrews.

Be sure to align the rod caps on rods with capscrews. First snug the capscrews, then align the cap and rod edges by tapping with a hammer. Final-torque these to 32 ft-lb.

Don't move the connecting rod around the crankshaft, or you'll smear the Plastigage. It's easiest not to bump the connecting rod when it is hanging straight down, which is why mounting the crankshaft horizontally is preferred.

Once the rod nuts are torqued, remove the rod from the crankshaft. This exposes the Plastigage strip that was squeezed flat. Hold the scale printed on the Plastigage wrapper against the flattened strip to determine the oil clearance. The wider the strip, the narrower the oil clearance.

How Much is Enough?—VW specifies 0.0008—0.0027-in. clearance, 0.006-in. wear

Tricky part about Plastigaging rod is not moving it while torquing. Letting rod hang straight down helps.

White grease is excellent bearing lubricant and won't run off before engine is started. Moly lube is another good choice. Motor oil is OK, but most of it drains off before first engine start.

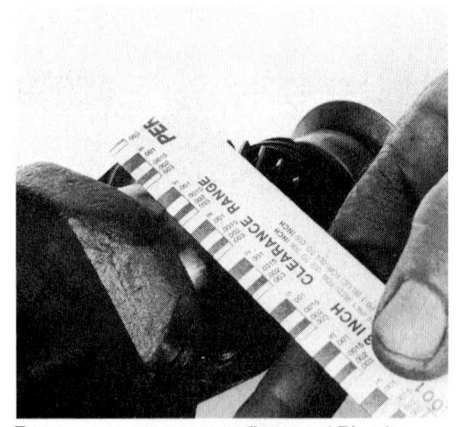

Remove cap to expose flattened Plastigage. Compare its width to scale on package. Optimum oil clearance is 0.002—0.0025 in.

Loctite is great insurance against loosening rod nuts. Threads must be completely oil-free for Loctite to grip.

Bumps on rods must all face *up* on Type 1—3 engines or rod offset will be backwards. Type 4 rods have symmetrical small ends and can go on either way.

limit. I would be happiest with 0.002—0.0025-in. clearance. You can adjust the clearance by grinding or polishing the crankshaft, or trying other bearings.

Practically speaking, it's a pain to keep swapping bearings or bringing the crank back for another 30 seconds of polishing, so we usually accept what the machine shop gives us. And 99% of the time they've done good work and are right on. Nevertheless, don't accept extreme clearances because that's what you received. If the clearances are barely 0.0008 in. or pushing 0.003 in., you have a legitimate gripe with the machine shop. Go back and get another crank or set of bearings.

Install Rods—Once you are satisfied with the connecting-rod bearing oil clearance, install the connecting rods for good. Lube the rod journals and bearings with white grease. Also wipe some along the sides of the rod's big end. This will help cushion any side loads the rod is exposed to until plenty of oil reaches this area.

Now fit the rods to the crankshaft, put a couple of drops of Loctite on the bolt threads and install the caps. Use the same torque sequence given above at **Plastigage Rods**.

Remember that the tangs in the bearing inserts butt against each other when the rod and cap are properly oriented. The three-digit numbers stamped in the rod and cap will be on the same side as well. Also, the bump in the rod's shank faces up when the crank is horizontal, as it would be in the engine. If the rods don't have the bump in the shank, orient the rods so the bearing tangs are at the bottom of the journals when the crank is horizontal.

Up to late 1972, rod nuts had a section to be peened over to lock the nut. Since then, rod nuts have been changed so no lock is needed. All rod nuts should be changed anyway, but if you have an earlier engine with peened-over rod nuts, you must replace them. You cannot repeen the early nuts.

Side-Clearance—Check the rod side-clearance with feeler gages. Push the rod against one end of its journal, then find the feeler gage that just slips between the rod and crankshaft at the other end. The specification is 0.004—0.016 in. (0.10—0.40mm) with a wear limit of 0.027 in. (0.70mm). That's a fairly large tolerance, so there is usually no problem staying within it. If the clearance is excessive, you will need new rods or a crankshaft to cure it.

Rotate the connecting rods around their jour-

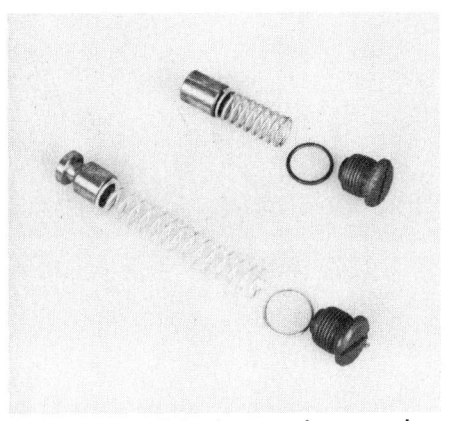

Oil pressure-relief valve uses longer spring, oil-control valve uses shorter.

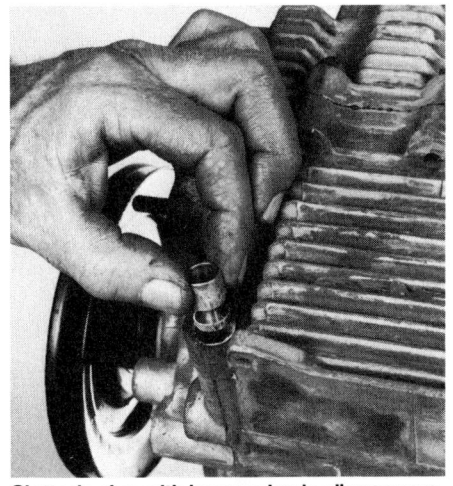

Slotted valve with long spring is oil pressure-relief valve. Place in hole closest to crankshaft pulley or cooling fan. Apply white grease to bore and valve before installation. Follow valve with spring, seal and cap.

Cinch caps with largest screwdriver and added leverage, or judiciously use a hand-held impact tool. But once tight, don't brutally force caps or you can *break the case*.

nals, feeling for any hangups or rough spots. If the journals are out of round, or a piece of grit got between the bearing and journal, you'll feel the rod jerk as you pass over that spot. Another check you can do when the crankshaft is horizontal is to hold all the rods level, then let them fall simultaneously. The white grease will slow their fall, and you can check that all the rods fall at the same rate. This is just a double-check that all the rod clearances are relatively close together, and not critically tight or loose.

CRANKCASE ASSEMBLY

Set the crank assembly aside for the moment and turn your attention to the case. Mount the left case half to the engine stand. You'll need one bolt for the upper bellhousing hole and one nut on the lower bellhousing stud. Make the case half secure on the stand as it will save wasted motion later, plus those two attachment points will have to hold the entire engine before you are finished.

Pressure-Relief Valves—After determining that the pressure-relief valves move freely in their bores, apply white grease to the bores and valves and slide the valves into position. Then drop in the springs, make sure there is a copper seal under the caps and thread in the caps. Cinch down the caps with your largest screwdriver as shown. A hand impact-tool is an acceptable method for tightening these, but too much force can *crack the case*. If you have a single-relief case, you will only have one valve to install, of course.

For some reason, all VW engine gasket sets have only one new seal for the pressure-relief valve cap. This is probably a throwback to the days before dual-relief cases. That's why I had you save one of the old seals; reuse it now.

Sealing Rings—Type 1—3 engines use rubber seal rings on the main-bearing studs. The old sealing rings may still be at the base of the studs on the left case half. If they are, remove them. Then install new rings from the gasket kit. There is no need to lube these sealing rings.

Distributor Drive Shaft—Installing the distributor drive shaft now makes setting its end play easy and eliminates the chance of dropping a washer into a sealed crankcase. The procedure may sound complicated, but is very easy and only takes a minute. Setting up the distributor with the cases joined takes much longer and has great aggravation potential: a washer dropped into the crankcase, the attendant screams of disbelief, and complete disassembly of the engine.

Gather up the distributor, the drive shaft and shims. If you have a Type 1—3, you need two shims. If you have a Type 4 there is only one. Remove the 0-ring from the distributor shaft so it will slide easily in and out of the case. Also remove the distributor hold-down clamp.

Now, without any lubrication, install the drive shaft and shims in the left case half. Next, place the distributor hold-down clamp on the engine. Tighten the clamp-to-case nut finger-tight, but leave the distributor cinch-bolt loose. Install the distributor, making sure it goes in all the way, seating against the hold-down clamp. Do not install the small coil spring between the drive shaft and distributor.

Now push up on the drive shaft until it touches the bottom of the distributor. Use feeler gages between the shims and case to determine end play.

You want a minimum of 0.020 in. (0.51mm) with up to 0.050 in. (1.27mm) acceptable. What you don't want is less than 0.020 in.

New main-bearing stud seals are in Type 1—3 gasket sets. Old seals should be out before now, but if not, pry them free. Install new seals dry.

clearance. It's much better to have too loose a clearance than not enough.

Adjust the clearance by changing shims. You can install any number of shims of different thicknesses to get the desired clearance as long as you have at least two shims in Type 1—3 engines and one in Type 4s. Shims are available at the dealer and VW parts houses.

Once the end play is set, remove all parts, apply white grease to them and reinstall. Before installing the distributor, fit a new O-ring from the gasket set to its shaft.

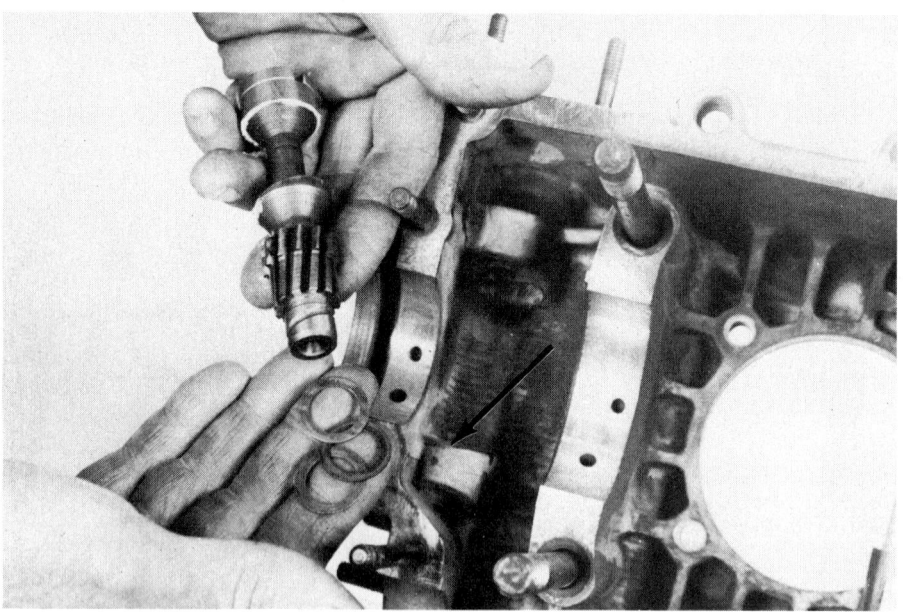

Adjust distributor drive end play with shims. Type 1—3 drives like this one need at least two shims, but Type 4s use only one. Extra shims are no problem in either engine. A factory-installed 1/4-in. thick spacer is sometimes pressed in where drive seats in case (arrow). It's a fix for case that was cut too deep.

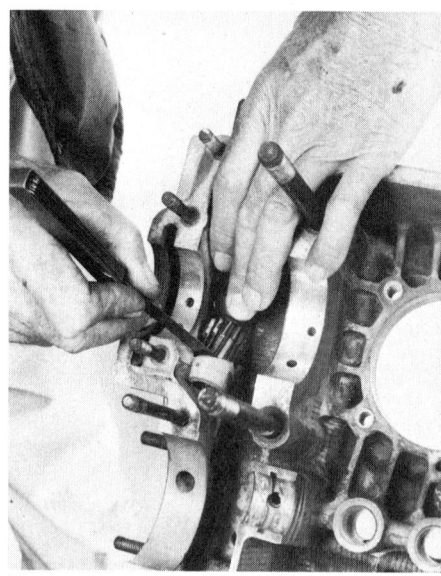

Measure end play while holding driveshaft up against distributor. Vary shims to get 0.020-in. minimum, 0.050-in. maximum. A little loose is much better than too tight on this measurement.

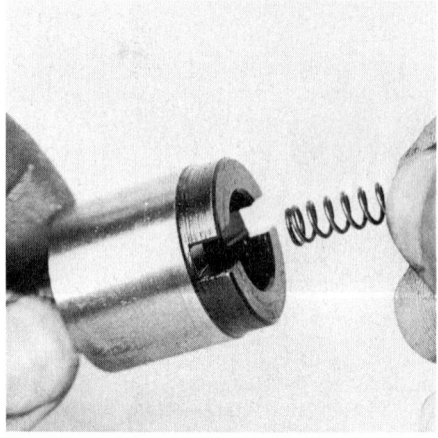

Don't forget: insert this small spring between drive shaft and distributor body when installing distributor. Note offset of slot in shaft; shaft must be aligned correctly when installing distributor.

After setting end play, and aligning drive shaft slot, lube driveshaft with white grease and install small coil spring and O-ring. Next, install distributor body (push stoutly) and rotate it to align #1 TDC mark with center of rotor's tip. Tighten clamp.

Before continuing, note the small line stamped on the edge of the distributor body which marks cylinder 1's position. Look straight down on the distributor to see the mark. When the distributor is correctly timed, the rotor will point at this mark. But first, you must orient the distributor body into its normal position for correct timing and so it won't foul equipment added later.

Install Distributor—We must align three points: the drive shaft slot, the line stamped on the distributor housing marking #1's position and the rotor.

When the shaft's slot is placed right, the matching tang on the housing's underside mates with it only one way. The rotor is then correctly positioned, and we turn the housing to align its timing mark with the rotor's tip. I explain the installation for Type 1s and early Type 2s here. The method is identical for the other engine types, and their slot and rotor orientation is specified on page 14.

Put the shims on the end of the shaft and install it. Face the engine flywheel and look in the distributor mounting hole. Turn the shaft until the slot is correctly angled and its offset (larger portion) faces the correct direction.

On Type 1s and pre-'72 Type 2s, place the slot perpendicular to the case parting line with the offset facing the flywheel. Install the coil spring between the drive shaft and body and then mate the distributor to the shaft. The vacuum advance can will point to the left-rear; the rotor should be pointing to 5 o'clock. Turn the distributor housing until its timing mark aligns with the center of the rotor's tip and tighten the clamp snug.

Install Cylinder Studs—If the cylinder studs were removed for some reason, install them now. Apply Permatex 3H to their threads to stop oil leaks. If you don't seal these threads, no matter what you do, some oil is going to eventually find its way around the threads and dribble down the outside of the case.

There is no need to overtighten these with a wrench, just run them in until snug. You may have a stud that is difficult to turn. This indicates a problem, i.e., the case-saver threads are dirty or damaged, or the stud is dirty or damaged. Remove the stud and investigate.

If there seems to be nothing more than a close tolerance between the stud and case threads, tighten two head nuts against each other at the other end of the stud. Then use a wrench on the

Coat cylinder-head stud threads with Permatex 3H before installation to stop oil leaks. Just finger-tighten studs. Don't bottom studs in case savers, or case-thread wear may result. Stud length varies among engine types; longer studs install in bottom row. Dual-port engine's two top middle studs are shorter; same length as all 40-HP studs.

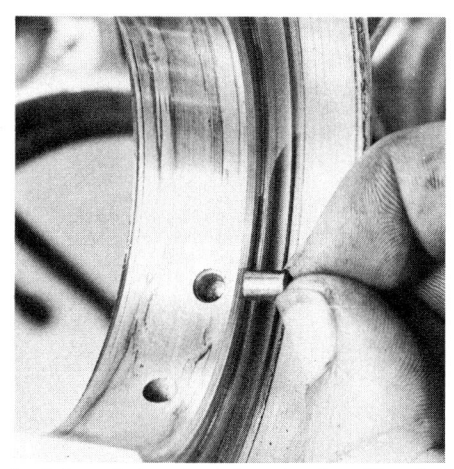

Install main-bearing dowels dry. Needle-nose pliers are handy for this job. Note: Case is drilled for six dowels, but only five are installed. Four in left case half and one in right at #2 main bearing. Right case half #3 bearing has a dowel hole drilled, but it isn't used.

outside nut to run the stud in. Excessive force should not be necessary. Make sure all the studs have threaded in a reasonable distance before continuing on. Don't forget to install the studs in the other case half.

Main Bearings & Lifters—Turn the case half so the parting line is parallel with the floor, cylinder studs pointing down. Also place the other case half so you can work inside it.

Install the small locating pins in the main-bearing bores (without any lubricant). Needle-nose pliers are the tool to use here. Now install the #2 main-bearing inserts in both case halves. Again, don't lube the backside of the bearing. Next, install all the camshaft bearing inserts. If you have an early 40-HP engine, you may not have cam bearings, so skip this step.

Prepare to install the lifters if working on a Type 1—3. If working on a Type 4 engine, install the lifters later, so ignore them now. Apply white grease to the lifters, lifter bores, cam bearings and main bearings, including #1 and #4. Also lube main-bearing journals #1 and #4. Then slip the two bearings onto their respective journals. Make sure the dowel holes in the bearing backsides are aligned toward the flywheel.

Type 1—3s are ready for lifter installation, so drop the lifters into place. If reusing old lifters, they must go into their original positions. New lifters can be installed anywhere.

If you have a set of hair-spring clips for holding the lifters in place, install them now. You really don't need them, however, as the white grease will hold the lifters in position.

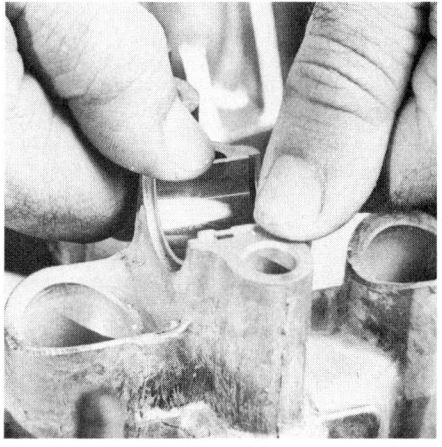

Cam bearings simply push in place once bore and bearing backside are wiped clean. Always start with tang aligned in its slot. Type 1—3 and Type 4 cam bearings look the same but are *not* interchangeable.

INSTALL CRANKSHAFT

Once the lifters are installed in Type 1—3 engines and you've double-checked distributor position, you are ready to install the crankshaft assembly. Grasp the assembly by two connecting rods, and lower it into the left case half; it's the one on the stand. Guarantee the crank is completely settled in the case by wiggling the main bearings back and forth. This will help the dowels and their bores find each other. Make sure the case is level and the crank drops straight in. Support the crank by holding the

Don't get carried away with white grease. A light coating on center main bearing, cam bearings, lifter bores and lifters works best. Too much grease will foul oil screen or filter. Note cam bearing installation. In all engines: #3 (thrust bearing) is widest, #2 is narrower, and #1 is narrowest.

small end of two rods, not by the crank's ends. Lower the crank into position with the rods perpendicular to the floor; if you don't, the ignition timing will be off.

It can be tricky to tell when the main bearings are fully seated. The best way of doing this is to rotate the crankshaft slowly while closely watching #2 main bearing. If the crank is fully seated, the bearing's edge will scrape white grease off its journal. If the crank is not fully seated, no grease will rub off. Keep wiggling

After lubing journal and bearing with white grease, slip on #1 and #4 main bearings. The dowel holes are offset; they are placed toward flywheel.

Lower crank assembly while holding at #1 and #2 connecting rods. If case parting line is level with floor, and crank dropped straight in, distributor timing is correct. Check that distributor rotor still points to distributor body mark, and timing marks on crank gear straddle case parting line.

Wiggle main bearings to make sure they seat on dowels. Expect a very slight rock even after bearing is properly seated.

Excellent check for main bearing seating is at #2 main. Rotate crank slightly. If grease scrapes off crank (arrow), main bearings are seated. Note: some bearing suppliers are making, in limited sizes, #2 main bearing without oil groove. It has a plain back and an oil hole that *must* align with oil hole in bearing bore. Don't match these oil holes and you'll cook #2 main.

Engage crank and cam gears while checking timing marks, then lay cam in bearings and maintain their timing relationship. Crank may need rotating while cam is lowered in place. Don't be afraid to do this step several times until you know it is right.

the main bearings until they are fully seated. If the bearings refuse to seat, remove the crankshaft and check that the bearings are not installed backward. Look for alignment between the dowel holes and the dowels.

When the crank is set, check ignition timing. The distributor rotor should be aligned with the mark on the distributor body. At the same time, the timing marks on the crankshaft's timing gear, two punched dots, should be aligned with the case parting line. They should face the bottom of the case, toward the camshaft.

Install Camshaft—Now apply white grease to the camshaft journals and lay the cam in the left case half. When installing the camshaft, line up the punched dot O on the camshaft gear *between* the two dots on the crankshaft gear. The alignment of these marks is *critical;* you are establishing the basic valve timing.

Camshaft Backlash—VW specifies 0.000—0.002-in. (0.000—0.050mm) backlash between the cam and crankshaft gears. This is a difficult measurement to take, requiring a dial indicator, and most mechanics don't actually measure this clearance. Instead, they use two tests that experience has shown to be accurate.

The first test is to rotate the *crankshaft* backward and see if the camshaft walks out of its bearings. If backlash is too tight, rotating the crank backward will lift the cam right out of its bearings, so watch for movement at the journals. Be sure to rotate the crankshaft, not the camshaft, or the test will not be valid.

The other test is to feel for backlash. Lightly rock the cam gear back and forth against the crank gear. You should either feel nothing, or a

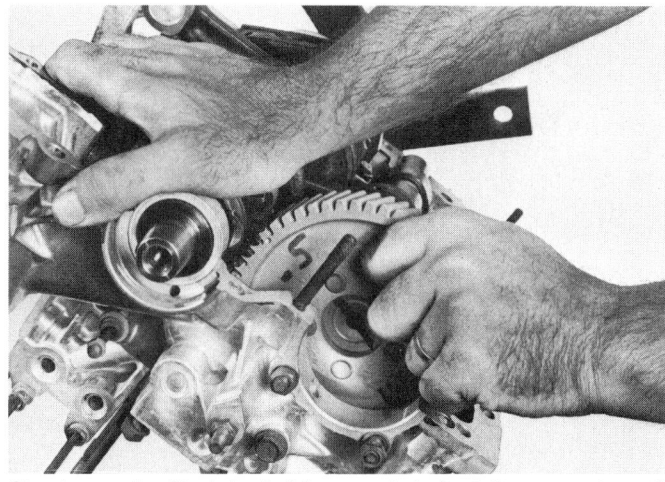

Check gear backlash by holding crankshaft stationary and gently rocking cam gear. Correct backlash is 0.000—0.002 in., so you should feel nothing or just a hint of teeth clinking together.

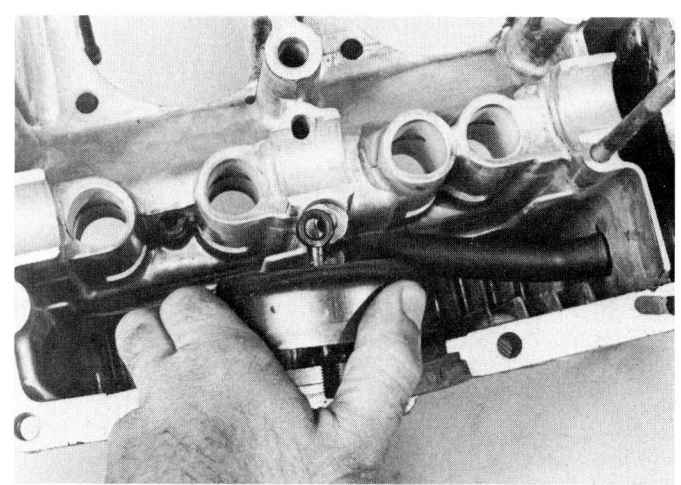

Install Type 4 oil pickup. Fit new 0-ring from gasket set and press in place. Loop on top of pickup fits over bolt hole in case.

Cam-gear tooth with large dot fits between teeth with smaller dots on crank gear. Crank-gear dots straddle case parting line as well. Sight across parting line; you should see only one crank-gear dot. Distributor rotor should be pointing to mark on distributor body, too.

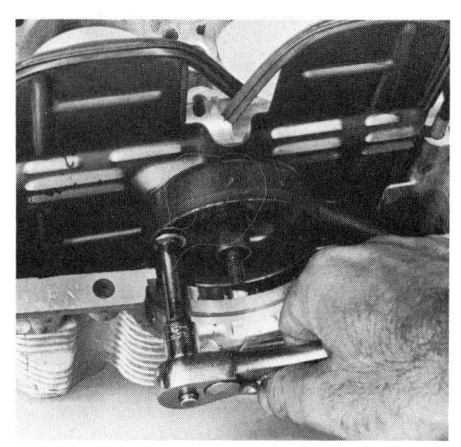

Rubber seals on Type 4 windage tray flop around until retained by case. Ensure their grooves fit over tray's edge right before joining case halves. Don't forget small bolt that secures pickup to windage tray.

Pass Type 4 case bolts through left case half, then snap small plastic cylinders over them. Slide cylinders down into recess in case parting line to hold bolts in place. Bolt heads and washers should be sealed with Permatex 3H before installation.

very small clunk.

So, if the cam doesn't walk out of its bearings, or go clunking back and forth, it's OK. If the cam lifts, or clunks excessively, get another camshaft-and-gear assembly.

Lube Lobes—After checking the backlash, coat all cam lobes with white grease or moly lube. You'll have to rotate the crank and camshaft assembly to get at all the lobes. That's fine, just don't absent-mindedly lift the camshaft and set it back down in the case again without resetting cam timing (aligning those punched dots). Give each lobe a medium coating of lube.

Cam Plug—Next install the cam plug at the flywheel end of the case. Seal the plug's circumference with Permatex 3H before setting it in the left case half. Install it with the open end facing into the case. On Auto-Stick engines, reverse this positioning. Install it with the open end facing away from the case.

Windage Tray—On Type 4 engines, slip the windage tray into position below the camshaft. Ever wonder what this piece of stamped metal is for? It stops excessive amounts of oil from being scooped out of the sump and whirling around with the crankshaft. Wind from the rotating crankshaft does the scooping, hence the term *windage* tray. Make sure the rubber edges are still intact, then slip it in place.

Oil Pump Pickup—On Type 1—3 engines the oil pump pickup is not removed for overhaul, but on Type 4s it is. Therefore, on the Type 4, fit a new O-ring to the pickup's end and slip it into its hole. Ensure that the small ring on top of the large dish lines up with the bolt hole in the case. Now install the small bolt that holds the windage tray to the oil pump pickup.

On Type 1—3 engines check for tightness between the pickup tube and case. Lightly try to wiggle the pickup tube where it joins the case. If the tube moves, peen the case material immediately around the pickup tube. This will displace some of the case material toward the tube, tightening it in the case.

Sealing Rings & Bolts—Type 1—3 engines use sealing rings around the main-bearing studs. These should have been changed when the left case half was mounted on the engine stand, and double-check them now.

Type 4 engines use long main-bearing bolts,

Apply sealer to right case half because it has fewer studs to get in the way. Permatex 3H or Gasgacinch work fine here, and you need seal only one case half. *Don't* use silicone sealant as adhesive for this joint.

Rotate case close to upright to keep lifters from falling out during installation. Use thumb to hold connecting rods up until they rest in their cylinder openings.

not studs. Seal the underside of the bolt heads and their washers with Permatex 3H, then pass the bolts through the left case half. Now snap the white plastic vibration dampers onto the bolt shanks where they protrude from the left case half. The engine may not have had these dampers originally, but you should install them.

The dampers can be snapped into position by pressing them against the bolt shank. Don't try to pass them over the bolt threads, then down the shank. They won't fit over the threads. Once snapped onto the shank, slide the dampers down until they butt against the left case half. They also help hold the bolts and keep them from falling out of the case.

Everything In Place?—The case halves are now ready to be joined together, so this is your last chance to check everything inside it. Did you torque the rod nuts? Are Type 1—3 lifters in place? Are the main bearings fully seated in the left case half? Are the cam and distributor timing correct? Is the oil pump pickup tight in the case? Is the cam plug installed? Have you lubed or sealed all the necessary pieces? Take a minute to closely examine the case until you are satisfied all components are correctly installed.

Clean Case Halves—For the case to seal correctly, the mating surfaces must be clean of all old sealer and dirt. Any foreign material between the case halves will force the case apart and give false torque wrench readings. You'll also get an oil leak, so inspect the case halves for cleanliness.

When removing old sealer or dirt, be very careful not to gouge or scratch the mating area, or you'll get oil seeping through the scratch.

Seal Case—Some sort of sealer is needed between the case halves; Gasgacinch and Permatex 3H are two. *Don't* use silicone sealer to mate the cases. It builds up 0.0015—0.0020 in. and will cause a loss of *bearing crush*. You only need to apply sealer to one case half, so put it on the right side. It is easier to work with because the main-bearing studs are not in the way, and you can easily move it around on the bench while applying the sealer. Run an even bead around the case, but don't goop it on. A light coat works best. Don't get any sealer into any oil passages or on the the crankshaft and camshaft bearings.

Now you can install the right case half onto

Sealant on case hardware may seem messy, but it will keep engine clean by stopping oil leaks later. Goop all threads, washers and the case area underneath them. Two smaller nuts in foreground are on the cam plug studs, and get torqued along with larger nuts. New style case nuts for main studs are self-sealing; put plastic edge of nut against case.

Once all main-bearing hardware is threaded on, snug it down with ratchet. This will draw case halves together, so use crisscross pattern. Once all are snug, torque to 25 ft-lb.

the left half. First, rotate the left half up from its completely horizontal position to one about two-thirds of the way upright; angle it about 30° to the left of straight up. Don't lean it upright so far that the crank and cam fall out, but enough so the right case half doesn't need to be turned completely upside down. That would only run the risk of the lifters falling out and possibly damaging the camshaft. Hold up connecting rods #1 and #2 and slip the right case onto the studs.

Install Hardware—Once the case halves are joined, install the main-bearing hardware. First coat all main-bearing stud or bolt threads with Permatex 3H. Then slip on the washers and apply Permatex to them. Thread on the nuts.

On Type 1—3 engines, install and seal with Permatex 3H the washers and nuts on the two 8mm studs by the cam plug. These studs are visible from the right case half, just below and toward the flywheel from the main-bearing studs.

On all engines, snug the nuts with a ratchet. Make sure the case halves have mated completely around the engine.

On Type 1—3 engines, torque the two cam plug studs 14 ft-lb. Then, starting in the center of the six main-bearing fasteners, torque all of them to 15 ft-lb.

Now retorque the main-bearing hardware at 20 ft-lb, then again at 25 ft-lb. Between each tightening step, check for crankshaft rotation. Just lightly rotate the crank to make sure it is still free. If the crank binds, something is wrong; don't continue tightening or you'll bend or break something. Remove all hardware and the right case half and investigate. The main bearings are probably not fully seated in the case, so make sure all are setting on their dowels.

Case Hardware—Now install the case parting-line hardware. On the Type 1—3, there are 10 studs and three bolts. One of the three bolts is longer than the others. This longer bolt is installed in the top of the case. The other two bolts are the same length. One goes under the oil pump, the other behind the flywheel. Because there are wave washers under all the hardware, there is no need to use Permatex 3H to seal them, so don't bother gooping them up.

There are 20 case parting-line fasteners on the Type 4. Three of them are particularly easy to overlook. Don't forget to install the one at the bottom-front, in the bellhousing area. Another is centered in the deep recess near the lifter bores in the left case half. And there is a fastener below the oil cooler, toward the fan, in the left case half. The others are roughly distributed as follows: seven across the top flange, five at the fan area, and five along the case's bottom.

On all engines, snug this hardware up, then torque to 15 ft-lb. See if anything loosened by going over the main-bearing nuts again at 25 ft-lb.

Avoid the temptation to torque any case hardware, especially the main-bearings pieces, more than specifications. Many people figure 25 ft-lb is some wimpy figure an engineer settled for so there wouldn't be any stripped hardware. Far from it. When the engine warms up, it expands. Because the magnesium or aluminum in the case expands more than the steel bolts or studs, the hardware is tightened by the expanding case. The effective torque is more like 60 ft-lb at operating temperatures—not 25 ft-lb. Therefore, if you think, "25 ft-lb isn't enough, I'll give 'em 40 ft-lb so this case can't leak," then you are really setting an effective torque of about 80 ft-lb when the engine is warm. That's way too much, and will distort the main-bearing bores and overstress the main-bearing hardware. Stay with the stock specifications.

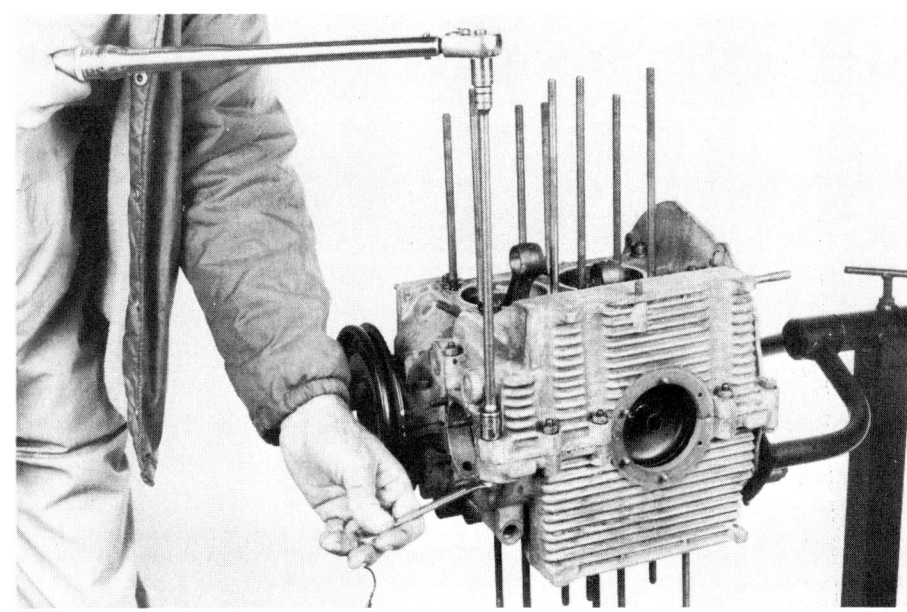

Case flange nuts and bolts need careful torquing. Hold head of bolts when torquing nuts.

Pry old O-ring from flywheels so equipped with small screwdriver or pick. Install new one from gasket set dry.

A few strokes with #220 emery cloth will clean and prep flywheel's oil seal surface. Rub until surface begins to shine, then stop and thoroughly remove all grinding grit. Surface should show light scratching when finished.

Type 4 TDC Sensor—Before moving on, install the TDC sensor in the crankcase bellhousing flange, if your '74 or later Type 4 engine has a TDC sensor. This sensor goes in from the flywheel side of the flange. Drive in the sensor, but do so carefully. Pay extra attention and don't hit the sensor's inner ring.

Check Flywheel—The next step is setting crankshaft end play. This requires some measurements using the flywheel, so make sure the flywheel is usable now. If you replace the flywheel, do so before setting crankshaft end play. If you must replace the flywheel after setting end play, reset it with the new flywheel.

Inspect the flywheel's ring gear teeth. If they are badly worn, consider installing a new flywheel. Check the area the clutch disc works against. If it is grooved or burned black in spots, it needs grinding or replacement. You should have no problems unless the clutch was badly worn or slipping.

The flywheel working surface will usually be coated with resin from the clutch disc. Remove it with abrasive paper or emery cloth. Use 220- or 320-grit paper. Sometimes a rag and a powerful solvent, like Berryman's B-12, will do the job. Be careful with such powerful cleaners because they can cause severe burns. Wear rubber gloves and eye protection.

Gland Nut/Pilot Bearing—This is also a good time to prep the gland nut or Type 4 pilot bearing.

First, check the gland nut threads. They should be clean of all residual Loctite, and completely free of any nicks or cross-threading damage. If the gland nut looks bad or marginal in any way, replace it.

Inside the gland nut is an end ring, felt ring and needle bearing. Remove these parts. If any of these parts are marginal (particularly the needle bearing) replace the complete gland nut. It's inexpensive and you'll have full service life. Clean the end ring, and place several drops of motor oil on the felt ring. Don't soak the felt, just wet it. Clean the needle bearing in new solvent and blow it dry. If the needle bearing looks somewhat clean, you can skip washing it. Once the bearing is dry, apply a light coat of multipurpose grease on the needles. White grease is fine. Replace the parts in the gland nut when finished.

Type 4s have the same parts, but they are housed in the flywheel's center, not the gland nut. Give them the same inspection and treatment. Remove and install the needle bearing in the flywheel with a socket and hammer. This is not a tight interference fit, so tap it in and out lightly.

Set Crankshaft End Play—For some reason, setting crankshaft end play has a lot of mechanics confused. It shouldn't, as it is straightforward. First, I'll outline the procedure using a dial indicator, which is the proper tool for the job. Then I'll explain how to do the job using feeler gages, because not everyone has a dial indicator.

I've presented end play and clutch assembly information here because it is the typical sequence followed by most mechanics. Its advantage is that any end play problems requiring engine disassembly (like installing a different #1 main bearing) will mean the minimum backtracking and disassembly possible. The disadvantage is your engine stand may be in the way of installing the flywheel and clutch. If so, either remove the engine from the stand and do the job on the bench, or leave flywheel and clutch installation for last. There's no reason

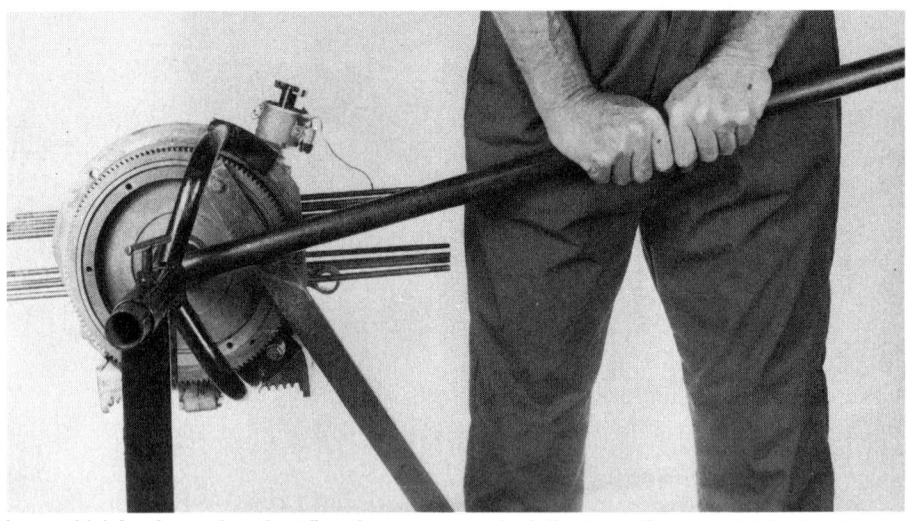

Long, thick bar braced against floor is easy way to lock flywheel. Commercial flywheel locks are another option, but be aware they can break case. If bar is used, tighten it firmly to flywheel. When using body-weight torquing method, don't bounce on bar. Merely press down until your toes just clear floor. Gravity is helping you apply torque, so don't expect a huge effort.

why you can't set end play and install the clutch right before the engine goes in the car. But if you discover a problem at that point, it may mean disassembling the entire engine.

The first step is to slip the three crankshaft shims onto the end of the crankshaft, then install the flywheel. This is only a mock-up for figuring end play, so don't install the oil seal. You're going to remove the flywheel after checking end play, so full tightening isn't required. On Type 1—3s, tighten the gland nut to 200 ft-lb. On Type 4s, tighten the five bolts to 80 ft-lb.

Type 1—3s use 215 ft-lb or more on the gland nut, depending on what VW information you read, for final installation of the flywheel. Believe it or not, however, gland nuts come loose when torqued to "only" 215 ft-lb. Most professional VW mechanics Loctite and torque gland nuts to 300 ft-lb to ensure they don't come loose. But, for the purpose of calculating end play, 200 ft-lb is fine.

Torquing Techniques—Accurately torquing to such high amounts requires special tools or some logic. The only completely accurate way of torquing any fastener is with a torque wrench. VW specialty mail-order houses sell a 300+ ft-lb torque wrench, and if you rebuild an occasional VW, you should have one.

If just doing one rebuild, apply the torque using your body weight. Let's say you weigh 180 lb and have rigged a socket and breaker bar tool to the gland nut. Now, grasp the bar 1 ft out from the gland nut and press down until your toes just touch the floor. You can safely estimate you've applied 180 ft-lb of torque to the gland nut.

If you grab the bar 2 ft from the gland nut and press down until all your body weight is on the bar, you've doubled the force because you doubled the lever arm. Thus, you've applied 360 ft-lb of torque, a little too much. And pressing 1.5 ft out on the bar would yield 270 ft-lb. Therefore, split the difference and press halfway between 1.5 ft and 2 ft, that is, 21 in. out along the bar. Then you'll be applying 315 ft-lb of torque.

Changing the body weight will make a difference. A 220-lb mechanic pressing down 1.5 ft out on the bar exerts 330 ft-lb, or 220 lb X 1.5 ft = 330 ft-lb. At the same distance, a 150-lb mechanic yields 225 ft-lb. You'll have to figure torque using your own bodyweight and bar length. Merely multiply your body weight (in pounds) by the length of the bar (in feet) to determine the approximate torque you and the tool will produce. No matter how much you weigh, don't bounce up and down on the bar or you'll apply too much torque. Just apply your weight to the bar in a steady motion.

Flywheel/Driveplate Locking—Besides calculating how much to torque the gland nut, you've got to keep the flywheel, or driveplate on automatic and Auto-Stick cars, from turning when applying all this force. Holding a flywheel immobile isn't too difficult. On the Type 4, you can hold a bar threaded between two clutch pressure-plate bolts against 80 ft-lb of torque. Type 1—3s and their 300 ft-lb are another exercise. No one can hold the flywheel

Dial indicator and magnetic mount attach on flywheel to directly measure end play. Aftermarket end play checkers are as accurate and less expensive, but are good only for checking VW end play. Specification is 0.003—0.005 in.

Once correct shims are selected, lube with white grease on all sides and install on end of crank. Six- and 12-volt flywheel shims don't mix. OD is same, but ID is different for step on crank.

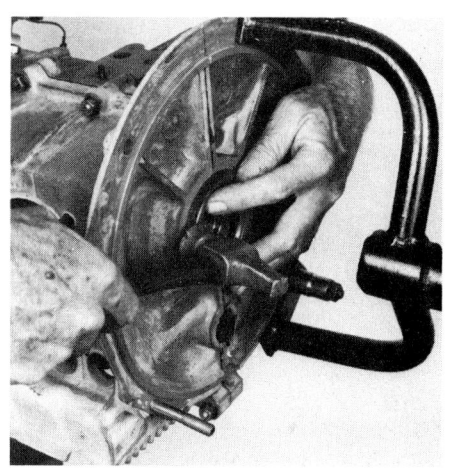

Now that end play is set and shims installed, oil seal can be installed. Alternate light taps from side to side to seat seal. Don't let seal cock in housing, then try and drive it square.

steady against that kind of torque. Here are two methods: fabricate a 5-ft bar for an anchor, or take the easy path and use a flywheel lock.

Fabricate the 5-ft anchor by drilling two holes in the bar's end so you can bolt it to the flywheel using pressure-plate bolts. The other end presses against the floor. This will give you a rock-steady flywheel and crankshaft to apply torque to. Use stout materials for this job. The stop must not fail while you apply all your weight to bear on the gland nut.

Of course, on all engines mated to manual transmissions, you can use an inexpensive flywheel lock to keep it fixed. Such locks bolt to the case bellhousing flange and have teeth that mate with those on the flywheel. They're easy to use, but be sure you have full engagement between the teeth. You don't want the flywheel to suddenly rotate.

Keep a driveplate immobile using the technique and tool specified on page 116. Then tighten the gland nut to 200 ft-lb.

Once the flywheel/driveplate is locked, the measurement method for checking crankshaft end play on each is the same. I use the term *flywheel* exclusively in the rest of the procedure. It's less cumbersome and the majority of cars have manual transaxles.

Mount Dial Indicator—After the flywheel is torqued, install a dial indicator. Use a magnetic base to attach it to the flywheel. Pull the flywheel away from the case, set the indicator's plunger against the case bellhousing lip, and zero the instrument. Now push the flywheel toward the case and read end play directly.

Make sure the end play is within specification: 0.003—0.005 in. (0.07—0.13mm) with a 0.006-in. (0.15mm) wear limit. If the case was not cut for end thrust, end play should be within limits. If end play is out of limits, get the necessary shim from a VW shop or dealer. Shims are available in a variety of thicknesses, so finding the right combination should be no problem. Thicker shims mean less end play; thinner shims mean more.

Of the three shims, two are usually quite thick and the third is thinner. Exchanging the smaller shim will usually set the proper end play.

Feeler Gage Method—If you don't have a dial indicator, you will need to rig a pointer or stop. Pass a bolt through one of the bellhousing bolt holes. You need a bolt with long threads to secure it with a nut tightly against the case. The remainder of the bolt should extend away from the case and past the flywheel. You also need two more nuts and a metal plate.

Now thread on one of the nuts, the metal plate and then the other nut. Pull the flywheel all the way away from the case and tighten the two nuts so the metal plate is just touching the flywheel. This is almost impossible with only two hands, so have a friend around to help. Once the plate is tight, push the flywheel back toward the case. You now have a space between the plate and the flywheel that is equal to crankshaft end play. Measure the distance between plate and flywheel with feeler gages. The gage that justs slips in is the amount of end play.

Special end play checking tools are available via mail order from many specialty firms, or you might find one in a VW parts house. These are handy, inexpensive ways of checking end play, so you might use one. A dial indicator is still the most accurate tool, and can do many other jobs besides measuring VW crankshaft end play, so I'd buy one instead if I could afford it. It all depends on how often you use such tools.

If end play was within limits, you can continue. Remove the flywheel so you can install the oil seal. If end play was not OK, get the correct shims, install them and check end play again. Once end play is correct, remove the flywheel to install the oil seal.

Crankshaft Oil Seal—Install the oil seal when end play is within specs. Otherwise, the seal must be removed to install the shims. Because removing the oil seal destroys it, it only makes sense to install it after end play is set.

First put white grease lube on the end-play shims and slide them on the crank. Then hold the oil seal to its bore in the case and tap it in. This is a large diameter seal, so tap around its circumference with a small hammer. VW has a special tool made out of a large plate and a bolt for drawing the seal into position, but you don't need it unless you install 10 or 20 seals a day.

The important point is to not cock the seal in its bore. If you do, gently remove the seal and start over. Drive in the seal until it is flush with the outer edge of the case. Most cases have room to drive in the seal a little farther, but by stopping a bit short, chances are the seal will bear against a fresh part of the crank. This is hardly critical, just one of those small touches, so don't worry if you bottom the seal in the case. After the seal is installed, wipe a light coat of white grease onto the sealing lip. This will lubricate the seal until oil reaches it.

127

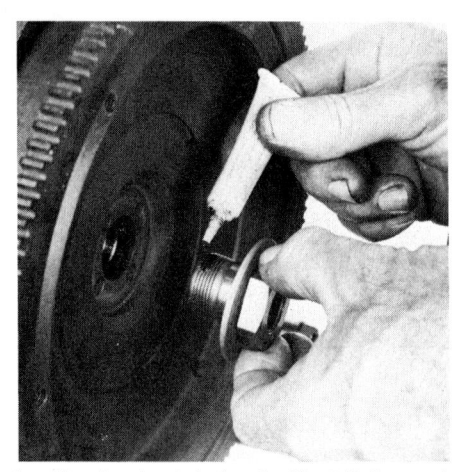

Loctite gland nut during final installation and torque to 300 ft-lb. That's a bunch, but necessary to stop nut from backing out. Large heavy washer is also necessary to distribute torque. Don't leave it off.

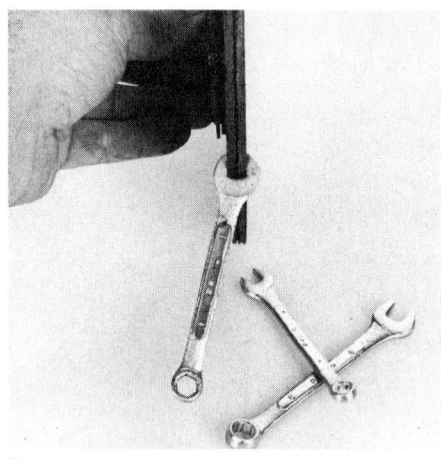

Quick and dirty clutch disc check is to fit 7mm open end over edge. This disc passes because an 8mm is still tight fit. Chassis where the engine is difficult to remove (late Bus and 914) should get a new clutch disc unless original is practically new.

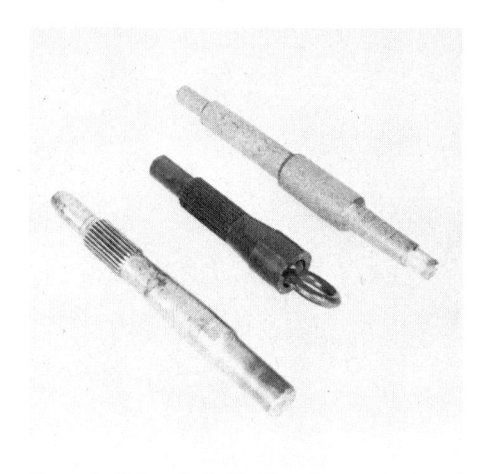

Sawed off input shaft, Kingsborne tool and wood dowel are three ways to center clutch disc. Kingsborne tool (center) is best choice in most cases, after balancing availability, cost and longevity. Plus, sawing tough input shaft is work!

The seal in the photos has Permatex 3H on its outer circumference. This is not necessary, but does aid seal installation and might help stop a leak.

Prep Flywheel—Get the flywheel ready for installation by changing the small O-ring and polishing the sealing surface with 220-grit paper. The O-ring is inside the lip behind the sealing surface on most flywheels. It's easy to overlook if you don't know it's there. Some early flywheels have no O-ring. A small screwdriver or pick is handy for pulling out the old ring. Install the new ring dry. Some mechanics use Permatex 3H to seal the flywheel inside the well area the O-ring is in, and if you are an oil-leak fanatic, you should also. Use a very light coat of sealer here or you'll upset flywheel torque.

After prepping the flywheel, bolt it to the engine. Take care when fitting the flywheel as it passes through the oil seal. Don't bang into the seal, or let the weight of the flywheel rest against it. Once on, torque the gland nut 300 ft-lb on Type 1—3s, or the five bolts on the Type 4 to 80 ft-lb. Don't omit the washer under the gland nut or those under the bolts. The gland nut washer is relatively thick, but the Type 4 washers are quite thin. Both are important because they spread bolt torque over a wide area.

Clutch Inspection—Because Beetle engines are so easy to remove, many VW owners reinstall old clutch components on a rebuilt engine. It's not bad logic, if you don't mind pulling the engine again. That way, you are sure to get the longest possible life out of that clutch. But, if R&Ring an engine is not your idea of Saturday morning fun, or your chassis doesn't feature easy engine removal (i.e., the 914 or late

If you don't have flywheel lock, use screwdriver braced into ring gear teeth while torquing pressure plate. Torque bolts in crisscross pattern to 18 ft-lb.

Buses), install a new clutch disc and pressure plate.

Clutch Disc—If unsure of how worn the clutch disc is, inspect the depth of the rivet heads. A 1/16-in. (2mm) depth should remain. Another check for the Type 1—3 disc is to place a 7mm open-end wrench at the edge of it. If the wrench slips on easily, the disc is too thin. Replace it. If the wrench can't fit over the edge, the disc is usable.

Clean a usable clutch disc with lacquer thinner and handle it like a record—by the edges.

Pressure Plate—Check the pressure plate for general wear by shaking it. If the large, diaphragm spring rattles, replace the pressure plate. Also look for worn pivots. The working surface should be smooth and free of dark spots. Dark spots indicate clutch slippage, and the attendant high heat. Lay the pressure plate on the bench and sight across the release levers

Lay on gasket, oil screen, then second gasket. Don't use sealer on these or gaskets will stick and tear at first oil change. Note: Types 1—3 have two screen styles. First was used until '68 and has smaller hole in center. From '69 on, look for larger mounting hole, and sleeve around edge of hole to keep screen from collapsing.

Acorn nuts on Type 1—3 oil strainers use flat washers. Make sure you start these small nuts by hand to avoid stripping their fine threads. Take it easy when tightening these, too. Once they cinch, that's it.

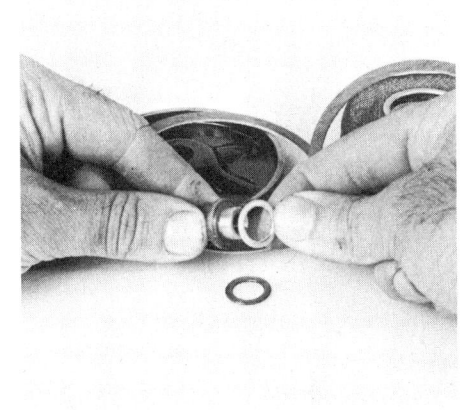

New sealing washer from gasket set will help single Type 4 oil strainer nut stop leaks. Torque is only 9 ft-lb—any more risks pulling stud out of case.

or diaphragm-spring *fingers*. They must all be at the same height. If not, replace the pressure plate. Also eyeball the release levers or spring fingers where the release bearing pushes on them. If the release bearing has worn a groove into the fingers, replace the pressure plate. I've seen fingers almost worn through from this, so it can be a problem. It's the result of improper clutch adjustment, causing the release bearing to ride continuously against the pressure plate.

Clean the pressure-plate working surface with lacquer thinner and keep fingers off it.

Release Bearing—Inspect the release bearing (also called *throwout* bearing). Spin it by hand. If you feel roughness, like it's full of grit, replace it. Do not wash the bearing in solvent, or you'll remove the factory-sealed lubricant. Then you'll guarantee that it growls. It is economical in the long run to install a new release bearing in almost every rebuilding job. It's fairly inexpensive, but the labor to R&R it can be substantial.

Install Clutch—To install the clutch, you need a *clutch centering tool*. It is used to align the clutch disc to the flywheel pilot bearing so the input shaft splines slide straight through into the bearing during installation. Most parts suppliers sell inexpensive and accurate Kingsborne alignment tools.

Place the clutch disc against the flywheel with the extended part of the disc's hub away from the flywheel. Install the centering tool through the disc and into the pilot bearing in the gland nut or flywheel. Now you can put the pressure plate into position and run its bolts in finger-tight.

Use a socket and ratchet to run the pressure-plate bolts down in a crisscross pattern. Tighten the first bolt one turn, then its opposite a turn, and so on until all bolts are wrench-tight. Don't tighten one bolt, then the next in a circular pattern. This will warp the pressure plate. Finish by torquing the bolts 18 ft-lb (2.5 kg-m).

A variation on this installation method is to install the clutch centering tool in position while the engine is on a stand. Most stands have hollow tubes supporting the engine on it. You can usually slide the centering tool down this tube. If not, leave clutch installation until after the engine is complete and off the stand.

Oil Screen—Roll the engine over on its side to install the oil screen. On Type 1—3s, put in a new gasket, the screen, another gasket, the cover plate, new copper washers and the acorn nuts. Tighten the acorn nuts to 5 ft-lb (0.7 kg-m) in a crisscross pattern. Finally, install the oil-drain plug in the center of the plate and wrench-tighten to 25 ft-lb (3.5 kg-m).

Installation on a Type 4 is similar. Install a new gasket, the screen, another gasket, the cover plate and, finally, the center bolt. Wrench-tighten the center bolt, but don't overdo it; torque is only 9 ft-lb. Overtorquing this bolt can pull its stud out of the case. Remember this value during future oil changes, too.

Oil Pump—Before installing the oil pump, check the oil pump stud length. This can be incorrect on Type 1—3 engines because of the many different oil pumps sold for these engines. The easy way to check stud length is to turn the oil pump body and cover plate backward and hold them up to the case. Note how far the studs extend past the pump. That's the length left for the oil pump nuts and washers. If it's enough, you're in business. If not, lengthening the studs by unthreading them from the case.

Pump Installation—Installing Type 1—3

Oil pump studs may be too short for pump and cover combination. If so, back studs out of case with two nuts jammed together. Test stud length by fitting pump and cover backward over one stud.

pumps begins with Permatex 3H sealer. Coat the gasket surface on the outside of the case. Don't get any sealer on the inside of the case. Install the thicker of the two oil pump gaskets. It has holes just for the four studs. Start the pump body into the case, making sure the gear-drive hole in the pump and the camshaft-drive slots lineup.

If the pump stops going in, the cam gear and tang are probably not lining up. Continue wiggling, pushing and tapping the pump into position while rotating the crankshaft. It takes two crankshaft revolutions to rotate the cam-

Sealer is fine on gaskets and case, but don't apply sealer to pump body where it slips into case. A little splashed up along the sides like this is okay, but avoid gooping pump body.

Start pump by hand and then tap in with hammer and piece of hardwood. Alternate tapping from top to bottom to avoid cocking pump in case.

Apply white grease to pump's interior, then install gears and lube them. Two large blobs on gears will help pump priming.

Sealer on pump studs stops oil leaks. Get all threads and both sides of washers. Aggressively sealing all possible leak sites like this will result in an oil tight VW, no matter what the non-believers may say.

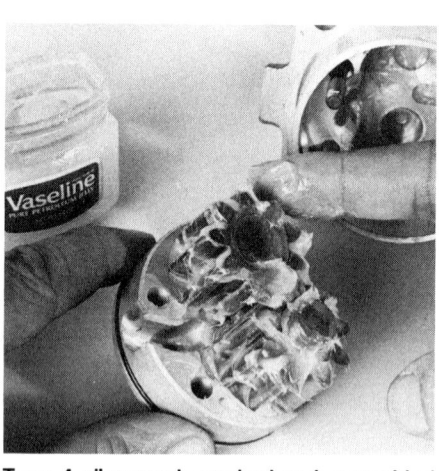

Type 4 oil pump is packed and assembled before installing in case. Petroleum jelly or white grease is fine for packing.

Always check oil-pump bore for excess sealer. Overlooking a blob like this can make pump installation difficult.

shaft once, so keep turning until the pump engages. Finish driving in the pump with a block of wood and a hammer.

Lube the pump body interior and gears with white grease. Install the drive gear so it engages the slot in the cam gear. Then install the driven gear. Pack some white grease into the pump. It will ensure oil system priming, and help build oil pressure rapidly upon engine start-up.

Finish the installation with the cover plate, gaskets, nuts and washers. This gasket is the thinner pump gasket, and has small bleed holes in it. Swab Permatex 3H onto the oil-pump studs before installing the washers and nuts, then coat the washers. Torque the nuts 14 ft-lb (2.0 kg-m) on 8mm studs and 9 ft-lb (1.3 kg-m) on 6mm studs.

Type 4 oil-pump installation is a little different. First, change the O-ring on the oil-pump bearing plate. Then put white lube on the pump gears and install them to the pump bearing plate. Lube the inside of the pump housing, and fit the bearing plate and gear assembly inside it. Complete the assembly by wrenching on the four small nuts and washers to 6 ft-lb.

Now the oil pump can be installed in the case. Apply Permatex 3H to both sides of the gasket, then slide the gasket onto the studs. The pump can now be installed in the case, but you must orient the pump-shaft tang with the slot in the camshaft gear. Line the two up as best you can by eye, then install the pump. A wood block and hammer may be necessary to drive the pump into position. Hammer as lightly as possible, especially when the pump is almost installed.

Once again, if the pump resists going in, the cam gear and tang are probably not lining up. Continue wiggling, pushing and lightly tapping the pump into position while rotating the crankshaft to change its slot's position.

Chances are slim you'll have any mismatch when installing a Type 4 oil pump, so don't be concerned. Besides, if the pump continues to hangup, you can always remove it and eyeball its placement.

Once it's in place, install the washers and nuts. Torque them 14 ft-lb (2.0 kg-m).

Type 1 & 2 Crankshaft Pulley—A common and frustrating mistake when building magnesium-case engines is forgetting to install

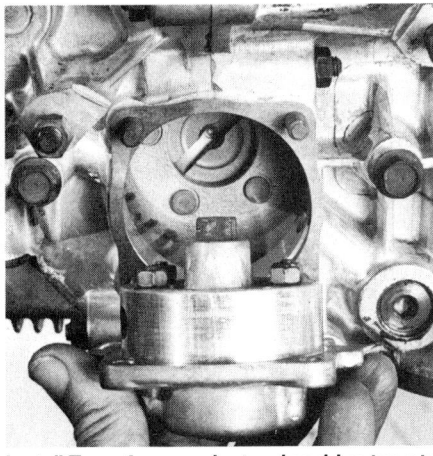

Install Type 4 pumps by turning drive tang to align with slot in camshaft timing gear. Tip: turn crankshaft until camshaft slot aligns with case parting line, then orient pump tang to same reference. Pump should slip right in place.

Don't forget to install this sheet metal while crank pulley is *off*. These two screws are the only fasteners, and are impossible to install once pulley is in place.

Some pulleys slip on easily by hand, but others need a few taps to seat fully.

Torque pulley bolt to 32 ft-lb, and use Loctite. Keep crank from turning by holding flywheel.

Depth gage may be elegant, but deck height can be measured with straightedge and feeler gages. Aim for 0.055—0.065 in., although an extra 0.015 in. either direction is within limits. Barrel shims (spacers) increase deck height but they offer more seams to leak and affect rocker geometry. Tip: increase deck height by machining 0.040—0.060 in. off piston top. It's inexpensive and doesn't affect rocker geometry or piston durability.

the curved piece of sheet metal behind the crank pulley. Simply mount the piece over the oil pump and install the two screws.

Now you can tap the crankshaft pulley into place with a hammer and block of wood. Apply Loctite to the pulley nut and torque it to 32 ft-lb (4.5 kg-m).

Set Deck Height—Because of the many variables in VW engines, you just can't slap on the pistons, cylinders and heads and know the pistons, cylinders and heads won't hit each other. You must ensure there is 0.040—0.080-in. (1.0—2.0mm) clearance between the pistons and cylinder head before bolting the two together.

This is done by mocking-up a cylinder, bringing the piston to TDC and measuring the distance from the piston crown to the top of the cylinder. When mocking-up the cylinder, install the cylinder-base gasket, piston and pin, and cylinder barrel. Don't use any sealer now. The pin circlips and piston rings don't matter here, so they don't have to be in place. Secure the cylinder with the head nuts. Use a socket or stack of washers on the stud to take up cylinder-head thickness. Snug the nuts wrench-tight.

Rotate the crankshaft until the piston is at the top of its bore. Now make the measurement. If you have a depth gage, great. Set the anvil across the cylinder's top and read the clearance directly.

No depth gage? Put a straightedge across the cylinder and find the feeler gage combination that fits between the straightedge and piston crown. As the photo shows, an extension makes a workable straightedge.

How Much?—Allowable deck height varies from 0.040—0.080 in. (1.0—2.0mm). Most cylinder and piston kits are designed to give 0.060—0.070 in. (1.524—1.778mm) on a stock case. Optimum is 0.055—0.065 in. (1.400—1.650mm), if you feel like tinkering.

If deck height is at or greater than specification you are in business. Well, if deck height is too much larger than specification, engine compression will be too low. For example, 0.150 in. (3.81mm) is excessive deck height, and should have you examining the crankcase or piston set you swapped to. Excessive deck height is usually rare, unless you goofed in a big way on crankshaft stroke or got the wrong pistons.

A more common error is too little deck height. This means the pistons will or might hit the cylinder head or valves, or compression will be too high. Again, the cylinder set could be at fault, as could the crank. This is likely if the problem is obvious, like pistons that protrude 1/16 in. out the cylinders. If your problem is more subtle, then the case-to-cylinder mating area has been milled excessively. This is typical of cases that have been rebuilt several times.

Adjusting Deck Height—A majority of the time, deck height is acceptable, and no adjustments are needed. If you need more deck height, you can install spacers between the cylinders and crankcase. These are called *barrel shims*. They are available in thicknesses of 0.010, 0.020, 0.030, 0.040, 0.060 and 0.090 in.. Past 0.090 in. they become spacers instead of shims and can be had in any dimen-

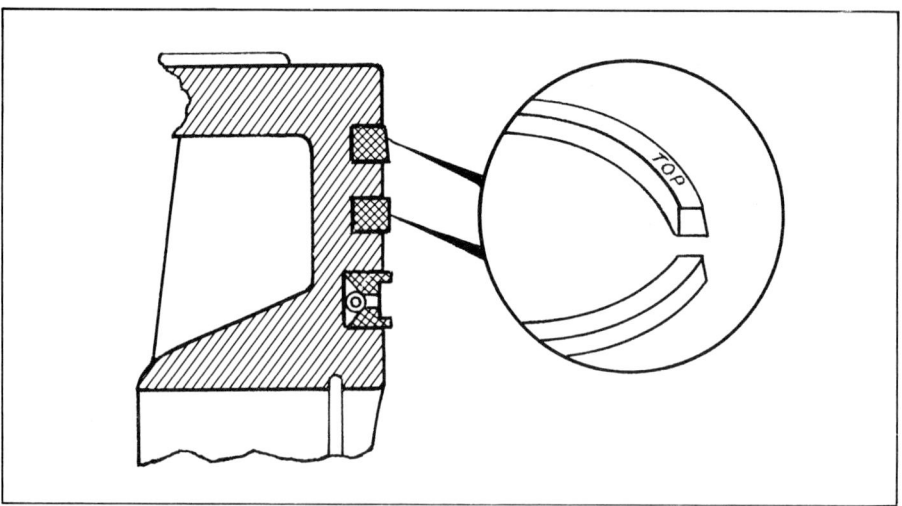

VW piston-ring installation. Other brands have different shapes and markings. Follow instructions packaged with rings.

sion a machinist can whittle out.

To decrease deck height, you'll have to remove metal or change parts. The easiest change is to eliminate the cylinder-base gasket. The engine will run fine without it, provided you apply Permatex 3H to the gasket area. The gasket's only function is to stop oil leaks, and sealer can do that. Removing the paper gasket will give only 0.008 in. less deck height. Significant reductions in deck height require the case be cut by a machine shop.

Minor changes in deck height can be gained by pairing pistons and connecting rods. By measuring rod length and piston pin bore-to-crown height with vernier calipers, you'll find that not all parts are exactly the same dimension. Pairing "short" pistons with "long" rods will help equalize deck-height changes among cylinders. This is a racing trick, and helps the most in performance engines where small differences matter. For street engines, it may not get you out of the woods, but it's free!

Pistons & Cylinders—Move over to the bench now and plan piston and cylinder installation. There are two methods you can use. The first is to install the pistons on the connecting rods, followed by the cylinders. The second is to install the pistons in the cylinders far enough to cover the rings, but leaving the piston pin exposed. Then the cylinders are brought to the engine, the pins pushed into the rods and then cylinders pushed onto the case. I prefer this method because it's simpler to install the pistons on the bench than with all the cylinder studs in the way. Also, it doesn't matter what type of ring compressor you use.

Clean Parts—While you are thinking about piston installation, clean the pistons and cylinders. Fill a wash tub or pan with hot water and liquid laundry soap. Wash each piston, pin and cylinder with a scrub brush until free of all oil and machine grit. You can also clean these parts in a dishwasher. Don't skip washing because you bought new pistons and cylinders, either. Such parts are shipped with dirt-collecting grease and oil, and are never cleaned at the factory with engine assembly in mind. That is your job. The same is true with rebuilt parts, like cases, cylinder heads and crankshafts. They are shipped with the intention the user will clean them before installing them.

It's best to clean these parts immediately before you assemble them, but sometimes that is not possible or convenient. In that case store the cleaned and oiled parts in sealed plastic bags. Be sure to wipe off all water with paper towels and spray the parts with water-dispersant oil, or they'll rust.

Check Rings—When you are ready for assembly, start with the piston rings. If the rings aren't on the pistons, install them now. You can either spread the rings by hand, or use a ring expander. Ring expanders can be simple or quite complex, and the prices can be cheap to very expensive. I recommend getting the inexpensive variety that holds rings at the gap. They make ring work much easier than by hand, and VW rings are not big enough or strong enough to warrant a fancy, high-dollar expander.

When installing rings, be sure to follow the directions enclosed with them. Be on the lookout for mixing top and second rings together. They look alike, and can be easily confused. They are different, however, and must be installed in the correct groove. Also, look for small dots or the word *TOP*. This side of the ring must face up (toward the piston crown). If the ring is installed upside down, *twist* built into the ring won't work and the ring will pass compression gases and oil.

Because VW rings are usually bought in a set with the cylinders, pistons and pins, they are normally already installed on the pistons. Then you don't have to work with the rings, even to clean the pistons. It's prudent to remove one top compression ring and check its end gap in a cylinder, though. This will check against the wrong rings being packaged with that set.

You should also check the piston-to-bore clearance of your new cylinders as explained in Chapter 5, page 90.

Once the rings are installed on the pistons, orient the gaps. Looking down on the piston crown, arrange the ring gaps so the compression ring gaps are about opposite each other, and not lined-up with the piston pin. The oil ring gaps should be staggered relative to each other on multi-piece oil rings. Also, oil-ring gaps should be oriented so they are *up* when installed in the engine. That is, the gaps shouldn't be at the bottom of the cylinder (toward the road).

While it is necessary to correctly arrange the ring gaps, don't sweat bullets trying to get the gaps lined up just so. When the engine's running the rings move around the piston on their own, so your carefully set spacing will change.

Piston-Pin Offset—VW piston pins are not centered in the pistons. Instead, they are offset slightly to one side. This helps load the piston against the cylinder wall, reduces noise and increases cylinder sealing. To ensure the pistons are correctly installed, an arrow is cast into the piston crown. This arrow points to the flywheel when the piston is installed. Keep this in mind when arranging ring gaps, installing pistons in cylinders and cylinders on the engine.

Prep Cylinders—Get the cylinders ready by spreading a coat of engine oil inside them. The best tool for spreading oil is your clean fingers. You don't need rivers of oil pouring out each end; a moderate film is fine. While you have the oil can out, give the pin bores in the piston and connecting rod a shot. Also, install the sealing ring and any shims at the base of the cylinder.

A coat of Permatex 3H will hold the gasket and shims during installation and help stop leaks. In fact, if you are missing the paper gasket and deck height was well within tolerances, just apply Permatex 3H to the cylinder and install it without the gasket. If you have the gaskets, however, use them along with the sealer.

Install Pistons—Now oil the piston rings, working the rings so oil gets behind them. Then clamp the ring compressor around the piston and rings, and feed the top of the piston into the bottom of the cylinder. Push the piston into the cylinder with a hammer handle butting against

Arrow on piston crown (arrows) points to flywheel when piston is installed on rod. This correctly orients pin-bore offset, reducing noise and piston wear.

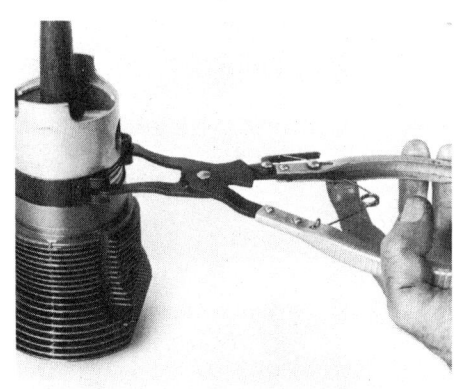

Install pistons in bores on bench. Drive just far enough to cover rings and leave pin bore in full view. If you meet stiff resistance when driving, STOP. A ring has popped loose and will break if more force is used.

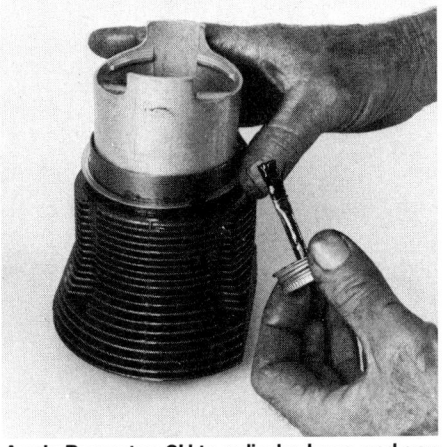

Apply Permatex 3H to cylinder base and any gaskets or shims that go there. If you choose not to use them, Permatex 3H or silicone sealer alone will effectively seal this joint.

Install piston and cylinder assembly on engine by sliding oiled piston pin into position.

the piston's underside. Stop inserting the piston before its pin is covered by the cylinder.

If you feel any solid resistance during this procedure, STOP! A ring has popped out from under the compressor and will damage the ring and piston if forced. Separate piston and cylinder barrel, refit the compressor and try again.

Next, install one of the circlips in the piston-pin bore. Look ahead to where the cylinder will be on the engine. You should install the circlip on the side of the piston which will be hard to get at. In other words, if the cylinder is going next to the bellhousing flange, put the circlip on the side of piston closest to the flange. This will save you from having to install the pin in close quarters.

Now take the assembly to the engine, and install the piston on the rod. Do this by sliding the cylinder onto its studs until the connecting rod is lined up with the piston-pin bore. Slide the pin through the piston and rod until it contacts the previously installed circlip. Now install the remaining circlip. Finish by pushing the cylinder completely on, until it butts against the case.

Continue with the remaining cylinders, keeping the arrow on each piston pointing toward the flywheel. If it isn't, remove the assembly, rotate the piston 180°, reset ring gaps and reinstall.

Type 1—3 Air Deflectors—Just as soon as you get the cylinders on the engine, install the lower air baffles. These are the sheet-metal pieces that snap into the cylinder-head studs, covering the space between cylinders. Sometimes they are yellow cadmium plated.

This is an important step, so don't forget it.

Make sure you install these baffles before taking a break, answering the phone, or standing back to admire your progress. Because if you do, you'll forget the baffles. Next thing you know, you'll have the cylinder heads and pushrod tubes installed, and you won't be able to get the baffles on. Then you'll be tempted to leave the baffles off—which is a **BIG MISTAKE**. The baffles are essential to route cooling air around the cylinders. Without them, the cylinders will overheat. So, if you do forget these baffles, *backtrack as much as necessary to install them.*

Install Cylinder Heads—The next major components are the cylinder heads. On Type 1—3 engines, the pushrod tubes must be installed first. New rubber seals from the gasket set should be installed on the pushrod tube ends. Both ends get the same seal, so don't look for any differences. Swab some RTV or Permatex 3H on the seals to prevent oil leaks, but they can be installed dry too.

On the Type 1—3, you need to hold four pushrod tubes in one hand, and the cylinder head in the other. Get the head on its studs, then place the pushrod tubes in position. Now slide

Installing outboard circlip is no problem, but as shown here, inside circlips must go on before assembly mounts on engine. After circlip is in, slide cylinder down until it butts against case.

These Type 1 and early Type 2 air baffles are critical for proper cylinder cooling. Clip them on right after installing cylinders so they aren't forgotten.

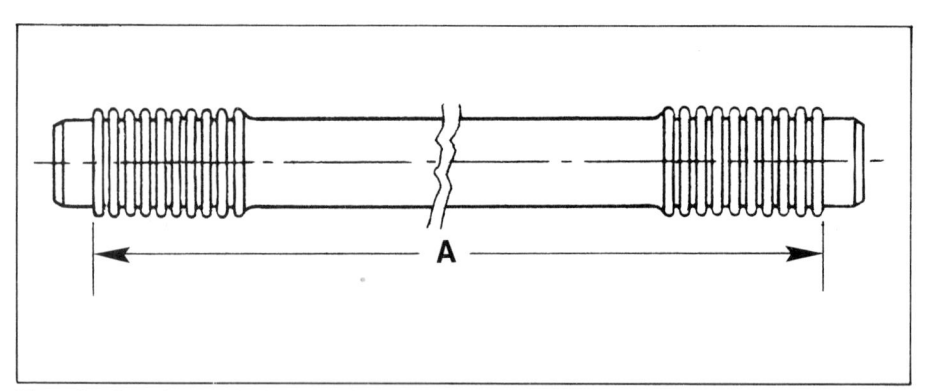

Type 1—3 pushrod tube length (A) must be 7-17/32—7-9/16 in. to get enough compression seal from cylinder heads. If short, lengthen them by gently pulling ribbed ends.

Pushrod tube seals can go on dry and are interchangeable from end to end.

the cylinder head down the studs, mating the pushrod tubes into the head as you do so. Place the seam on the pushrod tube "up," facing the cylinders. Then, if the pushrod tube seam splits, oil won't drain out. Granted, split pushrod tube seams are mighty rare, but it's a nice touch.

Install the washers and cylinder head nuts, then run them down snug with a ratchet. When finished with one side, roll the engine over and do the other. On Type 3 engines, don't forget to install the thermostat-linkage bracket on the right cylinder's upper center studs. The rod with the 90° bend in it passes through the center of the head to the thermostat. The less sharply bent rod runs to the fan end of the head. It goes toward the outside of the engine. If your Type 3 cooling system has been gutted, you won't have these parts.

Installing the heads on a Type 4 is easier. Place the metal sealing ring in the cylinder-head recess, then slide the head down its studs. Install the head hardware and snug it up with a ratchet.

On all engine types, make sure the cylinder head sealing surfaces are clean before installation. They must be free of dirt or nicks. This goes for the sealing surface of the cylinders, too. Wipe a finger across all sealing surfaces before installation. Remove any dirt you find.

Torque Cylinder Heads—Because of the flat cylinder arrangement and lightweight case construction, torquing cylinder heads is best done in an alternating pattern. This helps distribute stress evenly through the case. Follow the torque sequence in the drawings for each head and this order:

1. Torque one head to 15 ft-lb (2.07 kg-m).
   Torque the other head to 15 ft-lb.
2. Torque the first head to 20 ft-lb (2.75 kg-m).
   Torque the second head to 23 ft-lb (3.13 kg-m).
3. Torque the first head to 23 ft-lb
   Retorque the second head to 23 ft-lb.
4. Retorque the first head to 23 ft-lb.
5. After several minutes, repeat steps 3 and 4.

The previous sequence is for all Type 1—3 engines with 10mm studs, and all Type 4s. Type 1—3 engines from '73-on with 8mm studs should be final-torqued to only *18 ft-lb (2.5 kg-m)*. Over-torquing these engines will only result in stretched studs. Engines with 10mm studs and case savers can be safely final-torqued to *28 ft-lb (3.8 kg-m)*.

Valve Train—I'll cover Type 1—3s first,

134

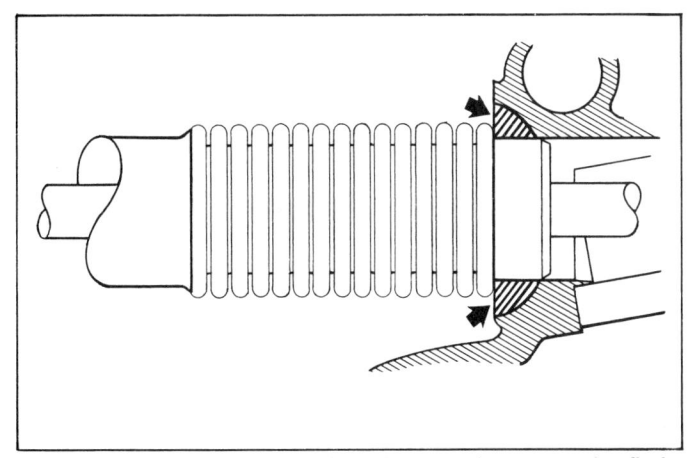

Here's how pushrod tube seals fit into cup in case and cylinder heads. After installing cylinder head, check seals for proper seating.

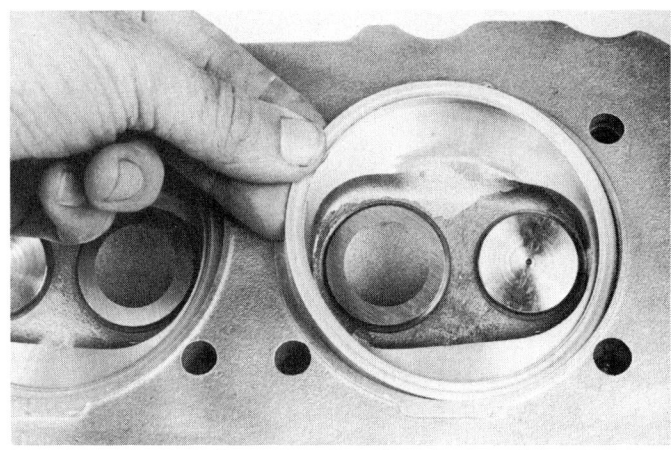

Don't forget to install metal sealing rings in Type 4 heads before sliding them on studs. No sealer is required on heads, nor is it necessary with Type 1—3 heads using shims.

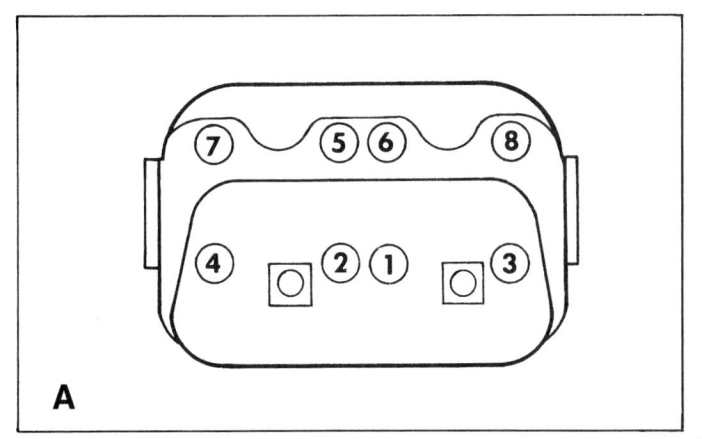

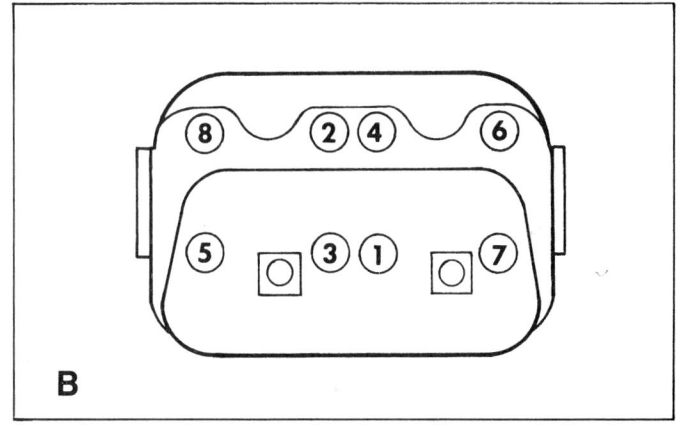

Begin torquing on all cylinder heads by snugging head nuts to 7 ft-lb using pattern A at left. It helps compress pushrod tubes so torque readings won't be artificially high. Final torque sequence at B is different from initial tightening pattern. Crisscrossing distributes torque evenly across head and pattern is same for all heads.

then discuss the Type 4. On both engines, it's nice to roll the engine so one head is somewhat higher. Then you can work on that side without fighting gravity until the rocker arms are on. Then you can roll the engine the other way and do the other side.

Because Type 1—3 engines already have their lifters installed, your next step is to slide the pushrods into position. Swab some white grease or moly lube on each end of the pushrods, then slip them down their tubes.

If you washed the rocker-arm assembly in solvent and didn't lube it with motor oil afterward, do so now. Squirt some oil onto the assembly and work it onto the shaft by sliding the rocker arms back and forth.

Install the rocker-arm assembly. Have the valve adjustment backed off all the way when installing and torquing the rocker-arm assembly because this reduces strain and binding of the rocker-arm shaft hardware. The rocker-arm stands have cutouts in them. The cutouts face up, toward the valve springs. The rocker arm stands also have one edge that is beveled. It must face away from the cylinder head, toward the rocker-arm cover.

Once the assembly is in place, slide on the wave washers and thread on the thin rocker-arm-stand nuts. Torque is 18 ft-lb (2.5 kg-m).

If you're installing swivel-foot rockers, make sure the feet are correctly positioned on the valve stems when mounting the assembly. You might have to look very carefully at the little balls in these adjusters to make sure the flat spot on them is against the valve stem.

Type 4 Valve Train—This valve train work begins with the cylinder bottom air deflector. This black painted sheet-metal piece fits under the cylinders and is secured in three spots by screws. Two go into the case, one into the cylinder head. Make sure you get this important cooling piece on now, or you'll have to backtrack later when it's discovered missing.

If the engine has mechanical lifters, all you need to do is lube their bottoms with moly or white grease and slide them into their bores. If you are reusing the cam and lifters, the lifters *must* go in their original bores.

Prep Hydraulic Lifters—Hydraulic lifters need a little more preparation. First, fill a small tin, like a loaf pan, with a quart of new engine oil. Completely submerge the lifter in oil. VW recommends completely disassembling the lifter in the oil, so oil reaches all parts. This means depressing the check valve with a thin piece of wire while compressing the plunger during reassembly. This is difficult, messy and not

135

Torquing heads should follow the head-to-head sequence outlined in text. Use a smooth, slow pull on the torque wrench to get accurate readings, and double-check the final torque after several minutes.

Apply white grease or moly lubing to pushrod ends and slide them into position. Give each a little wiggle to make sure it's seated in lifter.

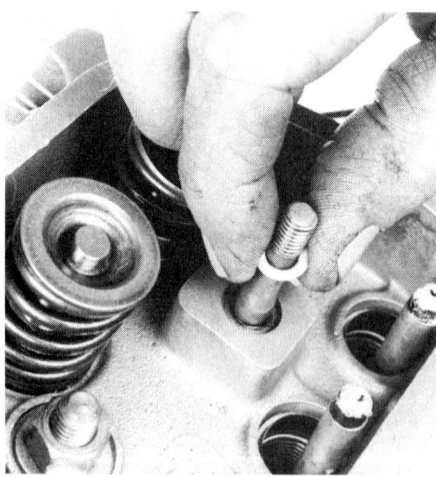

Check rocker-arm studs for O-rings. Use new replacements from gasket set, and install them dry.

Torque Type 1–3 rocker-shaft nuts to 18 ft-lb. Install assembly with valve adjustment backed off. Orient gap in rocker shaft stands to point toward valve springs; beveled edge sets away from head.

really necessary.

Instead, take one of the pushrods and depress the plunger. It's the center portion of the lifter where the pushrod normally rests. Continue depressing the plunger while the lifter is completely submerged. Air will bleed from the small hole on the lifter's side.

You'll find it difficult to determine when the lifter is actually depressing because it goes down so slowly. Therefore, continue holding pressure for 30 seconds. Then release, and start compressing again. The lifter will gain resistance as it fills with oil, so the second compression will be more difficult. Shortly, the lifter will fill and be impossible to compress. But keep holding pressure against it. Air can still bubble out, long after you thought there was none left inside. After three pressure and release cycles, the lifter should be filled. Remove it from the oil and go on to the next one. Store the lifter upright so oil won't drain out.

Your hand will, or should, get tired from pressing against the pushrod. Wrapping the pushrod end in a rag helps, as does switching hands. Apply just about all the pressure you can to the lifter, so securely support the tray of oil. Watch that it doesn't slip out from under you while you lean all your upper body weight against it.

This is the only way aftermarket hydraulic lifters can be bled. There is no hole provided for opening the check valve with a wire.

Squirting oil through the bleed hole with an oil can will make an oily mess in your hand, but won't do much for bleeding the lifter.

After bleeding, put moly lube on the lifter base, and insert it into its bore.

Pushrod tubes are next. Change the rubber sealing rings at each end. Each end uses the same type ring, even if they are sometimes different colors in the gasket set. A coat of Permatex 3H or RTV around the seals will prevent oil leaks, although these parts can be installed dry. Installing the pushrod tubes is easy. Just pass them through the holes in the head and press them into the case.

Pushrods are next. Dab each end with moly or white grease, then slide them into the pushrod tubes.

Next mount the rocker arms. The pedestals have a cutout section that must face *downward* and away from the valve springs when installed. Also, one end of the pedestal is beveled. It should face outward and away from the center of the engine. After checking for correct orientation, install the washers and nuts. Completely loosen the valve adjustment before tightening the rocker assemblies. Tighten them to 18 ft-lb (2.5 kg-m).

Valve Adjustment—All valves are adjusted when their cylinder is at TDC, but the exact procedure depends on whether you have mechanical or hydraulic lifters.

The first step, regardless of lifter type, is to set the adjusting screws until they just touch the valves. Rotate the crankshaft until you reach TDC on cylinder 1. The engine is at TDC on #1 when both valves for it are closed, the timing marks on the front pulley are lined up to the TDC mark or notch, and the distributor rotor is pointing at cylinder 1. You may want to mock-

Type 4 air deflector has three screws: one in head and two in case. There's no disgrace in plugging open lifter bores with paper towels while handling tiny air deflector screws. Fumbling a screw or washer into crankcase is a character-building experience.

After pre-lubing Type 4 lifters with white grease, slide them into their bores. Prime all hydraulic lifters before installation.

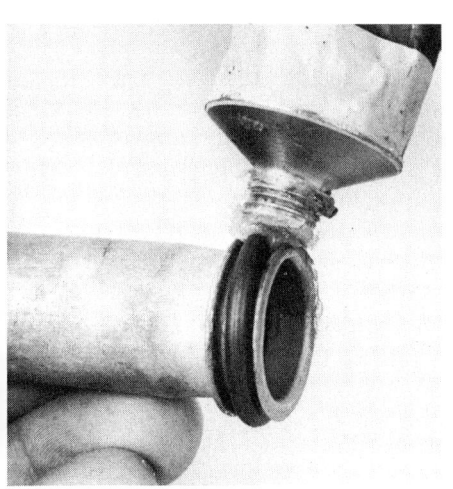

New pushrod tube O-rings are part of every gasket set. Apply coat of Permatex 3H or RTV silicone sealer when installing to prevent oil leaks.

Slide Type 4 pushrod tubes through head. Push them into case until they butt against shoulder in lifter bore. Use no tools other than fingers, or you'll crush tubes.

Install pushrods next. These have moly lube applied, white grease will also work. Jiggle pushrods to check their seating in lifters.

up the distributor cap and wires on the engine to see where the rotor should point. Most Bosch distributors have a small mark on the distributor body denoting the #1 position. You can see this mark on the edge of the distributor body when looking straight down at it.

Don't simply rotate the crank until the pulley timing marks line up. These marks line up twice, once when cylinder #1 is on TDC of the compression stroke, and once at TDC on the overlap stroke. The difference is the valves are both closed on the compression stroke, and both are open on the overlap stroke. You must check the valves to see they are closed by noting the valve spring is fully extended.

As for which cylinder is #1, stand facing the engine at the crankshaft pulley end. Number 1 is closest to the flywheel on your right, #2 is closest to you on the right, #3 is near the flywheel on the left side, and #4 is closest to you on the left.

Now proceed once #1 is at TDC. On engines with mechanical lifters, select a 0.006-in. feeler gage and adjust cylinder 1's valves. They are correctly adjusted when the feeler gage slips between the valve and adjusting screw with a slight drag. Tighten the jam nut and retry the feeler gage. If the adjustment loosened considerably (try a 0.007-in. feeler gage to see) the jam nut is stretching. Either you are tightening the jam nut too much, or the adjuster is worn out. I'll bet the adjuster is worn. Replacement is the cure.

When you're finished setting #1's valve adjustment, rotate the crankshaft 180° and adjust #4's valves. Then rotate the crank another 180° and set #3's, and then a final 180° to finish at #2.

Engines with hydraulic lifters don't require setting the valves with feeler gages. Begin again at #1 with it placed at TDC. Tighten the adjusting screw until it just touches the valve

137

Mount rocker-arm assembly. Gap in rocker-arm stands faces pushrods, and beveled end faces away from center of head. Torque nuts to 18 ft-lb; completely loosen valve adjustment before tightening.

Install Type 4 retaining wire (arrow) after rocker arms are in. Wire butts against casting at each end, fits into open part of rocker-arm stands and rides against pushrod tube ends. Check that wire is against tubes after installation. If not, use darning hook to pull them into place.

Adjust valves activated with mechanical lifters to 0.006 in. Use gages on each side of 0.006 in. To test: 0.007-in. shouldn't fit, and 0.005 in. should slip right through.

stem, then screw-in the adjuster *two additional turns.* Tighten the jam nut. When #1 is finished, move through the firing order as described in the paragraph above.

Rocker Covers—After you're finished adjusting the valves, install the rocker covers. Lay in the new rocker cover gaskets, but don't cement them in place with sealer. Just lay in the gaskets, install the covers and snap on the wire bales.

You can cement the gaskets in place if you wish after the engine has been started and the valves readjusted. I leave the Permatex 3H off, unless they leak. That way it's easy to change gaskets without having to scrape off all the old sealer.

When you've installed the rocker cover gaskets, you have just completed the basic engine. Add the external accessories and then it's finished.

EXTERNAL ACCESSORIES

To save a lot of confusion, I've separated this part of engine assembly into three sections. The first is for the Type 1, and applies to all Beetle, Karmann Ghia, Thing and early Bus engines. Second is for the Type 3: Squarebacks, Fastbacks and Notchbacks. Last is the Type 4 section, for all you '72 and later Bus, 914 and 411/412 owners. Therefore, you can jump forward to the section that concerns you. While each section stands alone, it's a good idea to look at all the photos. They supply important information, even if they don't exactly follow your engine.

In these sections there are many references to the "front" and "rear" of the engine. Remember, FRONT is the flywheel end and REAR is

---

SETTING VALVES

Here's another method for setting valves. You're going to use a characteristic of the VW engine to your advantage. When a valve on one cylinder is fully open, the corresponding valve in the opposing cylinder is fully closed. For example, when #1's exhaust valve is fully open, #3's exhaust valve is fully closed, and it's ready to be adjusted.

First, remove the rocker covers. Loosen the jam nuts on all valves and turn in their adjusting screws until they just touch the valve stem tips. Turn the crank clockwise until #1's exhaust valve is fully open. You can tell when this happens because its rocker-arm tip will be at the *bottom* of its travel and pushing the valve stem completely down. Practice turning the crank until sure you recognize when the rocker-arm tip has reached its lowest point (is nearest the rocker-arm-box floor). The valve is then fully open and its lifter is riding on the top of the cam's toe.

Now, when #1's exhaust valve is fully open, #3's exhaust valve is fully closed. Its lifter is resting on the cam's heel; the valve clearance is ready to be set. So set #3's exhaust valve clearance to 0.006 in. (0.15mm), which is, conveniently enough, the clearance for all the others. Tighten the jam nut.

Continue to set a closed valve on the #3/#4 side of the engine by turning the crank until the corresponding valve in the #1/#2 side is fully open. You've just set the #3 exhaust (#1 exhaust opposite and open). Next, from front to back, is #3 intake (#1 intake opposite), then #4 intake (#2 intake opposite), and finally #4 exhaust (#2 exhaust opposite).

The valve clearance in cylinders 3 and 4 is correct, so now just change the starting point. Turn the crank until a #3/#4 valve is fully open and its corresponding #1/#2 valve is fully closed. Follow the order specified above. For example, start with #3's exhaust fully open, and set #1's exhaust that is fully closed. Continue until cylinder 1 and 2 are in spec and then install the valve covers.

---

the pulley or fan end. Right and left orientations are based on looking at the engine from the crankshaft pulley or fan end.

TYPE 1

Two easy accessories are the sparkplugs and oil-pressure sending unit. Gap the plugs to 0.025 in. Sparingly apply anti-seize compound to the upper two-thirds of the threads to lubricate them. Don't get any anti-seize on the plug electrode or it will foul. Then thread them by hand into the heads and torque to 25 ft-lb (3.5 kg-m), or snug and one-half turn.

Apply a light coat of Permatex 3H to the oil-pressure sending unit threads. Then the unit can be threaded into the case, behind the distributor. Don't overtighten the sender or it will split the case. Finger-tight, plus maybe one full additional turn with a wrench is enough.

Oil Cooler—There are multiple oil seals in most gasket sets. See the accompanying photo for the correct ones for your engine. Install the

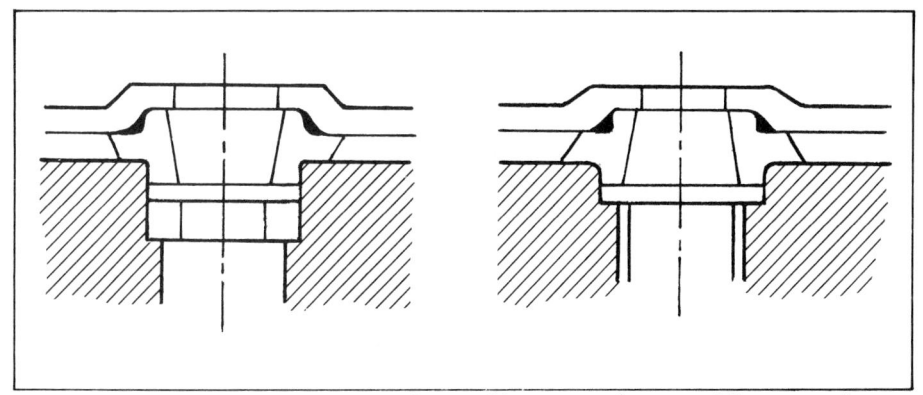

Grommets with tapered ID adapt new coolers to old cases, or vice versa. Oil passages in case and coolers were enlarged from 8mm to 10mm in '70, hence early and late designations. Grommets are supplied with three spacers for late case and early cooler applications.

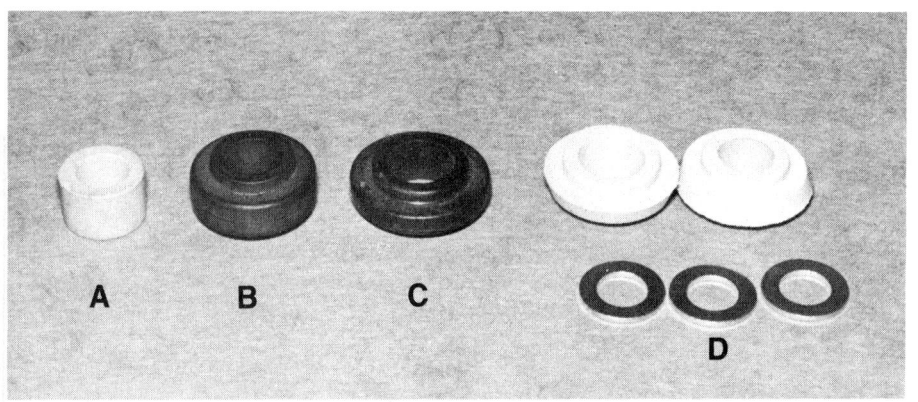

Four oil cooler grommets have been used. (A) is Type 1 and early Type 2 cylinder style. (B) is Type 3 grommet used through '69. (C) is '70 and later grommet (late style) for all Types. Install four of these on '70—79 Type 1 and '70—71 Type 2 to seal cooler and adapter on *doghouse* coolers. At right (D) are the two grommets used as tapered ID "adapters" and spacers, which are detailed in above drawing. Use no sealer when installing grommets because it can clog oil cooler.

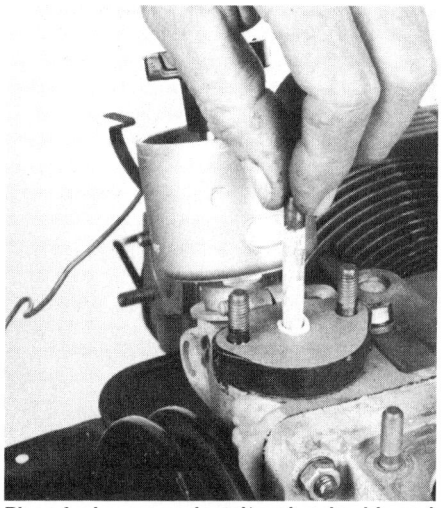

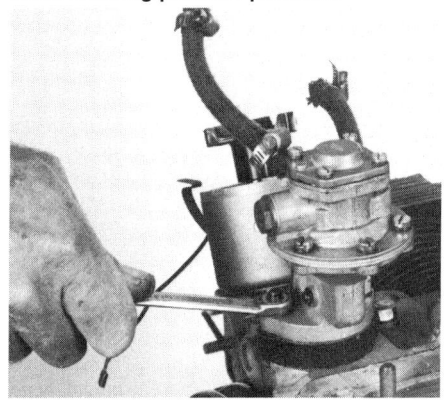

Coat fuel pump mounting area with sealer, then lay open-centered gasket over studs. Fuel pump gasket with open center is the one to use here.

Place fuel pump pedestal/pushrod guide and gasket with center hole are on studs, apply white grease to fuel pump pushrod and slide it into position (top photo). Pump will be easier to install if crankshaft is rotated so pushrod is at its lowest point. Apply sealer to bottom of fuel pump and set it on pedestal and wrench down its nuts (bottom photo). Snug nuts alternately from side to side to avoid cracking phenolic pedestal.

seals dry. Look for bumps on the cooler bottom before setting it in position. If there are bumps, tap them down with a hammer until flat. Such bumps can make the cooler stand too tall, causing the case mounting ears to break off.

It takes some fumbling to get the bottom two washers and nuts started, but the upper stud poses no problems. Wrench-down the three nuts to 5 ft-lb (0.7 kg-m). That's not much, only 60 in-lb, so don't overdo it.

Fuel Pump—Apply sealant to the case where the fuel pump insulator block sets, then lay on the lower gasket. Spread sealant on the insulator block's underside, then set it in position. Another gasket goes on top of the insulator block. Apply white grease to the fuel pump pushrod, then slide it into the insulator block, pointed end down. Seal the fuel pump bottom, then wrench it down on top of the insulator block.

Generator Pedestal—Next to the fuel pump is the square generator-pedestal mounting flange. Spread Permatex 3H over the square, then set the metal louvered piece over the studs, louvers down and facing the flywheel. Spread more Permatex 3H on the pedestal's base, then set it on its studs. You'll have to snake the *road draft tube* (crankcase vent) past the sheet-metal piece behind the crankshaft pulley on the way down. Finish with the four wave washers and nuts and tighten wrench-tight.

Thermostat—Install the thermostat and bracket to the right lower case. Align the bracket with the scribe marks you made in the case during teardown. If you didn't scribe the case, or have replaced it, leave the bracket loose on its stud. Also loosen the thermostat in the bracket.

From this point on, attach all parts loosely. Go ahead and thread-on all fasteners, but don't

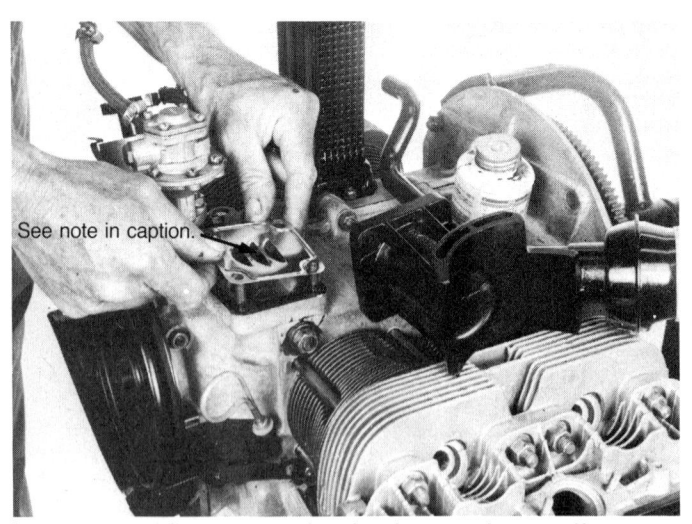

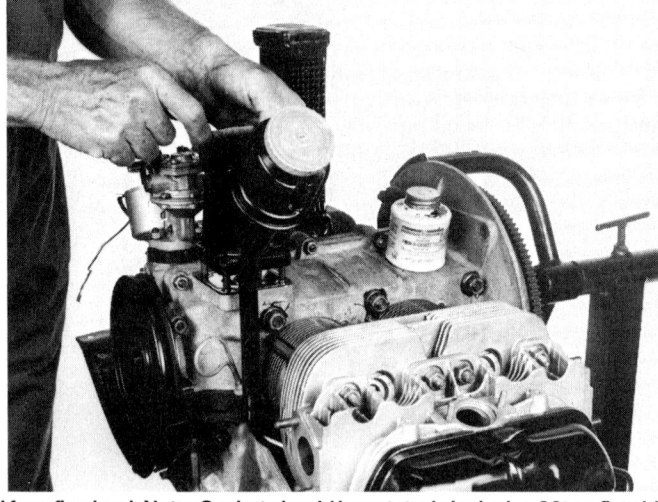

Louvers on metal generator pedestal gasket open downward into case and face flywheel. Note: Gasket should be rotated clockwise 90° so flat side of louvers face cylinders that are in view. Apply Permatex 3H to case sealing surfaces and pedestal bottom (left photo). Then, snake pedestal tube through sheet metal when setting pedestal on case. Install four nuts and washers wrench tight (right photo).

Put single-port intake manifold sealing rings in before upper cylinder sheet metal is installed. Lay rings in place dry. These copper washers come in large and small diameters. Use proper size for engine's manifold.

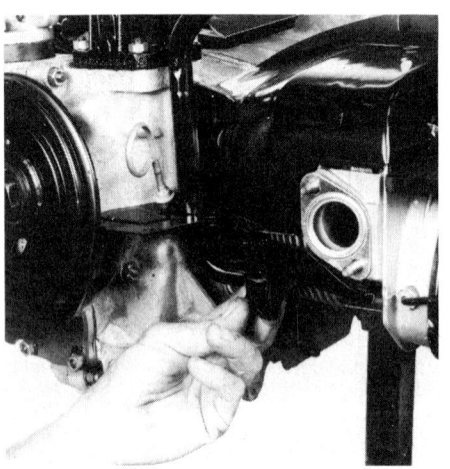

Upper cylinder covers and lower rear covers can go in place now. Upper and lower covers shown screw together at mating flange.

Install new gaskets before hanging heat exchangers. Gaskets go on dry. Keep gaskets straight while passing over studs to avoid tearing edges on threads.

wrench any of them down tightly. Only after all cooling-air tin, manifolds and the fan housing are in place should anything be tightened. This gives you some slack when fitting these pieces, and allows the parts to find their natural placement. If each piece is tightened as it goes on, misalignments are practically unavoidable.

Sheet Metal—Before covering the cylinders, install the copper sealing ring on single-port engines, and the gasket and cast-aluminum section of the intake manifold on dual-port engines. Then you can set on the large upper cylinder cover. Then add the small sheet-metal piece at the crankshaft-pulley end of the cylinder head. This piece screws to the upper cylinder covers. Install the pieces on both sides of the engine.

Below the cylinders go the two sheet-metal pieces with the large curves. The spooned end attaches at the flywheel end of the engine. There's also another attachment at the lower crank-pulley end of the case. Leave the right side piece off if installing the thermostat. Install this piece now if you choose not to install the thermostat.

Exhaust Heat Exchangers—Fit new gaskets over the exhaust port studs and don't use any sealer on them. Now hang the heat exchangers from the flywheel-end exhaust ports. The big swoopy end of the heat exchanger goes at the flywheel end. Install the nuts, but don't wrench-tighten them.

From the exhaust hardware kit, fit the new steel and asbestos rings to the heat-exchanger muffler flanges and crank-pulley end of the cylinder heads. Install the muffler loosely. You may have to straighten or reshape the muffler-to-heat exchanger mating area if it was twisted or knocked out of shape during engine removal.

Stovepipe—This is the sheet-metal tube that ducts hot air to the carburetor. Snake it to the right of the crank pulley, between the muffler, cylinder head and heat exchanger. It has two screw attachment points. One is behind the shrouding at the crank pulley, the other is the

After suspending heat exchangers from front exhaust-port studs, slip on new sealing rings. Leave exhaust system hardware loose until after entire system is installed.

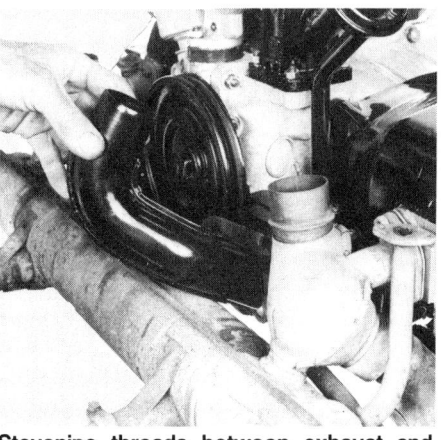

Stovepipe threads between exhaust and cylinders. It attaches at lower exhaust-port stud and on sheet metal near crank pulley.

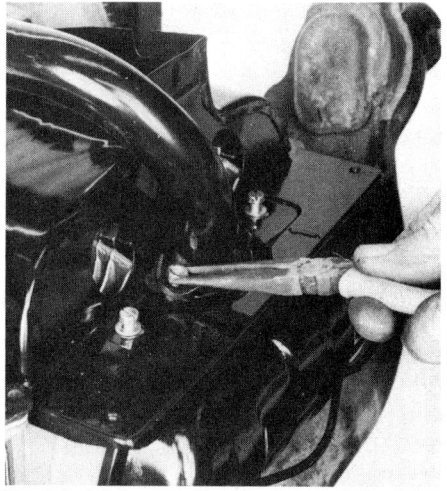

Bend in intake manifold makes it difficult to put on tiny hardware. Flex socket or needle-nose pliers help, as do a pick or magnet for the washers. Don't tighten this connection just yet, except on dual-port engines.

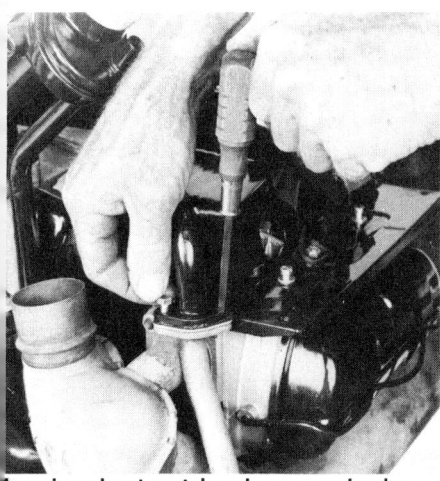

Leaving sheet metal and accessories loose pays off when trying to start manifold pre-heater bolts. A light pry easily lines these flanges up. Supporting muffler weight by hand helps too.

Fan housing flanges fit inside upper cylinder covers, requiring some eyeballing when installing. Also watch generator clamp. If it slides rearward, it will foul generator pedestal and prevent fan housing from mounting completely.

lower exhaust pipe-to-cylinder head flange stud.

Intake Manifold—Now install the intake manifold. Be sure to get gaskets onto the manifold heat-riser connections. Attach all hardware loosely, except for dual-port manifolds at the rubber joint. The rubber has plenty of compliance, so those hose clamps can be cinched.

Tighten Exhaust & Sheet Metal—Now begin tightening all the pieces together. Start at the exhaust pipe flanges at the cylinder heads. On single-port engines, bring the nuts down to just snug, not tight. This will bring the manifold preheat flanges closer together, but you don't want to tighten them yet.

Once the exhaust is snugged up, and thus fairly well aligned, tighten the intake manifold at the cylinder heads. Go ahead and bring these nuts wrench-tight to 14 ft-lb (2.0 kg-m). With the intake manifold tight, align the pre-heat flanges with the intake manifold. Install fresh gaskets and flange bolts, then tighten the flanges to 7 ft-lb (1.0 kg-m). Aligning the flanges can be troublesome. Use a drift punch or round-shanked screwdriver to help align these parts.

On dual-port engines the intake manifold is already fully installed, so when the exhaust system is tightened, the preheat flanges will start to mate. Make sure there is a gasket on each flange and install and tighten the preheat flange hardware wrench-tight.

Fan Housing—You are now ready for the fan housing. The housing fits inside the upper cylinder sheet metal. It will take a little jockeying to get the fan housing installed. A flat-bladed screwdriver makes a good shoehorn for this

141

Shoehorn fan housing into upper cylinder covers with a screwdriver. Ridges formed in fan housing will touch upper cylinder covers when housing is fully seated.

After fan housing is completely seated, slide generator clamp rearward and tighten. Bottom of clamp fits over generator pedestal, securing generator and fan housing.

Temporarily fit rear sheet-metal panel to align heater air inlets, but remove it to install engine. Large diameter hose nipples must be aligned with cutouts in rear panel, then panel can be removed. Once inlet nipple is centered in opening, tighten clamp.

Heater air inlets can be adjusted by hand once their clamps are loose. Clamp is large, flat one connecting the inlet to the heat exchanger.

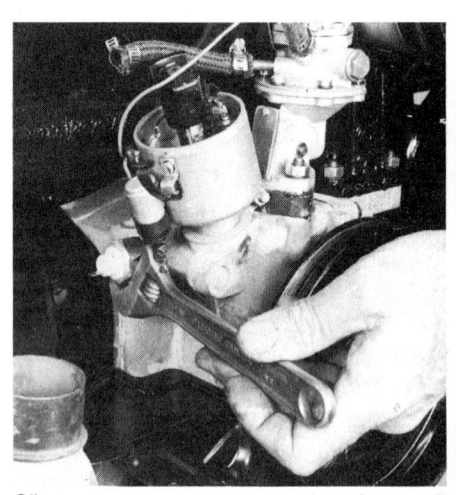

Oil-pressure sender uses pipe (tapered) threads. Coat threads with a dab of Permatex 3H and thread in hand-tight. Then lightly wrench-tighten. Too much force can crack case.

operation. Connect the fan housing in three places—once at each upper cylinder air shroud and once at the generator pedestal.

Final Tightening—Now go over the engine and tighten all exhaust and cooling tin, plus the generator clamp. If you haven't installed any new muffler clamp hardware yet, do so now.

Connect Thermostat—Join the thermostat to the rod coming down from the fan housing. Loosen the thermostat on its bracket, then thread it onto the rod. Pull the thermostat down and bolt it to its bracket. This is all you have to do if you scribed the bracket position on the case during teardown.

If there is no scribe mark to go by, adjust the thermostat. First, connect the thermostat and rod as outlined above. Loosen the bracket-to-case nut, and leave the thermostat loose from its bracket. Now raise the thermostat until the flaps in the fan housing are completely open. Move the rod up and down a couple of times to get a feel for when the flaps are open and shut. The flaps are open when the rod is up and closed when the rod is down. Now, put the flaps in the fully-open position: rod up. Hold the thermostat and rod motionless in that position. Bring the thermostat bracket down until the top part of the bracket just touches the top of the thermostat. Tighten the bracket nut.

That's all there is to it. It helps if the thermostat bracket nut is loose enough so you can move the bracket, but tight enough so the bracket doesn't flop around on its own. When you are finished, install the lower splash pan sheet metal. It screws to the upper cylinder cover at the front of the engine, and to the case at the bottom of the pan.

Rear Panel—For alignment purposes, you need to temporarily install the rear panel. Note the heater air-inlet tin where it pokes through the rear panel. The heater air-inlets are those round sheet-metal globes attached to the rear exhaust pipe. When the engine is complete, the fresh-air tubes from the fan housing mate with these, conducting air to the heat exchangers and finally the car interior.

By loosening the air inlet-to-heat exchanger clamps, you can move the inlets around. Adjust

Slot in front generator pulley is for holding pulley stationary with screwdriver. Fish around with screwdriver until it jams tight, then wrench-down pulley nut.

Type 3 oil coolers use a spacer on each stud or bolt. They look like small washers, and keep cooler hardware from over-tightening. Leaving them off may cause grommets to compress too much, shutting off oil flow. Install spacers between cooler and case.

Black stand at left is breather and multiple ground connection is mounted on stud closest to center. Between breather and distributor is Temperature Sensor II, and between it and oil cooler is thermo switch.

them so they are centered in the rear panel holes. Then tighten the clamps.

You can remove the rear panel now, or leave it on as convenient storage. But, it will have to come off for engine installation.

Distributor Cap—Take a second to see if the distributor needs new points. If so, put a set in, or at least make a note on your shopping list. Then you are ready to install the distributor cap.

It's easiest to completely remove the two longest wires—the ones that go to the right-hand cylinders. Then snap the cap onto the distributor. Starting outboard of the right side of the engine, thread the two long wires behind and under the generator and around the intake manifold downpipe to the distributor. There are clamps near the outer edge of the fan housing to hold the wires.

The cylinder-1 tower is on the right front and distributor rotor rotation clockwise. Firing order is 1-4-3-2. Therefore, the right front ignition wire at the distributor should go to cylinder 1, which is the cylinder closest to the flywheel on the right-hand side. Going clockwise on the distributor cap, the next wire should go to #4; it's farthest from the flywheel on the left side. Next is #3's wire. It is nearest the flywheel on the left side. Last is #2, farthest from the flywheel on the right side. See page 14 for a drawing showing firing order, distributor rotor rotation, and cylinder position.

Install Coil—If the coil was removed from the fan housing, replace it now. Hook up the high-tension lead (the heavily insulated wire) between the coil and distributor cap. Also connect the coil-ground wire from the negative side of the coil to the distributor. This is the small, lightly insulated wire.

Generator Belt—Unless the old generator belt was practically new, install a new one. It is cheap insurance for your new engine. Remember to put spacers between the pulleys to gain slack and remove pulleys to take up slack. Store unused spacers between the nut and outer pulley half.

Fuel Line—Check the fuel line from the fuel pump. If the metal section that runs around the left side of the fan housing is not in place, install it now. Be sure to use good clamps and fresh hose. Don't burn your car just to save a dollar or some time.

Lower Engine—If you built the engine on a stand, take it off now. If you are working on the bench, turn the engine so you can work on the flywheel end.

Front Sheet Metal—Install the sheet-metal piece that sits vertically above the bellhousing flange. It clips over the flange in two spots, and there are two screws as well. Pass the fuel line and throttle cable tube through it. Also, the throttle cable guide can be passed through the fan housing and front panel now.

That's It—Your engine is ready for installation. Go ahead to Chapter 8 and get the chassis ready for its new powerplant.

TYPE 3

As with other air-cooled VWs, the general idea when *dressing* a Type 3 is to start on the

---

DISABLE THE COOLING SYSTEM?

This is a controversial action. Some mechanics believe that removing the thermostat and air-control flaps in the fan housing does no harm. The primary advantage: the thermostat can't fail or the flaps can never stick closed and cause engine overheating.

The differing view is that keeping the cooling system intact means the thermostat manipulates the airflaps to keep the engine running at a correct operating temperature under various driving conditions. I agree.

Why? I want the engine to reach its operating temperature as quickly as possible and *maintain* it. Take out the flaps and thermostat and you increase warm-up time, eliminate any regulation of cooling air during severe loads, and essentially don't keep the engine operating at peak efficiency.

If warm-up time is increased, the oil stays cooler longer. This has two serious effects. Too much oil pressure can result and split the seams in the oil cooler. Plus, cooler oil doesn't lubricate as well as warm oil. The pistons have to overcome more friction, and cylinder wear is increased because combustion products condense on the cylinder walls. Effective engine life is shortened. That's not the point of all our work in this book.

How about higher temperatures? As the engine heats up when driven under severe conditions, the flaps open wider and let more cooling air flow. But more cooling air will be available if the flaps are completely removed, right? Wrong. There is no extra air capacity gained by removing the flaps; when they're fully open they flow enough air to maintain the optimum operating temperature and increase engine life.

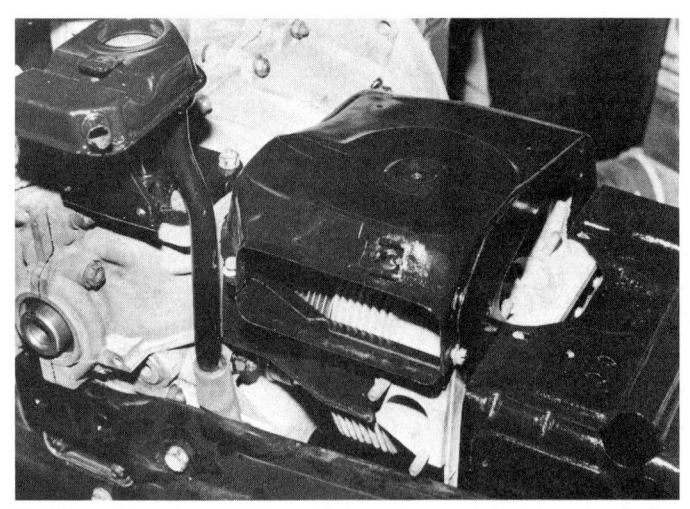

This Type 3 is going together without a thermostat, so upper cylinder covers and remaining sheet metal have been joined without thermostat linkage.

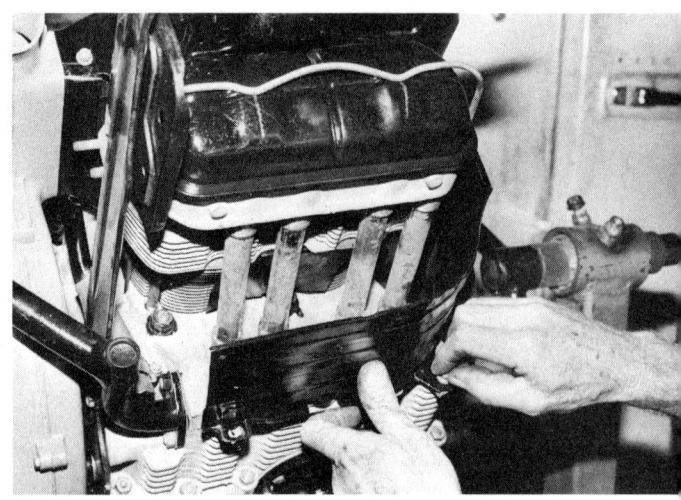

Lower sheet metal with big curve attaches to front and bottom of engine. Leave screws loose until all sheet metal, heat exchangers and mufflers are installed.

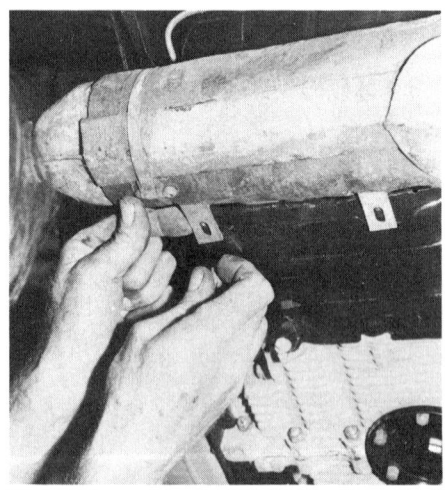

Roll engine over on stand to work on heat exchanger-to-lower sheet metal connection. Just start these screws. Tightening them now will only bind up remaining hardware.

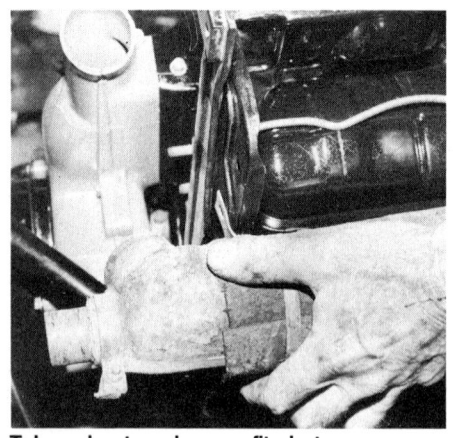

Tab on heat exchanger fits between upper cylinder cover and crossmember. This is connection that can be wrench-tightened now, because crossmember installation eliminates access. Engine here proves accessories can be installed in almost any sequence, as crossmember is already installed (sigh). Sequence in text is more efficient—and enjoyable.

Crossmember bolts aren't accessible in chassis, so Loctite them now. Upper center section bends out to clear oil pump, and threaded blocks at outboard ends face away from engine.

top and sides and finish with the rear.

Oil Cooler—Select the oil cooler seals from the gasket kit and lay them in place without sealant. Read the Type 1 section above, page 138, on the oil cooler to make sure you get the right ones. Slide the spacers over the stud or studs. Install the cooler over the single or triple studs. With a single stud, pass the two long bolts through the cooler.

Slip the spacers between the bolt and case ears, then thread on the through-bolt nuts and run them down finger-tight. This job takes small fingers and lots of patience, so take your time. Install the stud's nut and run it down. Now you can tighten all three fasteners in an alternating pattern. Don't tighten one side all the way, then the other. The case ears will snap off. The torque here is 5 ft-lb or (0.7 kg-m).

If you removed the oil-pressure sending unit from atop the cooler, replace it. Coat the threads with Permatex 3H before installation. You must wrench-tighten the sender, but don't expect it to bottom or cinch up tight. It uses pipe threads, which are tapered. So once it starts to wrench-tighten, that's it, don't overtighten it. Any more and you'll split the cooler.

Breather—Seal the breather baffle with Permatex 3H and set it on the case over its four studs. The louvers open down, into the case. Then put sealer on the bottom of the breather box and set it on the studs. Wrench down the nuts and washers. Install the multiple-ground connector that fits under the nut closest to the center of the case and at the flywheel end. Make sure this connector is clean, shiny and forming a solid connection so it can do its job.

Temperature Sensor II—Install this temperature sensor (found on fuel-injected engines) in the case near the distributor on '69 and earlier engines. Use a new gasket with Permatex 3H, and don't forget the throttle return spring stand goes under the rear bolt. That's the one closest to the fan housing.

Thermo Switch—Remove the center case bolt and install the thermo switch on it. Put the bolt back in and torque to 14 ft-lb (2.0 kg-m).

Oil Filler Tube—Get the filler tube gasket from the set and apply Permatex 3H to hold it to

Centering front fan housing half is just like centering a clutch disc by eye. Mate cylinder covers and fan housing first. Tighten hardware to just snug so housing can be adjusted with light taps, then eyeball distance between hole in housing with crankshaft snout. Tighten bolts when distance is the same around hole.

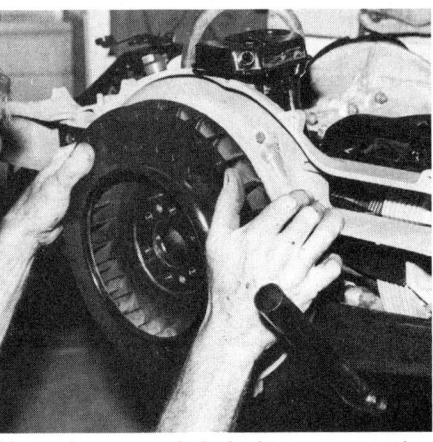

Mount fan on crankshaft, then rotate engine at flywheel and listen for scraping between fan housing. If fan scrapes, realign housing.

Dowels between front and rear fan housing halves require carefully applied force to mate. Housing is lightweight aluminum and cracks easily if abused. Gradually tighten perimeter bolts and squeeze dowel pin bosses in a crisscross pattern until both halves fully join.

the tube. Then bolt the tube to the case while slipping on the hose from the breather box. Clamp the hose, then wrench down the bolts. You'll probably have to grind down a cheap box-end wrench to get to the one bolt.

Thermostat—Install the thermostat and bracket to the lower right case half. If you marked the thermostat's placement before disassembly, then installation is simplified. Thread the thermostat onto the rod coming down from the linkage on the right cylinder head. Then nut the thermostat bracket to the case and the thermostat to the bracket.

Upper Cylinder Covers—Install the upper cylinder cooling tin. Start by laying on the large pieces that cover the cylinders. Add to these the pieces that complete the openings the fan housing mates with. They screw to the upper cylinder covers at each side. You can install both sides now if the thermostat and control linkage are missing. Otherwise, leave the right cover off until after the fan housing is installed.

At this point, leave all cooling system sheet-metal screws loose. This will let the various pieces adjust to each other as you add more parts of the system. After the engine is complete, come back and tighten all the screws.

Lower Sheet Metal—Roll the engine over on one side and attach the large, curved, lower sheet metal. There's one piece for each side. Each screws to the case in two spots and the front of the upper cylinder covers in two spots. Again, leave the screws loose.

Heat Exchangers—Install new gaskets on the front exhaust-port studs. Use no sealer. Now hang the heat exchangers from the forward studs and run down the nuts. Don't wrench-tighten the nuts yet, just run them on enough to support the weight of the heat exchangers so

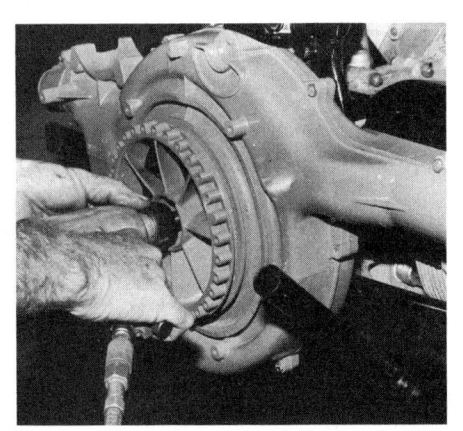

Next are the pulley and nut. Torque is 94—108 ft-lb, but many shops use an experienced zap from air wrench. Loctite and a torque wrench are more accurate.

they don't dangle from the studs.

At the rear of the heat exchangers, bolt the heat exchanger tab to the upper cylinder cover. The upper cover bends down at this point to join with the heat exchanger. This is one connection you can wrench-tighten as you are going to cover it up in a minute.

Crossmember—Hold up the crossmember to the rear of the engine and get it oriented. The relieved section bulges to the rear to clear the oil cooler. Also, the reinforcing blocks placed out at the ends face away from the case. Now bolt the crossmember to the case and torque to 30 ft-lb (4.0 kg-m). Use Loctite on the three attaching bolts. There's no way of getting to these bolts in the car, and it would be inconvenient to have to tear the rear of the engine apart to tighten them.

Fan Housing—Four bolts hold the front of the fan housing to the engine. Two thread into the case; two into the breather box. You need to get the housing half up and mated with the upper cylinder covers with all bolts started. Run in the bolts until snug, leaving enough slack so the fan housing will still move, but tight enough so it can't slip down from its own weight.

Using your best calibrated sight, gage the distance from the fan housing to the crankshaft snout. Adjust the housing until the hole in the housing is equidistant from the crank snout on all sides. Tighten the four bolts.

Check your alignment work by installing the fan on the crank and rotating the engine through one revolution. Listen carefully for scraping. You'll hear noise from the rings sliding in the cylinders, but not from the fan against the housing if it's correctly installed.

If the fan clears the housing, great. Complete the fan installation and move on. If not, remove the fan and readjust the fan housing until the fan turns without interference.

Fan—The fan should slide fairly easily onto the crankshaft. If it doesn't and you start looking to a hammer for help, check the fan housing alignment.

Check the *oil return thread* on the fan hub. It should be free of debris or peened over sections. The thread must be entirely clear to work correctly. If the thread is broken, use a triangular file to restore the groove.

Rear Fan Housing Half—Complete the fan housing by installing the rear half. The housing is located by two solid dowels at the 9- and 3-o'clock positions. Use slip-joint pliers to gently squeeze the housing together at these points. Be careful. Don't muscle a misfitting housing together with the pliers—you'll only

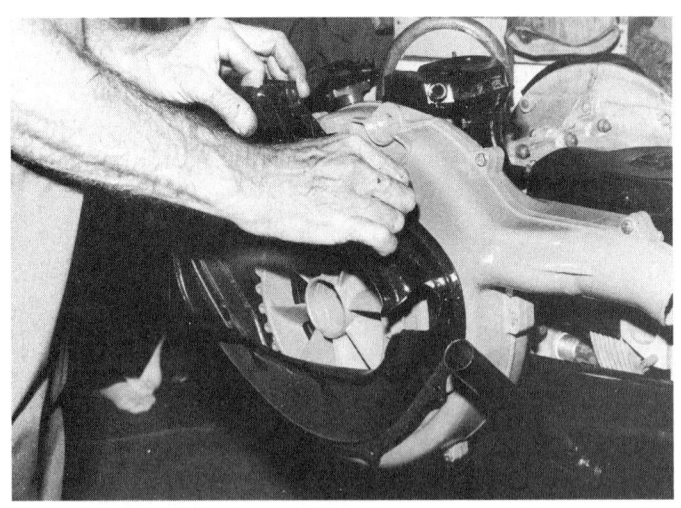

Sheet-metal air intake is next. Six bolts go around intake perimeter, with a seventh hidden inside generator-belt extension, blocked here by mechanic's wrist. This cover has been modified for an air conditioning bracket, hence cutout at 3 o'clock.

Muffler assembly attaches to rear exhaust-port studs, heater inlets and heat exchangers just like a Type 1, except components look a little different. Use new hardware at muffler-to-exchanger joint, and don't completely tighten any connections, until all are started together.

Generator must be positioned so index mark (arrow) lines up with dot on hold-down strap. Otherwise cooling holes in bottom of generator and cooling tin won't line up and it will overheat.

Fan housing-to-heater inlets are joined on Type 3s with curved sheet-metal tubes. Rubber joints and double hose clamps complete connection.

crack the housing. Instead, take the rear half off and check it for dings or burrs.

Install the seven or eight bolts around the housing circumference and wrench them tight.
Thermostat Linkage—Connect the thermostat linkage between the fan housing and right cylinder head. Lubricate all pivot points with white or moly grease.

To check adjustment, close the flaps in the fan housing. Both should close at the same time. If not, loosen the flap cinch bolt, rotate the flap on the shaft and try again. Then connect the spring between the fan housing and the cylinder-head bracket. The long part of the spring goes to the fan housing, the short part to the bracket. When the flaps are closed and spring installed, there must be a 1/32—1/16-in. (1—2mm) gap between the bottom of the flaps and the fan housing. Adjusting both flaps on the shaft will restore this gap if it is out of specification.

After connecting the linkage, install the upper cylinder cover. Don't forget the small piece that wraps under the bottom of the fan housing outlet.
Pulley—Always install the paper gasket between the pulley and the fan because it damps vibration between these parts. There should be a new gasket in the gasket set. There may also be a thin spacer between the pulley and fan. If so, use it. It is needed for setting pulley-to-housing clearance. If there was no spacer here when you took the engine apart, it shouldn't need one now.

With gasket and spacer in place, position the pulley on the fan with the dowel. Loctite the pulley bolt and torque it to 94—108 ft-lb (13—15 kg-m). That's substantial, so you'll need a flywheel lock or a helper holding the gland nut. A lot of garages just tighten these with a burst from a 1/2-in.-drive impact wrench and let it go at that. Snap on the plastic dome.
Air Intake—Next up is the air intake housing. The trick here is to remember to install the seventh bolt inside the housing. Reach up through the air inlet and to the left, under the generator area, to find the bolt hole. The other six bolts attach to the housing perimeter. Wrench-tighten all these bolts.
Generator—At the pulley end of the generator you'll find a locating line. The line should point at a dot on the generator hold-down strap when installed. So get the strap and generator and mock them up on the engine. When you've got them aligned, start the hold-down strap bolts. Don't tighten them yet.

The reason the generator must be aligned is more important than getting the terminals in position so the wires will fit. This procedure lines up a hole in the sheet metal with a cooling inlet hole in the generator. This lets cooling air from the fan blow through the generator. Without the holes lined up, the engine-mounted generator overheats.
Generator Belt—Insert the belt into the air-intake housing and loop it over the fan pulley. Now tilt the generator pulley end down. Roll the belt onto the generator pulley, then release the generator and tighten the hold-down strap bolts. Check the dot and line alignment.

This method allows fitting the belt without disassembling the generator pulley halves. This is good because it takes a special wrench, or

Mounting intake-air distributor requires removing some case seam hardware. It's easier to install if all runners and hoses are removed as shown, but label them all if you disassemble it.

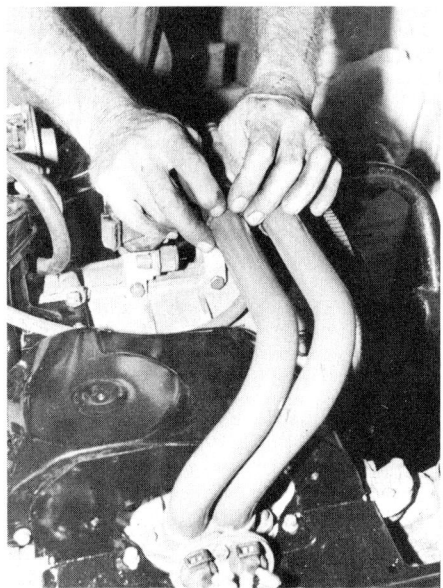

Center rubber joints between intake runners and intake-air distributor once runners are setting on engine. Rubber joints should move with moderate drag. If they slide easily on nipples or runners, they are too loose and need replacement.

slip-joint pliers and special language, to handle the pulley nut. You have to hold the domed washer stationary while wrenching the pulley nut; the domed washer has two flats for this.

Rebuilding is an excellent time to replace the belt, and you should give this cheap insurance to your engine. It's worth a little effort, therefore, to purchase a new belt of the same brand and length as the old one. That way you know the pulley won't have to come apart. Otherwise, you'll have to hassle with the pulley nut.

Muffler—Prepare to hang the muffler by installing new gaskets on the rear exhaust port studs (at the head). Inspect the slip joints on the muffler and heat exchangers for deformities. If they got pried out of shape during disassembly, bend them back into shape.

While supporting the muffler at both ends, guide it onto the heat exchangers and studs. There are three connections that must be made at each side. First, is the rear heat exchanger to head. The rear heat exchanger is the small, flat box attached to the muffler. Second is the rear heat exchanger to heat-exchanger joint. A broad clamp joins these two together. Third is the slip joint—muffler to heat exchanger. Get all these lined up, then install the two-piece clamps, band clamps and nuts. Use new two-piece clamps from a muffler hardware kit.

Intake-Air Distributor—Now go back to the top of the engine. A combination of three bolts and studs holds the intake-air distributor to the case. One bolt goes through the case "step" at the flywheel end. Going rearward, the next is a case through-bolt and finally the mid-case stud.

Completed Type 3 shows hose and wire routing.

Also attach the throttle-return spring between the throttle arm and its stand on Temperature Sensor II.

Phenolic Blocks—Lay a new gasket over the intake-port studs, then the phenolic block and then another gasket. You can use RTV sealant instead of the gaskets in a pinch.

Intake Runners—Check the fabric-covered rubber intake runner connecting hoses for looseness. Replace them if they don't fit snugly or are torn and worn. Now install the intake runners on the engine. Join the slip joints first at the intake-air distributor. Then slide the runner ends over the intake port studs at the head. Install and tighten the nuts to 14 ft-lb (2.0 kg-m), then go back and make sure the slip joints are centered.

Sparkplugs—Gap a new set of plugs to 0.024

147

Type 4 pulley (left) and thermostat (right) are simple stud and nut mounts. Lube pulley bushing and bolt with white grease, and align thermostat bracket with scribe marks on case.

Gasket and light coat of Permatex go between case and oil filter mount. Don't get wild with torque on nuts. Soft aluminum mount doesn't need lots of torque to seal.

Besides lubing oil filter gasket, pour filter full of oil before installing. Dry paper element will absorb oil, speeding priming at engine start.

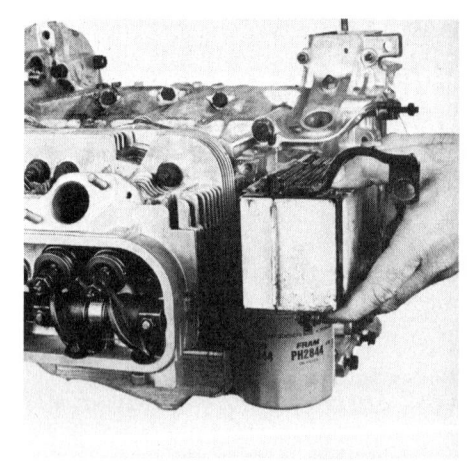

Trick to Type 4 oil cooler installation is keeping grommets in place. A *light* smear of RTV or Permatex will hold them, but don't put on so much it can squish into cooler passages.

in. (0.60mm). Put anti-seize compound on the upper two-thirds of their threads and install them at 22—29 ft-lb (3.0—4.0 kg-m). Use the lower end of the torque range. After all, someday you'll want to change them. As a reminder, when you've got plugs this tight in aluminum heads, always let the engine cool before attempting removal. Otherwise, you'll strip or gall the plug threads.

Now install the two cover plates over the plugs and around the intake-manifold runners.

CHT Sensor—If the engine has a CHT (Cylinder Head Temperature) sensor (also called Temperature Sensor II), install it now and tighten it snug. On '69 and earlier engines the sensor goes below the cylinder-4 exhaust port. On '70 and later models, the sensor threads into the head by #4's sparkplug. There's a hole in the upper cylinder cover for access and a rubber seal that closes this hole. If you're missing the seal, install a new one.

Fuel & Vacuum Lines—Using the numbering system you devised for disassembly, reinstall the fuel lines and vacuum hoses. There will be a fuel line running from the rear injector on one side to the rear injector on the other side. The front injectors will connect to the inlet and outlet lines when you install the engine; leave them loose until that step. The third nipple on the fuel manifold at the injectors goes to the intake-air distributor.

A vacuum line runs from the intake-air distributor to the vacuum advance on the ignition distributor. Other vacuum connections are on the Temperature Sensor II. There are two nipples, one vertical, the other horizontal. The vertical hose runs to the air filter, so it stays loose until engine installation. The horizontal hose goes to a nipple underneath the intake-air distributor.

Coil & Cap—Bolt on the ignition coil and connect its ground wire to the distributor. The coil goes to the left of the distributor on the fan housing or sheet metal. Install the distributor cap and wires. Connect the wires to the sparkplugs and insure the grommets are air tight. The firing order is 1-4-3-2; rotor rotation is clockwise. See page 14 for a drawing showing distributor position and cylinder location.

Front Sheet Metal—Lower the engine from the stand. Now you can attach the front sheet metal to the bellhousing flange. It clips over the flange and screws to the case in two spots.

TYPE 4

Begin attaching Type 4 accessories with the thermostat and pulley. Lightly lube the mount-

Dab a little Permatex on oil-pressure sender threads, then install wrench-tight. Sender's pipe threads will cinch up quickly after reaching finger-tight, so don't overdo it!

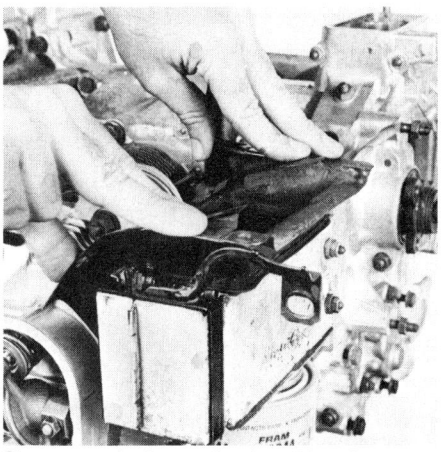

Sheet-metal piece simply slides between cooler and cylinder head. Give it a wiggle to make sure it is fully seated on cooler.

Another sender is the CHT up on left cylinder head. Because this is a blind hole, no sealer is required, just thread it in.

ing bolt and pulley ID with white grease, then mount them on the lower, right-rear case. The thermostat mounts on two studs just forward of the pulley, under cylinders 1 and 2. Let the cable hang free for now.

Sparkplugs & CHT—Two easily overlooked jobs are putting in the sparkplugs and the CHT sending unit. Gap the plugs to 0.024 in. and give the upper two-thirds of the threads a light coat of thread lubricant. Anti-seize is a popular choice for this job. Tighten to 22 ft-lb (3kg-m).

The CHT sending unit threads into the left cylinder head; wrench-tighten it.

Oil Filter—Carefully spread Peramatex 3H to the oil-filter mount gasket and attach it to the case. Install the filter mount and wrench down the two nuts.

Fill the oil filter before installing it to help build oil pressure more quickly when you start the engine for the first time. Pour a half quart of oil into the large hole in the filter's center, then swish the filter around. This helps the filter element absorb some of the oil. Lubricate the filter's gasket with fresh oil, then spin the filter onto its mount. Give it 2/3 of a turn after the gasket touches the mount. Do not wrench-tighten the filter, or it will be tough, if not downright difficult, to remove.

Oil Cooler—Get the oil cooler seals from the gasket set and fit them to cooler. A light dab of Permatex 3H or RTV will help hold the seals in place during installation, but don't slather the sealer on. It will only gush from under the seal and into the oil passages when you torque the cooler in position. The last problem you need is gasket sealer clogging the oil cooler.

Now turn the cooler on its side and pass it over the three mounting studs. Watch the seals as you install the cooler, they must not slip out of place. Thread on the three cooler-mounting nuts and torque them to 5 ft-lb (0.7 kg-m). That's not much torque, only 60 in-lb, so don't

Crankshaft oil seal can be installed with light hammer taps. Once in, lubricate seal ID with oil. This protects seal from overheating at engine start-up. White grease can be used if engine is going to sit for some time before running.

go overboard. You can crush the cooler if you aren't careful.

Sending Unit—Seal the oil pressure sending unit threads with Permatex 3H, then install it near the oil cooler. Wrench the unit in an additional turn after it's in finger-tight.

Now install the curved sheet-metal piece between the oil cooler and nearest cylinder. Just slide it into place. Right now there is nothing to bolt this piece to. Later, it will attach to the fan housing.

Crankshaft Seal—Check the crankshaft oil-seal bore for excess sealant. Wipe any that may have run or balled up at the case parting line in the seal bore. Then install the oil seal. You can use Permatex 3H on the seal's OD, or leave it dry. Carefully begin driving in the seal with light hammering about its edge. Don't let the

Install engine mounts before engine is covered by fan housing. Replace any spongy, torn or oil-soaked mounts.

seal cock in the bore. If it does, very carefully extract it and start again. It usually takes several times before the seal will begin. Drive in the seal flush with the case by tapping around the seal's circumference. Lubricate the seal lip with white grease or oil.

Oil Filler—If you have a Bus or 411/412 engine, bolt on the oil filler tube now. Use a fresh gasket from the kit, and apply Permatex 3H to both sides of it.

Engine Mounts—The 914 engine shown in the photo has two round engine mounts. Buses from '72—79 and 411/412 mounts are similar, and all must be installed before the fan housing is put in place. Installing new engine mounts are a good idea, particularly if the engine was shaking noticeably before you pulled it. Unfortunately, these mounts are surprisingly expensive, so you might be tempted to reuse the old ones. On Buses and 411/412s, attach the

Bolt in center secures fan hub to crankshaft and gets 23 ft-lb. Special large area washer going on is needed for correct fan spacing, and to stop vibration and wear on hub and fan.

Check fan housing fit to case before tightening these four nuts. Washers left on studs can cock housing, possibly cause a crack when housing is torqued down.

Left end of fan housing accepts bolt from sheet metal bracket at oil-cooler side of engine. Unlike Type 1—3 engines, these attachments can be wrench-tightened immediately.

Beware of small flaps and their unsecured hinges at bottom of Type 4 fan housings. Mounting housing and rolling engine on stand can cause hinge pin and flap to fall out. Pin is secured by ducting in chassis.

It's easy to forget this sheet-metal piece until after thermostat cable is neatly attached. But cable (in left hand) must be threaded through hole in sheet metal, requiring one free cable end.

Thermostat cable is held and adjusted at this small nut and bolt on air-flap shaft. Spring pressure is surprisingly strong, so use a helper to hold flaps shut while attaching cable.

crossmember to the studs on the engine mounts. Torque to 14 ft-lb (2.0 kg-m).

Buses from '80—83 have a different procedure. First, bolt the engine support with its four bolts to the engine and torque to 14 ft-lb (2.0 kg-m). The cutout goes over the oil pump. The support is the cast-aluminum part of the engine mount assembly. If you replaced the rubber mounts, bolt them to the engine support. Otherwise, they should still be attached. The carrier is the long bar that bolts to the chassis at each end. Bolt it to the support using the rubber mounts. Note the arrows on the carrier where it attaches to the chassis. These arrows point toward the engine when it's correctly installed.

Fan Hub—After lining up its keyway with the Woodruff key on the crankshaft, press the fan hub onto the crank. Use no sealer or locking compound. Thread in the bolt and special large area washer. The tapered attachment between hub and crank doesn't require a lot of torque to seat, 23 ft-lb (3.2 kg-m) does the job.

Fan Housing—Install the fan housing on its case studs. Be extra sure there are no washers left behind on the studs when fitting the housing to the case. This will cock the housing to one side, and could crack it when the nuts are torqued. The housing is retained by four nuts in the fan recess, plus a bolt at the upper corner of the oil cooler. Wrench-tighten these.

Right-Rear Air Duct—Before attaching the thermostat cable, install the sheet-metal piece between the right cylinder head and where the alternator goes. Just slide the duct into position for now, but be sure to pass the thermostat cable through its hole in the duct.

Adjust Thermostat—Double-check the threading of the thermostat cable as it goes around its pulley and up to the control-flap shaft arm. Press the control flap closed by hand. The flap is closed when the shaft's arm is pushed down against spring pressure. Now attach the cable to the shaft's arm using the screw and square-headed nut to pinch the cable to the arm.

Sheet metal at flywheel end of heads just slips into place until upper cylinder covers go on.

Upper cylinder covers screw to case in two spots, plus join with sheet metal at flywheel end of heads. Thermostat cable and 914 dipstick tube require some jockeying to get this side on.

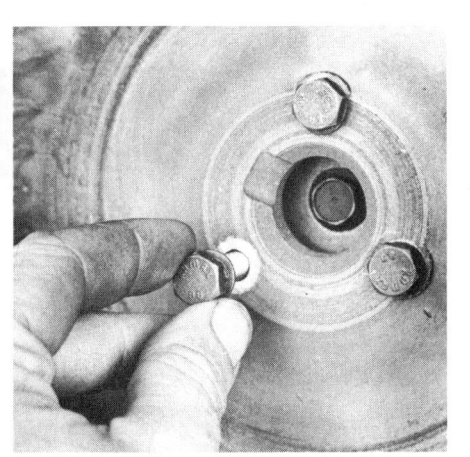

Three fan bolts get 14 ft-lb of torque. Wipe fan and fan hub mating surfaces clean before joining them. Dirt between them can cause imbalance and misalignment.

Hang alternator loosely on its mounts to start its installation. Lower pivot bolt also retains sheet-metal cover, so just stick it in place without cover for now.

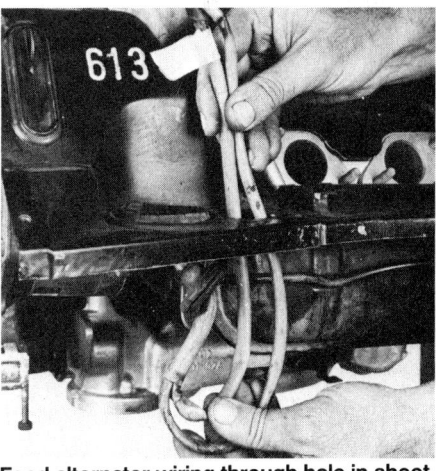

Feed alternator wiring through hole in sheet metal, then work grommet into place. Replace grommet if it's cracked or missing.

You need three hands to do this, so a helper is useful here.

Don't leave a little slack in the cable hoping to make the flaps open sooner. First of all, they won't. The thermostat controls when the flaps open; the cable adjustment controls how far. If you leave the cable slack the flaps will never close. Engine warm-up will take longer, increasing sludge formation and oil contamination. Besides, Type 4 engines don't melt like the overworked early Type 2, so no cooling system tricks are necessary.

Sheet Metal—It's time to cover the engine in its cooling tin. Start with the front ducts at the flywheel ends of the cylinder heads. Now add the upper cylinder covers, which screw into the upper case in two places as well as attaching to the front ducts. The right cover must be threaded around the dipstick tube on 914 engines. The left cover has a grommet for the CHT sensor wire, so you'll have to thread it through. There is also a grommet for the oil-pressure sender which should be installed now. If any of these grommets are old and cracked, replace them. Whatever you do, don't leave them off. That only lets cooling air escape without carrying its share of engine heat with it.

Install Fan—Get the fan's large area washer onto the fan hub. Then fit the fan onto the fan hub. The hub is doweled, so you don't have to worry about getting the timing marks off. Then bolt the fan up with its three bolts. If the engine is air injected, install the extension shaft with these three bolts. Torque on all engines is 14 ft-lb (2.0 kg-m). If you are installing the extension shaft, don't forget to install the timing scale under the shaft support-to-fan housing bolts.

Alternator—It's helpful to get all the alternator pieces organized before assembly. You need the alternator, upper and lower mounting bolts, cover plate, grommet and cooling duct.

Put the alternator into position and attach the the upper adjusting bolt. Leave it loose for now. Install the lower mounting bolt. Feed the wires through the hole in the right sheet metal and install their grommet. Attach the cooling-air duct at the rear of the alternator. It's bother-

Attach alternator cooling duct. Unit is mounted right next to cylinder head and exhaust, so it absorbs tremendous heat that must be carried away.

Adjust belt tension so it flexes 5/8-in. at mid-span with firm pressure. This belt is a tad tight, but being new, will loosen slightly after running awhile. Remove and replace lower cover for this step.

After fitting a new cork gasket to breather box bottom, sit in place and snap on its wire bale. Before installing, inspect inside for crusty oil deposits. If present, wash it out with clean solvent, then dry throughly.

some to have to connect this duct in the confined area behind the alternator, but it sure beats buying new alternators. Then attach the alternator sealing ring and cover.

Put on the belt, adjusting tension so it flexes 5/8 in. (15mm) when firmly pushed at its center. This is fairly tight, and going any tighter will only wear out the alternator bearings. After belt adjustment, install the dipstick grommet on Bus engines, and the oblong cover plate on all engines. 914 engines have a simple grommet where the Bus dipstick grommet goes.

On engines with a fan grille, bolt on the grille and timing scale now.

Bus engines from '80—83 require a modified procedure because of the heater blower. Here the bracketry attaches to the upper end and middle-rear of the fan housing. The alternator bolts extend through the plastic blower housing, alternator ears and brackets. The upper bolt comes from the front of the engine, not the back as you'd expect. And you've got to remember to install the belt before hanging the blower housing.

Also on '80—83 Buses, install the dipstick tube. Use a new grommet at the engine end, and bolt the tube to the top of the fan housing. Don't install the dipstick now or it will get in the way during engine installation.

Air Injection—If your '73 or '74 engine has an air pump, attach the pump and drive belt now. The pump and brackets should still be one unit. The pump mounting bracket bolts to the fan housing in two spots and the adjusting bracket in one. Fit the drive belt, and adjust it. When pressed at mid-span, the belt should deflect 3/16—5/16 in. (5—8mm). Complete the installation by hooking up the hoses to the pump and anti-backfire valve.

Cover Plates—Now that the crankshaft-pulley end of the engine is complete, attach the rear cover plates with the sheet-metal screws. These are the sheet-metal pieces that scoop out to clear the pulley, then form a flat tray at the rear of the engine. You may have removed the two cover plates as a unit, so it may appear as one large piece.

Breather Box—Install a new gasket and then snap the crankcase breather box into position. Depending on exactly how you removed hoses from the engine, you may be getting to the point in the assembly where those hoses will need to be attached. Because the exact hose routings have varied so much over 22 years of production, it would be a futile (if not monumental) task to specify them here. You have to rely on the numbering system and diagrams you made at disassembly.

Typically, however, you will have two breather hoses associated with the breather box. One hose each will come from the cylinder heads, joining at a valve. The valve attaches to the fan housing, next to the breather box. From the valve, a single hose runs to the breather box. Check all hoses for brittleness and replace them if they are cracked.

The breather on '80—83 Buses is a nipple that pushes into the pad on top of the crankcase. There is no breather box or wire bale to deal with.

Exhaust—If working on a 914 engine, skip ahead to "Prepare for Induction System," on the next page. You'll install the exhaust in the chassis. All other Type 4s get their exhaust attached now. With all Type 4 exhaust systems, the biggest headache is getting a wrench on the heat-exchanger nuts. Do your best to get these tight. A variety of wrenches, some modified by bending in a vise, will help. You may have bent some during engine removal, now you get to reuse them. Wrench-tighten all connections after the hardware is mated.

'72—74 Bus—The exhaust system for carbureted Buses is pretty simple. Using new copper sealing rings, bolt the heat exchanger to the head. The rear of the heat exchanger mates with the fan housing, and you need to screw the small covers on the top of these two connections. Also clamp the heater duct to the front of the heat exchanger if you removed it. Another upward rising duct goes at the rear of the exchanger. Finish the installation by installing the muffler using new gaskets.

'75—78 Bus—These fuel-injected engines have a more complex exhaust system. Start with the exhaust manifold/heat exchanger assembly. It nuts onto the exhaust-port studs and clamps over the lower, rectangular, fan housing outlets. Two blower ducts attach to this assembly, one front, one rear.

Now attach the crossover pipe/catalytic converter/muffler assembly using new gaskets to the heat exchangers.

'79 Bus—Use the above procedure, but add the catalytic converter and extra heat exchanger on the left side. Don't bump or bang the oxygen sensor probe when installing the exhaust.

'80—83 Bus—All Vanagons use a separate heater blower, so there are no lower fan-housing connections. They are replaced by the ducting running across the rear of the engine from blower fan to heat exchangers. First install the heat exchangers, then attach the crossover pipe/catalytic converter/muffler assembly to the exchangers. Leave the EGR filter-to-pipe connection off for now. You can install it after bolting on the induction system.

'81 California Bus—This system is like the '79 Bus. Install the right heat exchanger with its two heater ducts. Then bolt up the left manifold/catalytic converter/crossover pipe assembly. Then attach the left heat exchanger/muffler with its two heater ducts. Heater ducts are the round tubes running up and to the front or rear.

Vanagon Heater Ducting—On all Vanagon engines, attach the heater-ducting elbow to the fan housing and right heat exchanger. Then fit the long cross-engine duct to the elbow and left heat exchanger. This piece goes through the hole in the left-half rear engine cover plate.

411/412—Type 4 exhausts don't come much simpler than the 411/412 variety. Install the heat exchangers at the exhaust ports and at the fan housing. Attach the heater ducts to the exchangers is you removed them. Then, mount new gaskets, and bolt the muffler to the heat exchangers.

Prepare For Induction System—The next step is a big one, you'll install the carburetors or fuel injection. Take some time now to inspect

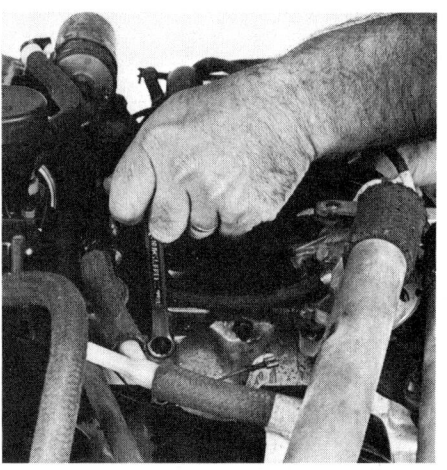

Setting fuel injection on engine crowds it in a hurry. Disconnect at one set of intake runners to get at case flange hardware that must come out for air distributor installation.

Once air distributor is in place, connect intake runners at both ends. Use new gaskets on both sides of the phenolic blocks between head and intake runners. Tighten nuts in alternating pattern.

the engine's topside. Go through your hardware boxes and look for small fittings that may have escaped installation with a larger component. For example, on fuel-injected engines, there is an electrical-ground fitting that mounts under the head of one of the case bolts. This fitting has three spades for electrical connectors. If you install the intake manifold without this piece in place, it's nearly impossible to install it with the injection in place.

This is also a good time to go over the hoses on the intake manifold, especially those on fuel-injected models. If you didn't replace cracked or deteriorated hoses earlier, do so now.

Install Manifold—Once you are ready, begin installing the fuel system. Start with the intake manifold spacers at the cylinder heads. Use new gaskets on both sides, then lay the spacers on their studs. Using gasket sealer doesn't hurt, but it isn't mandatory. Set the manifold in place. The intake tubes on the manifold for fuel-injected models will probably slip free from the air distributor, which is the black sheet-metal center section. That's fine, keep working from side to side until the manifold is in position, then reinstall the intake tubes to the air distributor.

On carbureted engines, it's straightforward to nut down the manifolds to the heads, then move on to hoses and linkage. If the engine uses dual carbs, and has the central idling system, attach it to the carbs now.

If you own a 914 with aftermarket car-

buretors, it's easier to leave the carb installation until after the engine is in the chassis. To stop dirt and stray hardware from falling down the intake ports, cover them with tape.

With fuel-injected engines, start at the air distributor. Using a selection of wrenches and ratchets, bolt the air distributor to the upper case through bolts and studs. This can be a real chore, depending on how large your hands are. Just take some time and save skinned knuckles. After installing the air distributor, nut down the manifold at the cylinder heads. Continue around the engine, attaching fuel system accessories. The fuel-pressure regulator can be attached to it with its bracket and hoses connected. The green and black vacuum hoses can be attached to the distributor. The green hose goes to the distributor side of the vacuum unit, the black hose to the outside. Use your number scheme for accuracy.

Make the electrical connections at the CHT sensor (Sensor II), oil-pressure sending unit, ground under the air distributor and to the ignition distributor. Again, follow your numbering scheme for accurate installation.

Install Coil & Distributor Cap—Bolt the coil to the slope on the left upper cylinder cover. Connect the coil ground wire from the distributor to the coil ground terminal. Look for the small + and − signs near the coil terminals if your numbering scheme fails you.

My favorite silly mistake at reassembly is putting the distributor cap on without installing the rotor. That error can frustrate your efforts to

TORQUE SPECIFICATIONS

Type 1—3	ft-lb	kg-m	Size (dia. X pitch)
Connecting-rod nut	24	3.3	M 9 X 1
Connecting-rod capscrew	32	4.4	M 9 X 1
Crankcase nut	14	2.0	M 8
Crankcase nut	25	3.5	M 12 X 1.5
Cylinder-head nut (8mm stud)	18	2.5	M 8
Cylinder-head nut (10mm stud)	23	3.2	M 10
Rocker-shaft nut	18	2.5	M 8
Heat exchangers at head	14	2.0	M 8
Muffler clamp bolts	7	1.0	M 6
Intake manifold nut	14	2.0	M 8
Preheat-flange bolt	7	1.0	M 6
Oil-pump nut	14	2.0	M 8
Oil-pump nut	9	1.3	M 6
Oil-drain plug	25	3.5	M 14 X 1.5
Oil-strainer nut	5	0.7	M 6
Oil-cooler nut	5	0.7	M 6
Oil-filler gland nut	40	5.5	
Flywheel gland nut	235	35.0	M 28 X 1.5
Clutch bolt	18	2.5	M 8 X 1.5
Sparkplug	25	3.5	M 14 X 1.25
Engine-to-transaxle nut	22	3.0	M 10
Crossmember bolt	18	2.5	M 8
Crossmember bolt	29	4.0	M 10
Generator-pulley nut	43	6.0	M 12 X 1.5
Fan nut	43	6.0	M 12 X 1.5
Crankshaft-pulley nut	32	4.5	M 20 X 1.5
Crossmember to body	18	2.5	M 8
Fan/Crankshaft pulley bolt—Type 3	94—108	13.0—15.0	M 20 X 1.5

Note: Flywheel gland-nut specifications vary among VW literature from 215—253 ft-lb (30—35 kg-m), but most machine shops install it at 300+ ft-lb (41+ kg-m) with Loctite to prevent loosening.

Type 4	ft-lb	kg-m	Size (dia. X pitch)
Connecting-rod nut	24	3.3	M 9 X 1
Crankcase nut	14	2.0	M 8
Crankcase-sealing nut	25	3.5	M 10 X 1.25
Cylinder-head nut (10mm stud)	23	3.2	M 10
Rocker-shaft nut	10	1.4	M 7
Heat exchanger/cyl. head	16	2.2	M 8
Muffler/Heat exchanger	14	2.0	M 8
Oil-drain plug	16	2.2	M 12 X 1.5
Oil-strainer nut	9	1.3	M 8
Driveplate-socket screw	65	9	M 12 X 1.5
Hub-to-Crankshaft bolt	23	3.2	M 8
Fan to hub	14	2.0	M 8
Pulley/Fan exten. shaft	14	2.0	M 8
Sparkplug	22	3.0	M 14 X 1.25
Engine-to-transaxle nut	22	3.0	M 10
Crossmember support to crankcase	22	3.0	M 8
Rubber engine mount to crossmember support or crankcase	14	2.0	M 8
Crossmember to chassis	14	2.0	M 8
Vanagon rubber mounts to crossmember	33	4.5	
Oil pump to crankcase	14	2.0	M 8
Oil-cooler nut	5	0.7	M 6
Flywheel bolt	80	11.0	M 12 X 1.5
Clutch bolt	18	2.5	M 8 X 20
Torque converter to driveplate	14	2.0	M 8

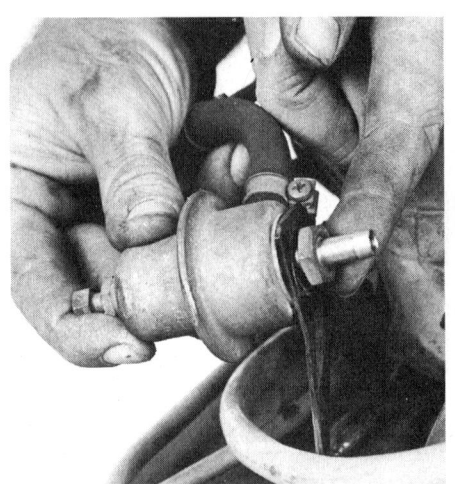

Fuel-pressure regulator mounts to its stand-off with nut. Nut and adjusting screw at other end of regulator adjusts fuel pressure, so leave these alone, at least until engine is back in chassis and running. Fuel pressure seldom needs adjusting.

start the engine. So after checking that the rotor is in place, snap on the distributor cap.

On fuel-injected engines, the plug wires are routed under the air distributor on their way to the right side cylinders. What a maze. Look for the ignition-wire clips on the upper cylinder covers when routing the plug wires. Check the sparkplug grommets for wear now. If the grommets that seal the plug wires as they pass through the sheet metal are old, cracked, warped out of shape or otherwise poor, replace them. This will help the engine cooling system be more efficient.

Air Filter Support—It's tempting to install center-mounted air filter supports now, but don't do it. The throttle linkage passes underneath this piece, so it must be left off until the engine is installed. Air filter supports that attach to left of center can be installed now.

Threading ignition wires takes a little patience, but if you look closely, you'll see clips on sheet metal for restraining wires. Wires to cylinders 1 and 2 shown here quickly dive toward case after leaving distributor.

Once engine leaves stand, front sheet metal can be screwed to upper cylinder covers and clipped to bellhousing flange.

CHAPTER 8

Engine Installation, Break-In & Tuneup

Before rushing to put the engine back into the chassis, perform some checks. First, is the car's battery fully charged? If you think it might be low, hook it up to a battery charger. Have you finished all you set out to do when you began this project? Perhaps you were going to change the starter motor and clutch, or detail the engine compartment? Now is the time.

Cleaning the engine compartment is always a nice touch and will make subsequent tuning more pleasant. Your engine is clean now, and stuffing it into a crusty old compartment won't make it any easier to work on or look at. Besides, what better time to reach all those nooks and crannies in the compartment than now? You can do a great cleaning job with the engine out of the way.

Scrub down all surfaces with warm soapy water. Liquid dish or laundry soap works well. A collection of nylon-bristle brushes will speed the process, particularly when cleaning in the corners and around wiring. Rinse out the soap with plenty of clear water. Blow dry with compressed air, if possible. Or soak up puddled water with a sponge. Surfaces can be dried with a towel or shop rags. Lubricate all hinges and unpainted surfaces to prevent rust. Spray-on oil like WD-40 will help remove water, also.

TRANSAXLE PREPARATION

Before covering the bellhousing area with the engine, perform some checks. First examine the input shaft or converter seal. If the bellhousing interior is covered with gear oil or ATF, the seal should be replaced. If the torque converter was removed to drain it, or accidentally fell out, replace the converter seal.

Manual Transaxle—Changing the seal isn't difficult, but may require resourcefulness when driving the new seal in. On manual transaxles, first remove the release bearing and release-bearing mechanism. The bearing can be freed by removing the two wire clips on either side of it. These clips are under tension, and are apt to go flying when released. Place a rag over the clips when removing them to at least keep them in the immediate vicinity.

Now slide the release bearing off the input shaft.

The bearing mechanism requires removing the clutch lever on the outside of the bellhousing. First remove the small circlip outboard of the lever, then slide the lever, wound spring and spring seat off of their shaft. Remove the shaft-bushing lock bolt. It's the bolt just inboard of the clutch lever. Shift the clutch operating shaft to the left, and the bushing, washer and spacer sleeve will come out. Shifting the lever to the right will free it. Check the condition of all bushings, washers and springs. Replace as necessary.

On '71 and later models, there is a *guide sleeve* over the oil seal. Remove the three bolts and take off the sleeve.

Pry the old oil seal out with a screwdriver. Be careful not to gouge the case. Prepare the new seal by applying Permatex 3H around its circumference. Set the seal in position after checking that the small coil spring inside the

Why is this man smiling? His rebuilt engine is ready to be installed and soon he'll be driving!

CLUTCH SERVICE

VW release (throw out) bearings and pressure plates have been changed over the production run. By now, many early clutch assemblies have been upgraded to later styles. Here is some general service information about the different pieces. If you're buying components, be sure to use your old ones for comparison.

Release Bearing—The early type (pre-'71) is used in a clutch housing *without* a supporting *guide sleeve*. This bearing is held to the clutch operating shaft's arms by two *retaining springs* that will fly off when their tension is released. Reinstall the springs with their bent ends fitting behind the hooked arms of the operating shaft.

The later bearing type ('71-on) is used with a guide sleeve that is bolted to the clutch housing. The bearing is held to the operating shaft's arms by two *spring clip retainers* and two *spring clips*. The retainers are extracted first, then the clips.

Inspect all hardware for cracks, wear and distortion; replace as necessary. The two bearing types are not interchangeable. Because they are inexpensive, it's worthwhile to install a new one to ensure full service life.

Clutch Operating Shaft—Inspect its two bushings for wear and replace if required. Lightly lube them with white grease. Closely examine where the two arms attach to the main shaft and the arms themselves. If the weld or the arms is cracked, replace the shaft.

Pressure Plate—These come in two diameters, 180 and 200mm, and two styles. The early one (pre-'73), has three *release arms* and the later one ('73-on) has *multiple fingers*. And there can be a ring joining these arms or fingers. The early type has pretensioning clips between the release arm and the assembly ring that *must be removed* before the plate will work. Install the plate first, then extract the clips.

Removing Clutch Assembly—Mark the pressure plate's position on the flywheel so you can install the clutch assembly in its exact place. Lock the flywheel and gradually loosen the six bolts a quarter turn at a time in a diagonal pattern to release the plate's pressure. Remove the unit and check the flywheel surface for wear, cracks and grooves. Inspect the pressure plate's (early type) release levers to see if they are in line; if not, replace. Check the multiple fingers on a later style pressure plate for cracks and wear; if any, replace.

Install Clutch Assembly—Use a clutch centering tool to align the clutch plate and then mount it against the flywheel. Install the pressure plate (match its position to the alignment marks you made earlier) and tighten the six bolts in a diagonal pattern until snug. Final torque to 18 ft-lb (2.5 kg-m). Remove centering tool.

---

seal lip is in position.

Driving the new seal means working around the input shaft. You can use a long drift and tap the seal in using a 12, 6, 3 and 9 o'clock pattern. Another method is to use a pipe that fits over the input shaft and butts against the seal. The pipe must be approximately the same size as the seal's OD, but you might not have one available that large.

Refit the guide sleeve (if applicable), release bearing mechanism and bearing. Remember the clutch operating shaft goes all the way to the left first, then the right. Coat all moving surfaces with white grease.

Release Bearing—Always check the release bearing for roughness. Press in lightly on the bearing's working surface while rotating it. If it moves freely and quietly, the bearing is OK. Any roughness, snags or gritty feeling is cause for replacement. Remember to beware of flying clips when removing the bearing.

The input shaft on manual transaxles needs lubrication, but not so much that it flies off and contaminates the clutch. Dry moly powder is the ultimate choice, but a light application of white grease will work. Also put a touch of white grease on the starter shaft.

Starter Pilot Bushing—Now is an opportune time to renew the *starter bushing* in the transmission bellhousing. If worn, the starter's pilot shaft can move and the starter will drag. Remove the starter and check the bushing. If suspect, extract it with a pilot bearing puller. Then drive a new bushing into place—don't cock it in the bore, the case will be damaged—until it's flush with the bore's edge. Replace the starter.

Automatic Transaxles—The converter seal is mounted on the bellhousing input shaft behind the torque converter. To extract it, remove the converter, and pry it out using a flat-blade screwdriver or hooked seal puller. To install a new seal, lube its inside with ATF and lightly tap its edge with a hammer until it seats.

Auto-Stick Transaxles—The converter seal on these transaxles is also located behind the torque converter. To extract it, remove the converter, and use a hooked pry bar at the seal's outer edge. Lube the inside of the new seal with ATF and place it over the input shaft. Lightly drive the seal in place with a large-diameter cylinder (about 2-in. dia.). On early models, once the seal is fully seated, peen its outer metal collar over in three spots. On later models, a serrated edge holds the seal without peening.

Engine Compartment Seal—Many mechanics still don't understand the importance of proper engine compartment sealing on air-cooled VWs. On all engines, air is drawn from the top of the engine and forced by the fan over the cylinders and heads out the bottom of the engine. Any hole between top and bottom of the engine lets cooling air escape and hot under-engine air enter the engine compartment. Because the cooling fan moves such massive amounts of air—as much as *900 liters per second* (1920 cfm)—there's hardly an excess of air waiting around.

The point is, seal the engine and compartment as well as possible. Specifically, the engine compartment seal fitted around the circumference of the hole the engine sets into may need replacement. Inspect it for cuts, tears, drying out and compressed sections. Go over the engine and double-check for missing spark-plug grommets or any other open hole between top and bottom.

Finally, double-check ignition timing and fuel-system connections. You don't want to excessively crank the engine on the initial start-up only to find the timing was out or no fuel was available. Extra cranking at such low rpm and oil pressure is likely to damage all those new bearings and surfaces.

ENGINE INSTALLATION

In general terms, you need the same tools for engine installation that you did for engine removal: floor jack, plywood and jack stands. Lay out all pieces to be installed, such as the rear sheet metal, air filter, throttle-cable guide, intake air sensor (on fuel-injected models) and so on.

Lift Car—Raise the chassis high enough to get the engine underneath the rear bodywork. Support the car with jack stands and block the front wheels. If your jack stands are not tall enough, you can probably get the engine under the engine compartment by tilting it. Lift the muffler end until the fan housing will clear the body. It takes two people to tilt and push the engine under the car. Then you continue tilting the engine enough to get the floor jack under it. Don't forget to put the protective piece of plywood on the jack pad.

Lifting the engine and mating it to the transaxle are straightforward actions on all chassis except the 914. If you are working on a 914, go ahead and read the following section on getting the engine in the chassis. It gives pertinent background information. Then move ahead to the specific section on 914s, page 166.

Orient Driveplate—On Auto-Stick and cars with automatic transaxles, turn the driveplate and torque converter so their bolt holes line up. And make sure one of the bolt-hole pairs is in front of the access hole. Then when you've got the engine in, you can install the first driveplate bolt without having to hunt for it through that

At this point, stop lifting and make sure all hoses and wires are correctly routed. Check throttle cable, fuel line and electrical wiring. Remove wheel to increase working room and light for bottom work.

Type 3 lower bellhousing nuts are out in the open. Fully mate engine and transaxle before wrenching these tight. Plywood on floor jack saddle protects soft magnesium case and is more secure base than engine against jack.

little access hole. On the Type 2 and 3, the hole is in the bottom of the transaxle. On the Type 4, it's in the bellhousing flange of the engine case, covered with a plastic plug. Although this may be obvious, don't forget to remove the bracket you installed to hold the torque converter in place while working on the engine. It's attached to the lower bellhousing stud.

Install Engine—While guiding the engine on the jack, raise it into the car. This is a two person job, and a third is helpful, so don't try it alone. When the engine gets close to the transaxle, stop for a minute and feed the throttle cable through the front sheet metal. Also arrange any wiring and fuel lines through the front sheet metal.

Now continue raising the engine until the lower studs are ready to slide through their holes in the bellhousing. Push the engine forward until the engine has fully mated with its transaxle. That's easy for me to say, you mutter under your breath. Yes, there are a couple of points to watch for.

First, make sure the gap between engine and transaxle is constant from top to bottom. This is your gage of the engine's vertical tilt. Look at the edge of the sheet metal vs. the bodywork for horizontal cocking. Unfortunately, any mismatch between clutch and input-shaft splines is invisible. You must take care not to damage the gland nut needle bearing and avoid bending the main shaft. If the engine goes most of the way forward, but then stops about an inch short, the splines might not be mating. Select any gear with the shifter, then turn one of the rear wheels while your helper holds the other one. This

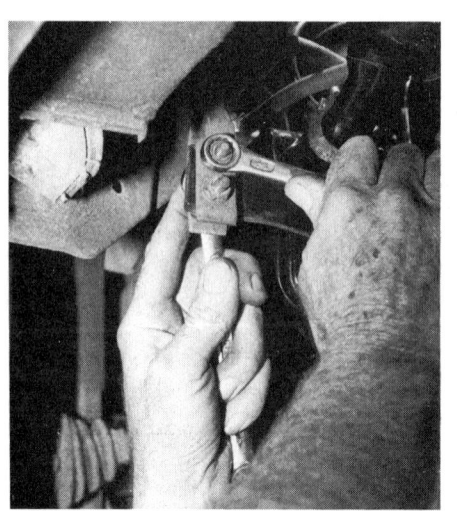

Bolt Type 2, 3 and 4 crossmembers to mounts and check engine centering in its compartment. Check engine centering by eyeballing gap between engine and compartment edge. If engine is cockeyed, straighten it by moving mounts. This shouldn't be necessary unless mounts were loosend earlier or car is out of line from body damage.

turns the input shaft, and should get the splines lined up. It really helps if a third person pushes the engine forward at the same time.

Another way to line up the splines is to put the transmission in first gear and set the parking brake. Put a wrench on the crankshaft-pulley or generator-pulley bolt. Then, if the splines don't line up, you can rotate the crankshaft via the

Pass solid end of heater cables through hole in heater-valve-arm clamp and tighten cinch bolt to hold cable. Head of cinch bolt faces ground when installed.

pulley bolt. This method is useful if you are short on help.

Getting the engine to mate with the transaxle seems to be one of those jobs that just demands a certain amount of grunting; then all parts match up in place. Keep at it, and don't get flustered. The engine will go in eventually. While jockeying the engine around, don't let the engine hang from the lower bellhousing studs. This will bend the studs and make installation almost impossible. Keep the engine supported.

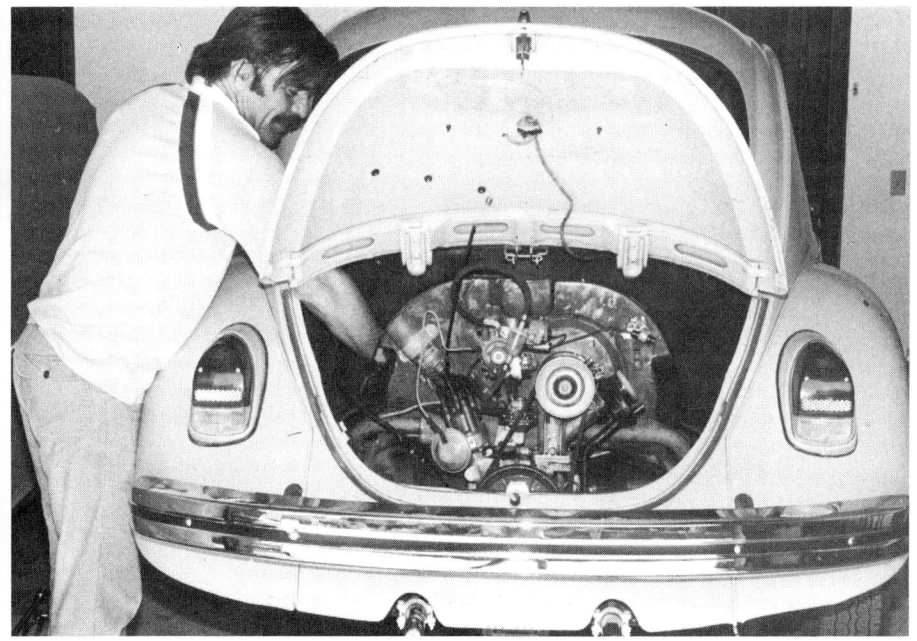

Tightening upper bellhousing bolts takes faith. A helper under car may be necessary to hold nut on other end.

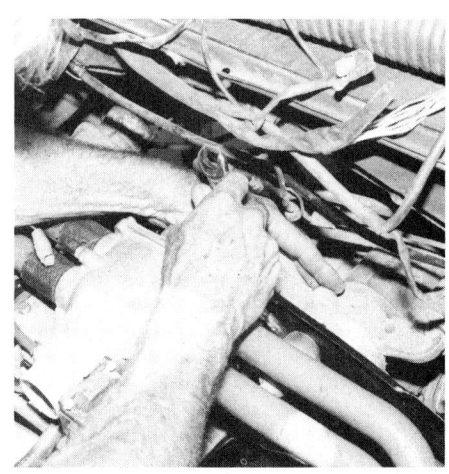

Install Type 3 upper bellhousing bolts from top while engine and transaxle hang low on floorjack. After all fuel and hose connections are made, raise engine into place.

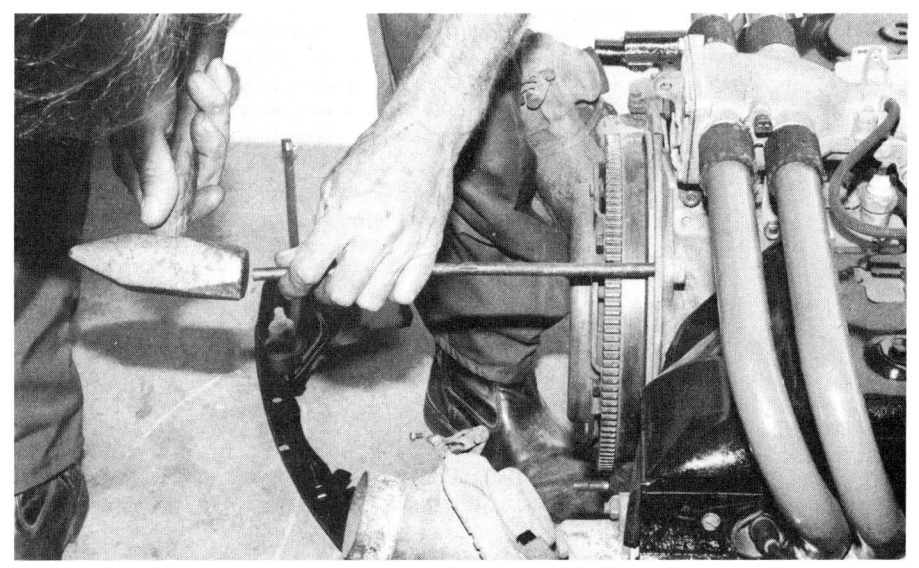

Threaded plug for left upper bellhousing bolt hole used in '71 and later Type 1 cases and universal replacement cases simply presses in. Tap it in or out as necessary.

If installing the engine behind an auto transmission, you've probably been chuckling. Because there is no input shaft to line up, the engine should slide right in.

Check the positioning of the engine compartment seal on the engine sheet metal. One lip should go under the sheet metal, another over it. Straighten out any folded or creased sections with pliers or hands.

Once the engine is in, double-check the stability of the jackstands, and go underneath and wrench-on the two lower 17mm bellhousing nuts and rear crossmember, if your chassis has one. Bellhousing nuts torque to 22 ft-lb (3.0 kg-m). Crossmember bolts torque to 18 ft-lb (2.5 kg-m). Crossmembers on '72 and later Buses require a different order of operations. Go to the Late Bus section, page 162, now. The 411/412 installation is different too, so go to the Type 4 section, page 163.

While underneath, connect the fuel line and heater cables. You may also be needed below to fit and secure the upper bellhousing bolts.

The fuel line is a slip connection. Ensure that you get 1-1/2 in. of mating between hose and metal pipe and fit hose clamps to these connections. You may have already made this connection earlier to stop fuel from leaking, but double-check it anyway.

Get the heater cables installed by removing the cylinder and cinch bolt from their ends. Feed the cylinder into the control-valve arm, thread the cable end and tighten the cinch bolt. When correctly oriented, the cinch bolt faces downward.

Connect the flexible heater ducting to the heat exchangers. Make sure the clamps are tight and the lock tabs are bent down.

On the Type 3, connect the two small fresh-air ducts at the right and left edges of the engine. On '72 Type 3s, connect the EGR wire to its transmission switch.

Upper Bellhousing Bolts—On '71 and later cars, install the left upper bellhousing bolt from below the car. The right side uses a nut and bolt combination. Fasten it like both nuts on pre-'71 cars, from inside the engine compartment. This can be a real chore, because the bolts tend to turn with the nut. If you have a helper, one of you can hold the bolt while the other turns the nut. If you are alone, try putting a wrench on the bolt head and letting it turn against the bellhousing. Another trick is to push some putty against the bolt heads so they'll stick. Sounds hokey, but it works. Window sealing putty works great.

Another tactic on Type 3s is to make all four bellhousing connections while the engine and

159

Guide rear panel into position, then screw it in place. Watch engine compartment seal. One lip fits under panel, other lip fits over it.

Smaller panel fits around crankshaft pulley. Lay it on panel just installed, and screw it down. You might have to remove screws installed during engine assembly to secure this piece.

Throttle cable attaches like heater cable. Use pliers to grip throttle arm and cinch bolt while tightening cinch bolt. Don't rely on cable tension to tighten bolt against, or you'll kink cable.

Labeling wires really pays off during installation. Lay wiring loom behind carburetor, letting wires fall in place. Make all connections, double-check them, and remove tape flags.

transaxle are tilted down. Then you can get the upper bellhousing bolts from the top.

Auto-Sticks & Automatics—Auto-Stick cars before '71 use two nuts on studs. Too bad they aren't all so easy. Also on Auto-Stick cars, install the four torque converter-to-driveplate bolts. Use a 12-point socket on these. Access is through the small rectangular hole in the bottom center of the bellhousing on fuel-injected engines. With carbureted ones, the hole is larger and on the side of the bellhousing. These bolts are torqued to 25 ft-lb (3.0 kg-m). Remember to unplug and reconnect the two ATF lines at each side of the engine on Auto-Stick cars.

Automatic transmissions have only three driveplate bolts. Install them now and torque to 18 ft-lb (2.5 kg-m). Connect the vacuum hose and kickdown-switch wire as well.

From this point on, I cover the specifics of engine installation separately for each chassis, so go to the section you need.

TYPE 1

That's it for all the underneath work, so slide out and dust yourself off. Get the floor jack and stands out of your way and lower the car.

Control Valve—More extra work awaits Auto-Stick owners topside. Reconnect the wiring and vacuum connections on the control valve at the left side of the engine. Use your disassembly numbering system to get the connections correct.

Rear Cover Plate—Guide the rear cover plate into position. On engines with emission controls, include the air-duct grommets. Also put in the crank-pulley cover plate and the heat-riser cover plates. After all three are in position, install all screws. Only after all screws are started should you tighten them. Fuel-injected engines don't have a pulley cover plate, so don't wonder where it went.

Throttle Cable—Pass the throttle cable through the fan housing and pull out all the slack. Insert the throttle cable guide over the cable and into the fan housing. Finally, attach the cable to the throttle lever on the carburetor. Insert the metal cable end into the throttle lever's cylinder and pull out the slack. Pull on the cable, hold the throttle lever in the closed position, then tighten the cinch bolt or screw.

Wiring—Bring the wiring harness around the top of the fan housing, letting the wires drape over the engine. Reconnect the wires to the

Remember to install fuse inside fuse holder before joining two halves. If you forget, you won't be the first to do so. More wires are around corner of fan housing, so don't forget them.

Cable for pre-heated intake air is connected with filter housing turned out so you can see it.

oil-pressure sending unit, coil, carburetor, generator, voltage regulator and backup-light fuse holder. Don't forget to put the fuse in the holder before joining the two halves. You may have another wire that runs in front of the fan housing, joining into a multiplug connector, near the backup-light wires. Check for the connector at the right front fan housing. Follow your disassembly numbering system when connecting wires for accurate unions.

Fuel-Injection Wiring—Now reconnect the injection wiring harness. Using your numbering system, connect the leads at the coil, injectors, TDC sensor, temperature sensor in the crankcase and the intake-air distributor.

To complete the injection connections, install the intake-air sensor and air filter. Start with the large rubber intake boot. Install it at the intake-air distributor. Then fit the air sensor into position, installing the upper end of the rubber boot and the two mounting bolts. Tighten the boot clamp and two bolts. Now you can install the wiring plug into the air sensor's side. Take care not to bend the wiring during installation or while pushing on the protective rubber boot. Finish the installation with the air filter. Use a new filter element, align the housing cover and snap the four clips in place.

Air Hoses—On Clean Air engines, install the two large air hoses between the fan housing and heat-exchanger fittings protruding from the rear cover plate. You may have already installed these, but I like to leave them until last so they don't get in the way. If they are torn, replace them, or at least patch them with duct tape.

Air Filter—On non-fuel-injected engines, the air filter must still be installed. To avoid oily messes, remember to keep all oil-bath filters level. Tighten the mounting clamp first, then connect the hot-air control-flap cable at the flap lever. Install the hot-air hose to the filter-housing snout. This is the flexible hose that passes through the rear engine cover plate. Slip on all vacuum hoses and use your numbering system to guarantee all are attached.

On filters using paper elements, install the bottom half first, then the element and upper half of the housing. Close the four clips.

TYPE 2

From here, I've separated '71 and earlier, and '72 and later installation into early and late sections to avoid confusion. Go straight to the appropriate section.

EARLY BUS (PRE-'72)

Installing these engines is about the same job as in a Beetle, so read the previous Type 1 section for details and additional help.

Lines, Cables & Ducts—From underneath, you should slip on the fuel line, heater ducting and heater control cables, and passed the throttle cable through the front sheet metal. Have a helper thread the cable through the fan housing at the same time. Attach the starter-solenoid connections, too.

Rear Cover Plate—That's it for the work underneath, so crawl out and install the rear cover plate. It uses a total of ten screws. Six are in the left and right forward corners of the plate. The other four are in the rear corners, mounted vertically on the plate's rear face.

Vanagon crossmembers have arrows on mounting pads. Arrows should point to front of vehicle.

Wiring—Get the throttle cable through the fan housing and install the cable guide. Attach the cable at the carburetor. Using your numbering system, connect the wires at the distributor, coil, generator, oil-pressure switch, carburetor and backup lights. Don't forget to join the wire running in front of the fan housing on some engines.

Air Filter—Install the air filter. Up to '68 models, just clamp the filter in place and connect the hot-air and breather hoses. From '68—70, there's the addition of the hot-air flap cable. The '71 has no cable.

161

LATE BUS (POST-'72)

Bellhousing & Crossmember—Once the engine and transaxle are together, loosely install the lower bellhousing nuts. Then install all rear-crossmember hardware. If a Vanagon crossmember doesn't seem to fit, you probably have it on backward. Check the arrows on the crossmember mounting pads, the part that goes against the chassis mount. The arrows must point forward, toward the engine. With the crossmember installed, the jack and transaxle support can be removed.

Now tighten the lower bellhousing nuts and install the upper bellhousing bolts. You'll need a helper again to turn the nuts from the engine compartment side while you hold the bolt heads from below. Torque these fasteners to 22 ft-lb (3.0 kg-m).

Look at the crossmember from directly below it. Sight along the gap between the crossmember and fan housing. This gap must be equal from side to side. If it is, tighten the crossmember bolts to 14 ft-lb (2.0 kg-m). If not, fiddle with the crossmember brackets at the chassis connection until the gap is equal. Then tighten the crossmember hardware.

Engine Compartment Seal—Before going any further, have a helper lean into the engine compartment and check the rubber seal that routes all the way around the engine compartment. It should have one lip over the engine sheet-metal panels, and another below it. Push out any kinks or creased sections.

Lines, Cables & Ducts—Connect the fuel lines in the area forward and right of the engine. On carbureted engines, the lines connect at the fuel pump; on injected engines, the connection is at the pressure regulator. Secure the connections thoroughly; you don't want any fuel leaks. Thread the throttle cable through its hole in the front engine sheet metal and push it in as far as it will go. Finish off by reconnecting the heater control cables and ducting. That's it for the bottom work, except for the rear gravel guard. Four bolts and it's back on.

Engine Compartment—Up in the engine compartment, install the rear cover plates. The right plate goes on first, then the left. Install the oil filler tube and bellows, then install the oil dipstick.

Torque Converter—If you have an automatic transmission, bolt the torque converter and driveplate together through the small hole in the bellhousing flange. These bolts are torqued to 14 ft-lb (2.0 kg-m). You must rotate the crankshaft to bring each bolt hole into view. If there's a guard over the fan, remove it. Then you can rotate the crank by grasping the fan. This is much easier work if the sparkplugs are out. When done, install the plastic plug over the hole in the bellhousing flange.

More Hookups & Ducts—Also on '74 automatics and all '72 and '73 engines, make all the vacuum and electrical connections at the

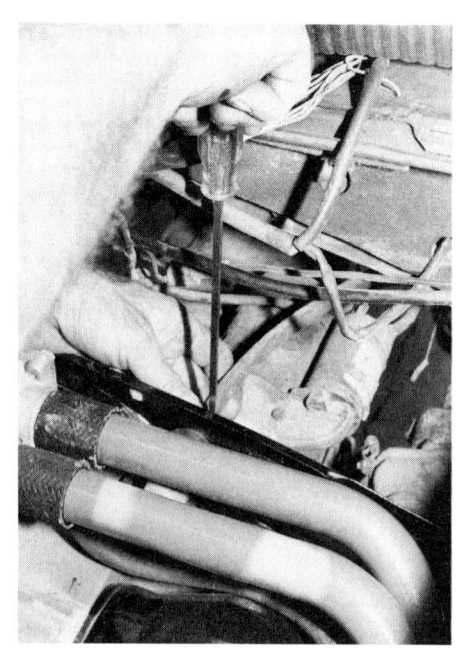

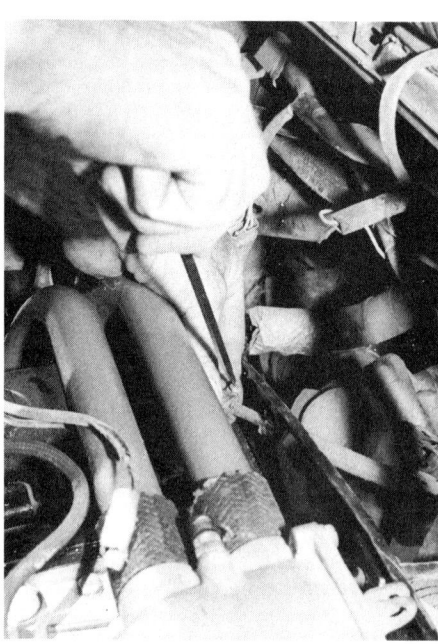

Connect Type 3 fuel lines at both sides of engine. Use new hose and clamps! Engine and transaxle are hanging down slightly in these photos.

vacuum-advance cutoff valve mounted near the blower motor. On all Buses, connect the blower-motor ducting. This step was done during engine assembly on Vanagons.

On the right ceiling of the engine bay is a charcoal filter. Connect its hose to the engine. There's also a hose to connect on the left carburetor, if those are installed.

The coil may need installation, and the backup-light fuse holder near the coil needs to be connected. If there's a temperature sensor mounted in the upper right side of the engine compartment, connect its wire.

Connect the throttle cable at the crossbar or throttle-body section of the intake-air distributor. A vacuum line from the automatic transmission can be joined to the air distributor, too.

Wiring Harness—Following your numbering system, reconnect the electrical leads at the distributor, alternator, regulator, oil-pressure switch, fuel injectors or carburetors as necessary. Placing the wiring harness across the engine and letting the wires drape where they want helps sort this procedure out.

Air Filter—The air filter is next. Set the bottom half across the carburetors and clip it on. Install the breather and hot-air hoses, then lay on the top half and secure its clips. Be careful not to tip the top half or you'll spill oil. In '73 and '74, paper filters were used, so you can tip them all you want.

On fuel-injected engines, the filter housing and intake-air sensor are mounted as a unit. They clamp to the large S shaped hose. Once the assembly is in place, connect the electric plug and carefully install its protective rubber boot. Then you can install the filter and clip on the cover. Reconnect any hoses to the cover.

TYPE 3

Fuel Lines—All bottom work should be complete, so start in on the engine compartment chores. Connect the fuel lines. Look in both forward corners of the engine compartment. If the right side is already together, double-check that you joined it to the line underneath the car. The line on this side is the return line. If it is left disconnected, the engine will start and run, but fuel will pour under the car. This could lead to a total disaster, so guarantee that the lines are connected.

Engine Mount—In the engine compartment, install the upper-rear engine mount on those chassis without a crossmember. This mount is often left off, apparently without causing any problems, so your car may have been missing this piece all along.

Throttle Cable—Connect the throttle cable. With a single carburetor, this is done at the carburetor. On dual-carbureted engines, make the connection at the throttle-linkage crossbar. Fuel-injection throttle linkage is connected at the intake-air distributor. If you removed the idle cutoff solenoids from your dual carburetors, reinstall them and their wires.

Electrical & Vacuum Connections—Using your numbering system, connect all electrical and vacuum leads. On a carbureted engine, these will be at the carburetors, oil-pressure switch, generator and coil. On a fuel-injected engine, lay the wiring harness across the engine. It will go to the: distributor, injectors, crankcase sensors and grounds, cylinder head temperature sensor, and intake-air distributor. Also connect the vacuum hose at the fuel-pressure sensor on the left wall of the engine compartment.

Cooling-Air Bellows—Clamp on the oil-filler tube and install the dipstick. The cooling-air bellows needs to be installed and clamped in place. Make sure it is in good condition, without rips, tears and holes. If the bellows isn't intact, warm air from below the engine will get sucked up by the fan and the engine *will overheat*. If nothing else, repair the bellows with duct tape until you can buy a new one.

Air Filter—Install a new air filter. On carbureted engines, set on the filter housing, close any clamps, hook up the air-supply and breather hoses and attach the wing nut or nuts. Snap the dual-carburetor throttle linkage onto the carburetors and center bellcrank.

Air filters on fuel-injected engines are even easier. Set the unit in place and install the wing nut. Hook up the intake elbow and breather hoses. And make sure the lid covering the generator belt is snapped in place.

TYPE 4

Install Engine—This installation is different because you probably have the transaxle crossmember loose. If so, get a jack under the transaxle and lower the bellhousing end slightly. If the crossmember isn't loose, the engine can still be installed, but it's more difficult.

Raise the engine into the chassis, watching for snags. The injectors and fuel hoses are real fish hooks during installation, so keep an eye on them. A helper is a must for this. As the engine joins the transaxle, have your helper install the upper bellhousing hardware.

Now raise the engine and transaxle combination into position and attach the transaxle and engine crossmembers. The transaxle crossmember may misalign on the way up and cause a blockage. Loosening all connections at the body will help them slide together. Your helper must get inside the car to hold the transmission crossmember bolts while you secure them from underneath. Torque to 18 ft-lb (2.5 kg-m).

WARNING: Don't get directly underneath the engine/transmission package until both crossmembers are installed.

Do this attachment work from the side instead. The engine crossmember is installed completely from underneath, without any help from above. After the crossmembers are in, remove the jacks and install the lower bellhousing nuts. Torque these to 22 ft-lb (3.0 kg-m).

Lines, Cables & Ducts—Connect the heater control cables, ducts and fuel lines. Push the throttle cable through the front engine sheet

Not much of Type 3 throttle cable shows until it reaches throttle arm. Lightly pull slack from cable before installing; drawing cable tight with pliers will only stretch cable. Don't forget throttle cable return spring.

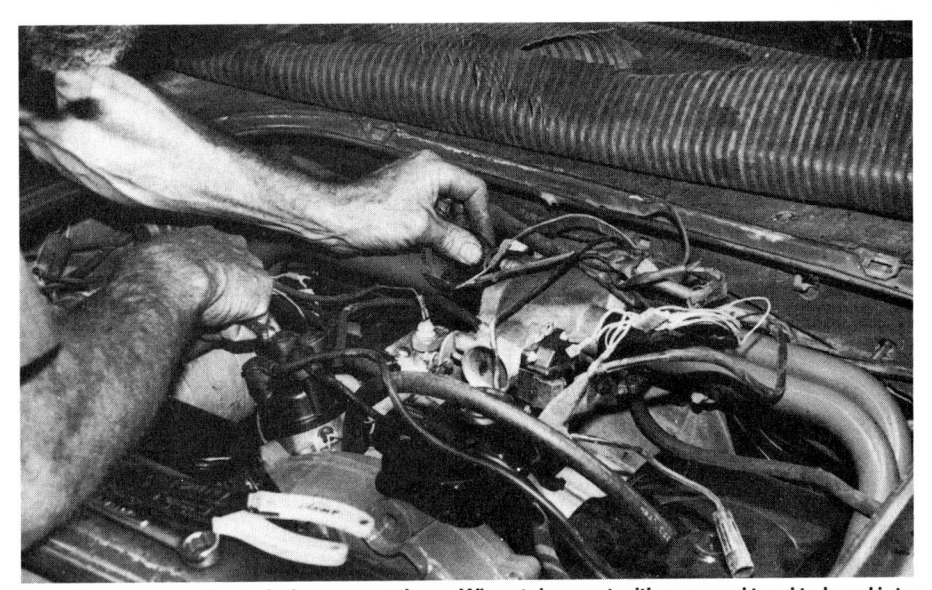

Lay wires over engine to help connect them. Wires take a set with age, and tend to bend into place when laid out. Still, there's no substitute for labeling all disconnections during engine removal.

metal. You can also attach the rear splash pan, which is the slotted body panel that mates to the rear of the body.

Manual Transaxle—From on top, reinstall the transaxle driveshaft on manual-transaxle cars. Push it until the splines meet, then rotate and push the shaft so the splines mesh. Reinstall the circlip and nut. Tighten the nut snug. Then install the cover and rear seat cushion.

Automatic Transaxle—Install the driveplate-to-torque converter bolts through the hole in the left side of the bellhousing flange. This is done from inside the engine compartment. Use the engine fan to rotate the bolt holes into position.

When you are finished, install the plastic plug in the bellhousing-flange hole.

Fuel-Injection Harness—Using your numbering system, reconnect all injection wires and hoses. Start by draping the wiring harness over the engine, then connect the system at the intake-air distributor, ignition distributor, injectors, case ground, plus the temperature sensors at the case and heads.

Electrics & Throttle—Also connect the multiwire connector at the voltage regulator in the right front side of the engine compartment. A mirror is helpful here. Pull the throttle cable through the front sheet metal and connect it at

163

D-JETRONIC

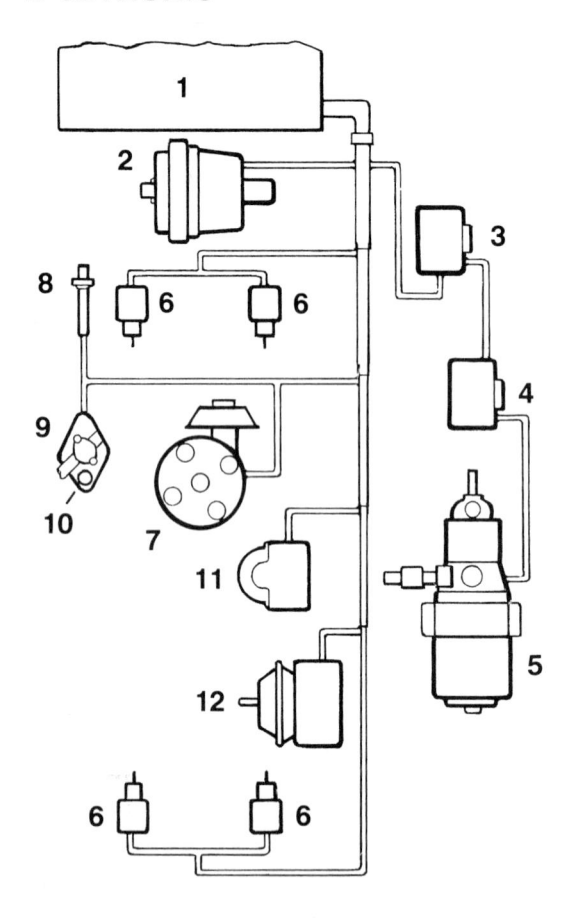

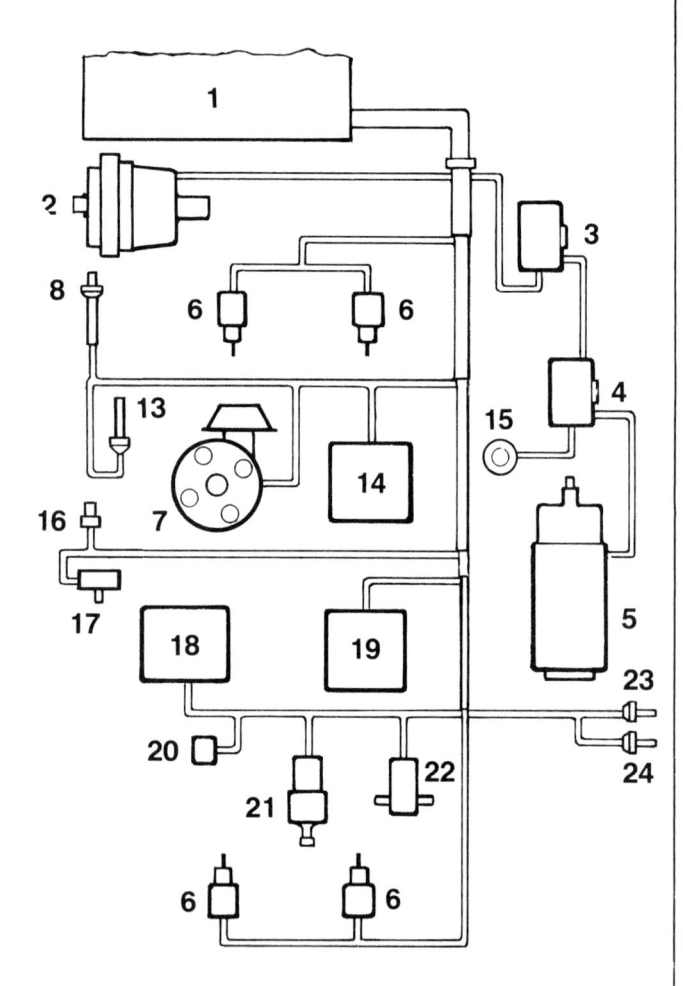

(1) Electronic Control Unit (ECU)
(2) Pressure sensor
(3) Main relay
(4) Fuel pump relay
(5) Electric fuel pump
(6) Fuel injector
(7) Distributor w/triggering contacts
(8) Temperature sensor II in cylinder head
(9) Auxiliary-air regulator
(10) Temperature sensor I in crankcase
(11) Throttle valve switch
(12) Pressure switch
(13) Temperature sensor I in intake manifold
(14) Engine speed relay, 411E, ('72 only)
(15) Auxiliary-air device, electrically heated, 1600 automatic
(16) Thermo-time switch
(17) Cold-start valve
(18) EGR switch, 1600 automatic
(19) Throttle valve switch
(20) 2-way valve, disconnects ignition vacuum advance ('72 only)
(21) EGR valve, 1600 automatic
(22) Solenoid-operated air valve 1600 & 411E automatic ('72 only)
(23) Oil-pressure switch in automatic transmission ('72 only)
(24) Thermo-switch for EGR

Basic fuel injection schematics for '67—July '69 Type 3 (left) and August '69—73 Type 3, '71—73 Type 4, and '70—73 1.7 and '73—75 2.0 liter 914 (right). Rely on your labeling for specific connections on your engine.

L-JETRONIC

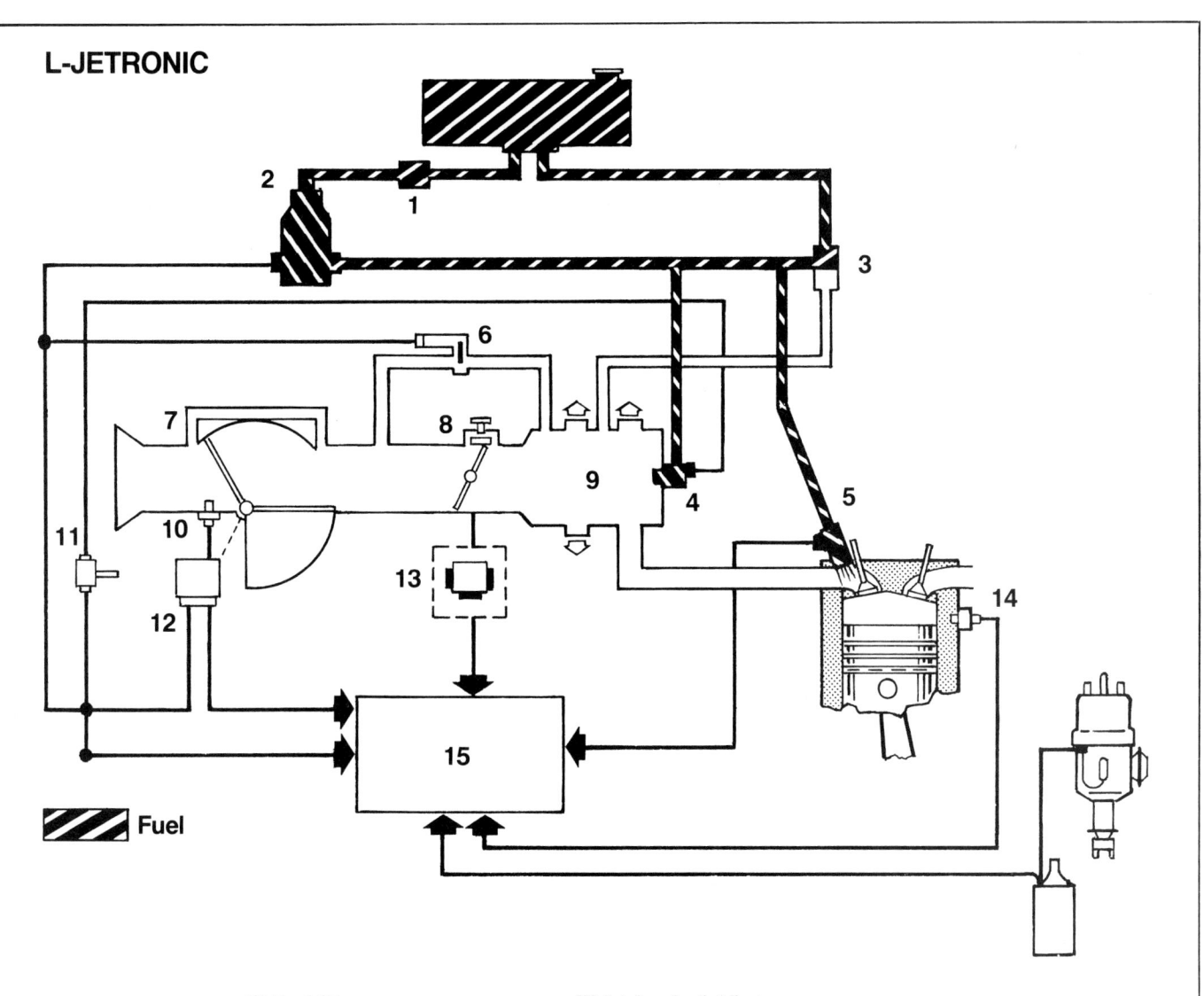

////// Fuel

(1) Fuel filter
(2) Electric fuel pump
(3) Fuel pressure regulator
(4) Cold-start valve
(5) Fuel injector
(6) Auxiliary air regulator
(7) Intake airflow sensor
(8) Throttle valve housing
(9) Intake-air distributor
(10) Temperature Sensor I ('76 and later)
(11) Thermo-time switch
(12) Potentiometer/fuel pump switch
(13) Throttle valve switch (except '78 Calif.)
(14) Temperature Sensor II
(15) Electronic Control Unit (ECU)

Basic fuel injection schematic for '75—79 Type 1, '74—79 Type 2, '80—83 Vanagon, '74 412, and '74—75 1.8 liter 914. Rely on your labeling for specific connections on your engine.

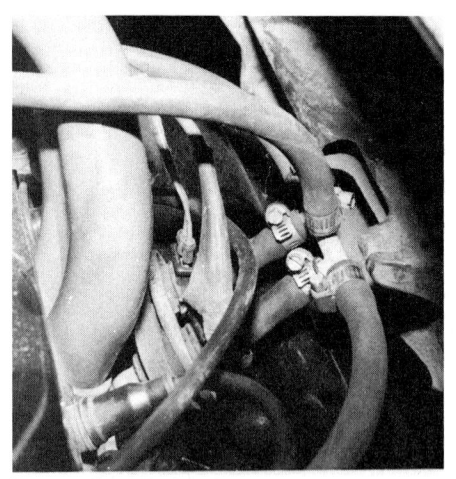

Except for missing boot, installation looks good. Fuel lines are new and clamped, spark-plug boots are fresh and tight.

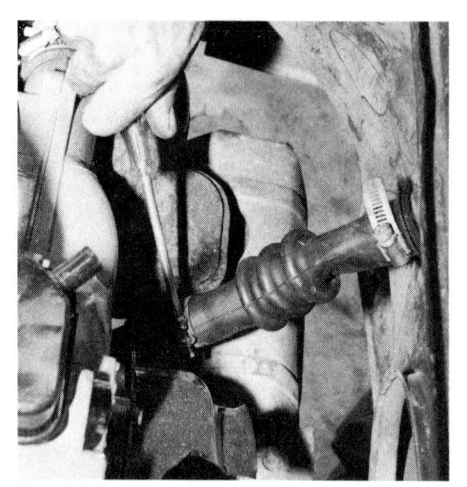

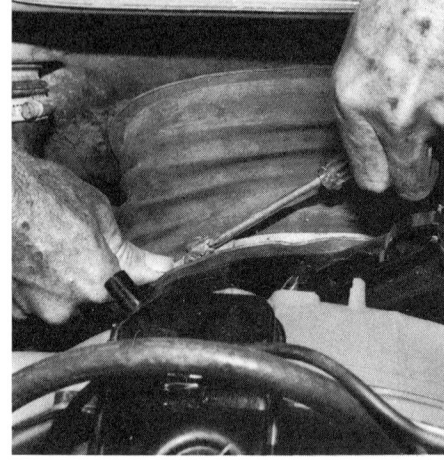

Dipstick bellows (left) and cooling air intake (right) go on with little bother. Check both for cracks or perforations, and repair or replace as necessary. A dirt-tight dipstick bellows is essential because it's a direct path for dirt into engine if open. Blowing grit through fan and onto engine isn't smart either.

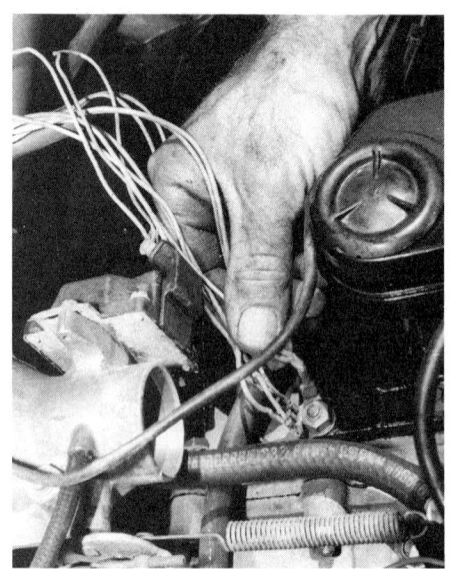

Multiple ground connection on breather pedestal nut is mounting point for this bundle of ground wires. A few tie wraps will dress up this installation.

Fuel-injected Type 3 air cleaner sets on pad at right. Tighten center wingnut and all hoses.

the front of the intake-air distributor. Make sure the throttle return spring is in place. Connect the leads to the oil-pressure sending unit and ignition coil. The coil may still be off the engine, install it now.

Air, Oil & Heat—Install the cooling-air bellows at the extreme rear of the engine. Install any heater blower ducting and install the oil-filler tube and dipstick. Install the air filter, and hook up the breather hoses.

914

Replacing 914 engines requires a different plan. I recommend installing the engine first, then the transaxle. You can bolt the two together on the floor, then install them together, but it sure makes for a heavy, unwieldy package. If you have a concrete floor and lots of helpers, go ahead and give that method a try. The balance point will be directly under the clutch.

If you choose to install the engine separately, the balance point is under the oil strainer.

Transaxle Preparation—Check for transmission fluid in the bellhousing area. This is a sign of a leaking transaxle seal. Also check the condition of the release bearing and release-bearing-arm pivot.

Check the release bearing by rotating it with your fingers. Noise or roughness calls for replacement. To remove it, unthread the Allen-

166

head bolt in the release-bearing arm. This will free a small nut and clip piece behind the arm, so be ready to catch it. Now the bearing and arm can be removed.

On the backside of the arm is a plastic cup. This cup can wear through and break, so clutch adjustment is upset to the point where under some conditions you can't adjust the clutch at all. So, replace the plastic now. Prying the old piece out is the hardest part. Icepicks and screwdrivers will do the job, but be careful. Drive the new one in with a hammer.

Under the release arm is a guide sleeve. Remove the guide sleeve bolts and the sleeve to expose the transmission oil seal. If it is leaking, pry the seal out with a screwdriver. Don't gouge the seal's bore. Coat the new seal OD with Permatex 3H and the ID with oil. Install it with its lip facing in. A drift or thick-walled tube is needed to transfer the hammer blows to the seal. Tap the seal lightly, and drive it straight into its bore.

Once the seal bottoms in its bore, replace the guide sleeve, release arm and bearing. Dab some grease (not too much) on the release-arm pivot.

Raise Car—Loosen the rear-wheel lugnuts, then lift the car and support it on jack stands. Test the stands by shaking the car. Adjust the stands until you are certain they are secure. Then remove the rear wheels for extra light and room.

Maneuvering the engine into position is straightforward, but isolate the greasy CV joints. You covered them with plastic bags during engine removal, didn't you? If you neglected this step, do yourself a favor and cover them now. Hold them to the axles with rubber bands. If you don't cover the joints, you and your engine are going to get covered in grease.

Raise Engine—When raising the engine, or engine and transmission, try to raise it straight up. Tight spots are at the suspension's trailing arms and the fuel lines to the injectors. Careful jockeying will get the engine past the suspension. The fuel lines will escape unscathed if you just keep an eye on them. The danger is they will get caught by the bodywork just as the engine fits into position. The lines are then bent back and pinched.

Place one person at the engine compartment, looking down on the engine, and another raising it with the jack. The person at the engine compartment can then guide the fuel lines with a screwdriver or stick. *Under no conditions stick your hand or fingers on or around the engine while it is going in.* That's a sure way to get them pinched, cut or crushed.

Front Crossmember—Once the engine is positioned, install the front crossmember.

WARNING: Don't get directly under the engine. Work from each side of the car.

This crossmember is heavy, so have a helper kneel down on one side and support the crossmember while you support and guide from the other side. Place the crossmember on the two engine-mount studs and attach the washers and nuts. Don't wrench-tighten them yet. Then thread the two large bolts at each end of the crossmember upward into the chassis. Wrench-tighten the two large bolts with a breaker bar. Then tighten the engine-mount studs to 14 ft-lb (2.0 kg-m)

Raise Transaxle—Support the rear of the engine with wooden blocks, then remove the floor jack. Mount the transaxle on the floor jack and slide it into position. Raise the transaxle and mate it to the engine. Install the lower bellhousing nuts with 22 ft-lb (3.0 kg-m) of torque. Then connect the transaxle mounts at the rear of the unit. Now you can remove the floor jack and blocks from under the engine.

Upper Bellhousing Bolts—With the jack out of the way, you can install the upper bellhousing bolts with 18 ft-lb (2.5 kg-m) of torque. The left bolt doubles as the upper starter fastener. You'll need a person on each side of the upper bolts to keep them from turning while installing the nuts. That means one person leaning into the engine compartment and another underneath the car.

Also make all electrical connections to the starter while you are there.

Transaxle Connections—After the biggest pieces are in and secure, I attach the transaxle. The very first connection is the axles. Use new gaskets between the CV joints and the transaxle flanges. Have a helper step on the brakes, or hold the axle by threading a bar through two partially installed lug nuts.

Once the axles are bolted-up, you won't be pushing them out of the way or dragging your hair through the grease. Apply Loctite and tighten the four CV-joint bolts in a criss-cross pattern to 28—30 ft-lb (3.8—4 kg-m).

Connect these next: shift linkage, ground strap, backup-light wiring, speedometer and clutch cables. The shift-linkage bar must be threaded through the crossmember, then connected front and rear to slip couplings. Don't tighten the Allen-head lock bolts until both ends are installed. Cover the front joint with its rubber boot and the rear joint with the plastic cover. Carefully rewind the sardine-can type band around the plastic cover.

Clean the ground-strap mating surfaces to guarantee a good ground. Everyone seems to forget to install the ground strap until the stop lights won't work and the running lights flash off and on. If you have electrical problems later, make sure you connected the ground strap.

The backup-light wires just plug into the backup-light sending unit on the left side of the transaxle.

The speedometer and clutch cables are banded together. Install the speedo cable first, and it will help support the clutch cable as you work with it. The speedo cable nut should be finger-tight only. Don't use a wrench on it.

If you haven't cleaned the metal plate the

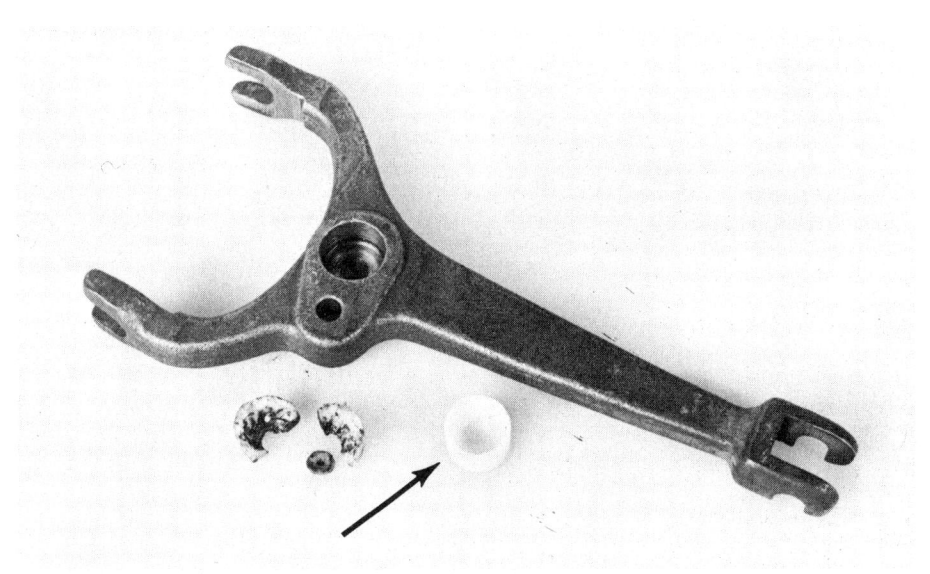

914 release-bearing arm pivots on plastic cup. When bottom of cup breaks through, it can be impossible to set correct clutch adjustment with thick, new clutch. Pry out old cup and tap in new (arrow).

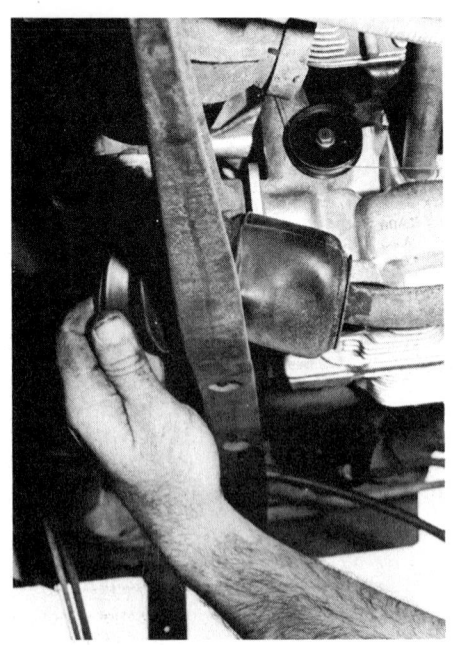

Push/pull front shift-rod boot through hole in crossmember. Get this end of the shift rod installed . . .

. . . then move to rear and connect transaxle end. When finished, pop plastic cover back in place and tighten sardine-can clamp.

Throttle cable runs across intake-air distributor and cable housing anchors with two nuts. Throttle arm connection has usual cinch bolt.

clutch pivot is on, do so now. Wrap the cable around the plastic pivot wheel, grease the wheel's ID metal sleeve and slide the assembly onto the metal plate. Apply Loctite and install and tighten the retaining screw. At the release arm, lay the cable's block into the release arm's ears, then run down the adjusting nut. Adjust the clutch so there is 1 in. of *freeplay* at the clutch pedal. Freeplay is the distance you can press the clutch before meeting any resistance. Always measure freeplay by hand, never by foot. Your foot just isn't as accurate.

Make the freeplay adjustment back at the transaxle underneath the car. Hold the clutch cable stationary with pliers while turning the adjusting nut. Facing the rear of the car, turn the nut clockwise to reduce freeplay.

The throttle cable has probably been in your way during all this work. Solve this problem by sticking it straight up through its grommet on the right-side (passenger-side) engine sheet metal.

Exhaust—Now wiggle forward to the engine and attach the exhaust system. Use new copper sealing washers from the engine gasket set. Hang one heat exchanger and start as many nuts as you easily can. Then do your best to get the remaining nuts on. It will take your best efforts to get some of these nuts tight to 14 ft-lb (2.0 kg-m). Use sockets and wrenches to do this job. It might be worth getting an inexpensive wrench and bending it to fit past the pipes.

Heating Controls—Once both heat exchangers are fitted, install the heater control cables, warm air guides and associated ducting. Start with the ducting. It runs between the fan housing where the little flap doors are and the heat exchangers. Screw it in place at the tabs. Then attach the heater control valves. The fabric ducting between the control valve and bodywork has probably come loose, so reclamp it to the valve is necessary. Install the heater control cable if you removed it.

Now you can screw on the warm-air guides. You'll probably have to remove a screw or two from the metal ducting as some screws do double duty. By now, you're probably lightheaded from working upside down on your back, so crawl out from underneath and start the topside work.

Lines, Cables & Wires—Now you can't say I didn't warn you at disassembly, but here is the final test of the numbering method you used: reconnect all hoses and wires.

Start with the throttle cable. It runs across the top of the engine to the throttle at the left rear. The throttle-cable housing fits into a notch in the top of the intake-air distributor where it is secured by two clamping nuts. The cable attaches to the throttle with the usual cylinder and cinch bolt. Hold some light tension on the cable to remove all slack while tightening the bolt.

Once the throttle cable is on, attach the air-filter support. On most 914s this three-legged bracket covers the intake-air distributor and screws to the center of the case way down there past all those hoses and wires. There's nothing to do except stand on your head and install the screws. A magnet or screwstarter are real aids on this job.

Left-side mounted supports should be on already.

ECU—If you are feeling like a surgeon, what with your meticulous care when putting in those tiny screws, then great. You get to install a brain next. The ECU mounts in front of the battery. First attach the long and narrow electrical plug to one end of it. This plug is secured by screws, so don't forget them. Then slide on the plastic cover and install the wiring-harness clamp. Slide the assembly into the space between the battery box and bodywork, then attach the mounting hardware.

Fuel Lines—These are attached at the right side of the engine compartment. There are slip-on connections there, most likely at the pressure regulator or sensors. Install hose clamps at these slip-on connections. Several vacuum lines attach in this general area as well. Follow your numbering system.

Relay Board—On the other side, attach the wiring to the *relay board*. This is the platform with the domed plastic cover that is used for mounting the voltage regulator. Remove the knurled center screw and cover, then attach the

various electrical plugs. Reattach the cover.
Air Filter—Attach the large flexible heater ducts to the blower motor. Take one last look for loose hoses or wires, then install the air filter. Use a new filter element, and make sure the two filter-housing halves fit together.

FIRST START!

I'll bet you're eager to start your new engine, but there are a few more checks you must finish first. If you installed a new clutch, check clutch freeplay. Do this by depressing the clutch pedal by hand. It should travel 3/8—3/4 in. (10—20mm) before encountering resistance.

Change freeplay by turning the wingnut on the end of the clutch cable at the release-bearing arm. Facing the nut, use pliers to keep the clutch cable from rotating while you turn the wingnut clockwise to decrease freeplay. An assistant here saves a lot of effort; you don't have to keep crawling under and out from under the car.

Check for throttle return-spring attachment. Manually check this spring's action: does it quickly return the throttle to it's rest?

Lower the car before starting the engine, and clear all tools and materials away from the chassis.

Build Oil Pressure—Before starting your engine, you need to build its oil pressure, i.e., prime it before actually starting. This prevents high initial wear from metal-to-metal contact between close-fitting engine parts before oil pressure is sufficient. Of course, you must have oil in the crankcase to build oil pressure. Fill Type 1—3 engines with 2.65 qts, and Type 4s use 3.7 qts.

Build oil pressure by using the starter to crank the engine over. This is an accepted priming method because the engine's oil galleys are short, the cam gets plenty of lubrication by *oil splash* and the oil pump primes quickly.

First, install the battery. Clean the terminals and cable ends while you've got them apart. To keep the engine from starting while cranking it, disable the ignition. On breaker-point ignitions, disconnect the coil's negative terminal. On breakerless systems, disconnect the coil high-tension lead from the distributor cap and ground it. This lead is the heavily insulated wire in the cap's center. Don't disconnect the coil negative wire, or disconnect any ignition wire while the engine is running, or you could damage the ignition system.

Now turn on the ignition and crank the engine until the oil-pressure warning light on the dashbord goes off. This should take 30 seconds to one minute. Continue cranking for 15 seconds after the light has gone out to get oil to all parts of the engine. Warning lights are set at notoriously low values, typically around 6—8 psi, so just because the light has gone out doesn't mean oil has reached all parts of the engine.

Relay board on left engine compartment wall accepts multi-pin connectors from fuel injection and alternator. Because of connectors' shape, it's impossible to get them mixed up.

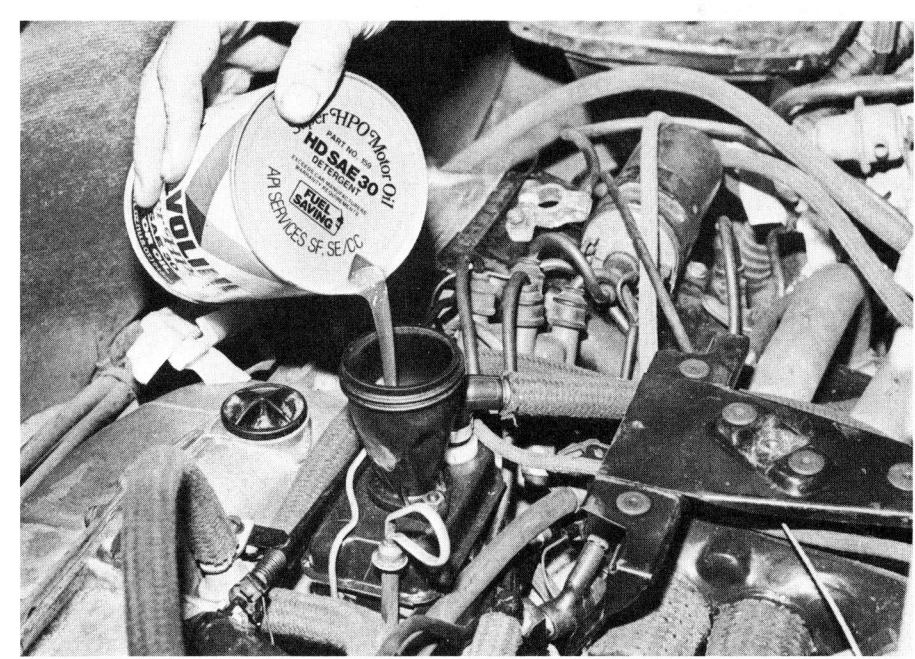

Fill engine with oil. Type 1—3 engines take 2.65 quarts, Type 4 engines use 3.7 quarts with new filter.

If the oil-pressure light doesn't extinguish, STOP! Determine the reason before continuing. Did you connect the oil-pressure sender? Is the sender working?

Start Engine—Once the engine is primed, reconnect the ignition and start the engine. Immediately raise the idle to 2000 rpm and leave it there. Don't rev the engine or let it idle at normal speed. Just set it to run at 2000 rpm for 20 minutes. You'll think the engine is making a lot of noise and that this treatment can't possibly be good for a new engine, but it is. It raises oil pressure and volume so all those new meshing parts in the engine are sure to get enough lubrication. This is especially true of the cam and lifters. The extremely high loads between cam and lifter demand plenty of oil during break-in.

You'll also think 20 minutes is forever when running-in the engine, but it really isn't. That's about what it takes to get the cam and lifters through their first critical break-in period. Use a clock when running-in or you'll stop at 10 minutes and swear it was 20 or more.

During this run-in period, you really should have a helper. Then one of you can start the engine, and the other turn in the idle screw until it holds 2000 rpm. Check the engine for leaks, especially at the fuel connections. On most Type 4 engines, you can see the cooling-flap arm where the cable from the thermostat joins it. As the engine heats up, this arm will move some, indicating the cooling system is functioning.

When the 20 minutes is up, back off the idle screw until a normal idle is reached: about 750 rpm. Double-check the ignition timing. If you have an inductive-pickup timing light, you can attach it now and check the timing. Otherwise, shut off the engine, hook-up the light and make the check. This also goes for early engines that are static-timed. Consult the owner's manual or underhood sticker for timing specifications.

Finally, closely examine the engine to check all exposed fasteners for proper torque, and for fluid or air leaks. Put a wrench on all of the hardware and a screwdriver on the clamps. Check all fuel and vacuum connections for leaks. When you are satisfied with the engine's condition, you're ready to drive it.

Professional tuneups are useful on later, more complex engines. Fuel-injection setup can require tuning experience and equipment to sort. Also, professional tuner can double-check your work.

BREAK-IN & TUNEUP

Now that you have run your new engine for the first time and established that it's working, it's time to break it in completely. Engine break-in is a subject of great interest to shop gossips, and you are bound to hear a lot of folklore about it. In reality, however, there are only a few steps to take, and none are as exotic as some of the "experts" would have you believe. Don't mistake me, it is a step you should take to begin the new engine on a long and useful life, but no magic is required.

What Is Break-In?—First, let's define what breaking-in is. If you stop and think about it, what's the primary difference between the engine you pulled out of your car and the one you put back in? *Clearance* is the answer. The biggest difference between an old worn engine and a new one is the clearance between all the close-tolerance parts. That's why you did all that measuring work, to determine if a part was worn out or not.

During break-in, the close-tolerance parts are going to machine themselves into perfect adjustment through their movements. Thinking back to the overhaul, which parts wore from motion with companion pieces? The valves and their guides did, so did the rings, pistons and cylinders. The camshaft and lifters also showed related wear patterns. Other parts may have been worn, but not in the same amounts as these, or from relative motion. Exhaust valves, for example, always show wear, but little of it comes from touching the valve seat. Most of it is erosion from hot gases.

Therefore, think of break-in as a period when the camshaft, lifters, pistons, rings, cylinders, valves and valve guides shape themselves to each other. The camshaft and lifters have already been taken care of. They are a good example of parts that once worn, will continue to wear in the same pattern. So the best tactic for longevity is to not start a wear pattern. That's why you ran the engine at high rpm for so long—to lube the cam and lifters so a wear pattern couldn't start.

The pistons, rings and cylinders take a little longer to lap together. An important concept to understand here is that rings do not necessarily seal between themselves and the cylinder wall. They seal mainly between the top and bottom of the piston-ring groove and the top and bottom of the ring. Also, realize that combustion pressure supplies the energy to force the rings against the cylinder wall, not ring tension. Therefore, if you don't increase combustion pressure, your rings will not seat as quickly as they could. In the meantime, they will be passing oil into the combustion chamber and combustion gases into the oil.

Ring Wear—So, what you want to do is increase pressure on both sides of the rings. Do this by accelerating from 20—55 mph at three-quarter throttle, then completely close the throttle and coast back down to 20 mph. Do this at least five times the first time you drive the car.

When accelerating at wider throttle openings, you are increasing combustion pressure and sealing one side of the rings. By closing the throttle at speed and coasting, you are increasing vacuum, which pulls the rings in the opposite direction. That seals the other side.

There is no real need to use full throttle, and it would only put undo stress on the rest of the engine, so don't use it for this exercise. Also, don't cruise at steady speeds—fast or slow—during the break-in period. That will not help ring sealing, except under those exact conditions, and you want rings that seal under all conditions.

The accelerate/deaccelerate exercise also helps the rings conform to minor variations in the cylinder walls. Increased pressures force the rings against the cylinders, where metal is removed as the high spots rub over each other.

Valve Wear—Valves and valve guides benefit from the proper valve clearance. This keeps the side loads on the valves to a minimum, and side loads are what wear the guide and stem. You set valve clearance during engine assembly, but once the engine has run some, the clearance will change. On the average, it takes 25 miles or so for the valves to completely seat. This is because of motion among the cylinder heads, cylinders, case, lifters, pushrods, and valve springs. Also, valve-grinding stones leave small ridges and grooves that are knocked off and pounded flat during the first hours of valve operation. Thus, the valve adjustment must be checked after the engine has been run.

Break-In Procedures—Putting all of this information together, how should you drive your car? Well, after the 20 minute run-in, check the timing and fastener torque as outlined in the last chapter. It's a good idea to pop the rocker covers and check valve clearance at this point, as well. The engine is hot, so you can't get an accurate reading, but you can tell if any one valve is too tight. That's the main danger at this point. Don't worry about subtleties, just run through the adjustment sequence looking for excessively tight or loose valves.

Don't worry about retorquing the cylinder heads. There is no way you can get at the nuts with the sheet metal in place.

Trial Drive—Now drive the car. Make the first trip a short one; several miles is fine. Do the accelerate/deaccelerate sequence, and otherwise drive as you would normally. Go home and check under the hood, or at least stop alongside the road and take a look before heading out on longer drives. If you see anything wrong—leaking fluids, missing part—fix it right away. All should go well, though, so continue to drive the car as much as you want. Again, drive normally, but avoid full throttle and steady-speed cruising.

Change Oil—VW says to change the oil at 300 miles, then at the standard intervals. I prefer to change the oil at 25 miles, then again at 300 miles. Change the filter each time, you Type 4, 914, and late Bus owners. The shorter interval between oil changes removes all that metal being ground off by tight-fitting parts. Use any modern, SF grade oil you prefer. Single or multi-grade is fine. There is no need to use non-detergent oil for break-in, so why bother? Just buy a case of your usual oil.

If you don't think that changing the oil at short intervals is necessary, drain some and look at it in the drain pan. Hold it out in sunlight and sight across the surface. It will look like metallic paint, with thousands of tiny shiny metal particles floating on the surface. Blow across the surface and you'll see clear oil underneath. Those shiny pieces of metal are from the rings, cylinder walls, pistons and so on.

When you change oil at 25 miles, set the valves. You must let the engine cool off for this. Overnight is best. Carefully go through the valve-adjusting sequence until all valves are at exactly 0.006 in. This is the last special valve adjustment, so apply Permatex 3H to the rocker-cover gaskets if you wish. Adjust the valves at least every 6000 miles hereafter.

If you installed hydraulic lifters, there is no further valve adjustment. The settings made during engine assembly should be fine. If you want to check, pull the rocker covers and go through the procedure again. The engine can be lukewarm for this, but it's more pleasant when cold.

How Long?—Most engines are broken-in at 300 miles. They may loosen up a little more over the next 2000 miles or so, but for practical purposes, 300 miles is it. Only when harder rings (chrome) are used, does the break-in period lengthen. Chrome rings may take 500—1000 miles before oil consumption drops, signaling ring seating. If you are having a problem, a leakdown test will help spot it.

The point is, if the engine is smoking, running rough or consuming measureable quantities of oil at 700 miles, something is wrong. A ring may be broken, the cylinder wall honing job too smooth or the mixture incredibly rich.

TUNEUP

After you have 300 miles or so on the new engine, take it to a tuneup specialist. Well, if you have one of those basic 40-HP engines and have been tuning it yourself for 20 years, go to it. All the rest of us should use professional tuners. They can use their exhaust-gas analyzer, oscilloscope and experience to give your VW a sharp tune.

It's a fact of our times that a professional tuneup is one of the best methods to tune today's cars. The oscilloscope makes the job so much easier, and the exhaust-gas analyzer is impossible to duplicate. With the analyzer, a tuneup mechanic can really optimize the mixture for maximum mileage and engine life, along with minimum emissions.

Another reason for a pro tuneup is as a double-check. Again, on earlier engines, this might not be such a big selling point, but with later engines and all their emissions hardware, it's a good idea to have a professional review your installation. Then again, if the engine has fuel injection, it really takes a sharp tuneup specialist to straighten out any driveability problems and adjustments.

There's one job most tuneup mechanics don't do, and that's adjust valves. Make sure they are set before taking the car in. Otherwise, any tuning effort will be wasted.

As for keeping your new engine running well as long as possible, frequent tuneups and oil changes are the most important factors. You went to a lot of trouble to get the inside of your engine clean during the rebuild, so keep it clean with regular oil and filter changes. Do these every 3000 miles and establish a regular tuneup schedule, too.

METRIC CUSTOMARY-UNIT EQUIVALENTS

Multiply:	by:	to get:	Multiply:	by:	to get:

LINEAR
inches	X 25.4	= millimeters(mm)	X 0.03937	= inches
miles	X 1.6093	= kilometers (km)	X 0.6214	= miles
inches	X 2.54	= centimeters (cm)	X 0.3937	= inches

AREA
| inches^2 | X 645.16 | = millimeters^2(mm^2) | X 0.00155 | = inches^2 |
| inches^2 | X 6.452 | = centimeters^2(cm^2) | X 0.155 | = inches^2 |

VOLUME
| quarts | X 0.94635 | = liters (l) | X 1.0567 | = quarts |
| fluid oz | X 29.57 | = milliliters (ml) | X 0.03381 | = fluid oz |

MASS
pounds (av)	X 0.4536	= kilograms (kg)	X 2.2046	= pounds (av)
tons (2000 lb)	X 907.18	= kilograms (kg)	X 0.001102	= tons (2000 lb)
tons (2000 lb)	X 0.90718	= metric tons (t)	X 1.1023	= tons (2000 lb)

FORCE
| pounds—f(av) | X 4.448 | = newtons (N) | X 0.2248 | = pounds—f(av) |
| kilograms—f | X 9.807 | = newtons (N) | X 0.10197 | = kilograms—f |

TEMPERATURE

Degrees Celsius (C) = 0.556 (F - 32)   Degree Fahrenheit (F) = (1.8C) + 32

°F -40   0   32 40   80 98.6 120   160   200 212 240   280   320 °F
°C -40 -20   0   20   40   60   80   100   120   140   160 °C

ENERGY OR WORK
| foot-pounds | X 1.3558 | = joules (J) | X 0.7376 | = foot-pounds |

FUEL ECONOMY & FUEL CONSUMPTION
| miles/gal | X 0.42514 | = kilometers/liter(km/l) | X 2.3522 | = miles/gal |

Note:
235.2/(mi/gal) = liters/100km
235.2/(liters/100km) = mi/gal

PRESSURE OR STRESS
inches Hg (60F)	X 3.377	= kilopascals (kPa)	X 0.2961	= inches Hg
pounds/sq in.	X 6.895	= kilopascals (kPa)	X 0.145	= pounds/sq in
pounds/sq ft	X 47.88	= pascals (Pa)	X 0.02088	= pounds/sq ft

POWER
| horsepower | X 0.746 | = kilowatts (kW) | X 1.34 | = horsepower |

TORQUE
pound-inches	X 0.11298	= newton-meters (N-m)	X 8.851	= pound-inches
pound-feet	X 1.3558	= newton-meters (N-m)	X 0.7376	= pound-feet
pound-inches	X 0.0115	= kilogram-meters (Kg-M)	X 87	= pound-inches
pound-feet	X 0.138	= kilogram-meters (Kg-M)	X 7.25	= pound-feet

VELOCITY
| miles/hour | X 1.6093 | = kilometers/hour(km/h) | X 0.6214 | = miles/hour |

Index

A
Abnormal fan noises, 11
Accessory removal—flat
  Air injection, 63
  Alternator, 64
  Breather hoses, 64
  Carburetors, 62
  Coil & distributor, 63
  Cylinder covers, 66
  Dual carburetors, 62
  Fan housing, 64
  Fuel injection, 62
  Fuel-pressure regulator, 62
  Fuel ring, 62
  Injectors, 63
  Intake-air distributor, 63
  Multiple ground connector, 63
  Oil breather, 63
  Oil breather box, 64
  Oil-pressure sending unit, 66
  Thermostat, 66
  Type 3 Fan housing, 64
  Type 3 Generator, 64
  Type 3 Intake Housing, 64
  Type 4 Fan housing, 65
Accessory removal—uprights
  Carburetor, 59
  Crankshaft pulley, 62
  Distributor, 59
  Dual-port manifolds, 59
  Exhaust, 59
  Fan shroud, 58
  Fuel pump, 59
  Generator pedestal, 61
  Heat exchangers, 61
  Intake manifold, 59
  Oil cooler, 61
  Oil-cooler grommets, 62
  Oil-pressure sending unit, 62
  Sheet metal, 61
  Thermostat, 58
Accumulated mileage, 5
Air deflectors, 133
Air injection, 152
Air intake, 146
Airflow sensor, 8
Alternator, 151
Alternator cooling duct, 152
Assembly lubrication, 116
Auto-stick transaxles, 157
Automatic transaxles, 157

B
Basic engine
  Air ducts, 67

Camshaft, 71
Camshaft bearings, 71
Camshaft plug, 71
Case parting-line fasteners, 70
Center main bearing, 71
Connecting-rod side clearance, 72
Crankshaft teardown, 72
Crankshaft, 71
Cylinder heads, 68
Cylinders, 70
Distributor driveshaft, 14, 71, 73
End-play shims, 73
Fan hub, 69
Front main bearing, 73
Lifters, 68, 73
Main-bearing dowels, 71, 73
Main-bearing studs, 73
#3 Main bearing, 73
#4 Main bearing, 73
Middle main bearing, 73
Oil pump, 68
Oil pump pickup, 72
Oil screen, 69, 71
Oil slinger, 73
Pistons, 70
Piston pins, 70
Pushrods, 67
Pushrod tubes, 67
Rod cap, 72
Rod nuts, 72
Splitting cases, 69—71
Valve train, 66—68
Windage tray, 74
Woodruff key, 73
Battery, remove, 19
Blowby, 6
Break-in, 170—171
  Clearance, 170
  Procedures, 171
  Ring wear, 170
  Trial drive, 171
  Valve wear, 170
Breather box, 152
Breather, 144

C
Camshaft
  Base circle, 97
  Cam wear, 96
  Camshaft gear backlash, 98
  Clean lifters, 98
  Closing ramps, 97
  End play, 97
  Four-rivet cam gear, 54
  Heel, 97
  Hydraulic lifter, 54, 98
  Inspect cam gear, 96

  Lifter inspection, 98
  Lobe lift, 97
  Mechanical lifter, 54
  Opening ramps, 97
  Pitting, 97
  Pushrods, 99
  Replacement, 97
  Runout, 97
  Surface hardened, 96
  Three-rivet cam gear, 54
  Thrust bearing, 97
  Toe, 97
Carbon deposits, 9
Cases
  1300, 36
  1500, 36
  1600, 36
  40 HP, 35
  Case leaks, 7
  Case savers, 9
  Changing cases, 37
  Cracking, 37
  Deep-stud, 37
  Doghouse cooling, 36
  Dual oil relief, 35
  Oil passages, 36
  Single-relief, 36
Casting number, 32
Chassis numbers, 32
CHT sensor, 148, 149
Clutch
  Centering, 116
  Clutch disc, 128
  Clutch operating shaft, 157
  Inspection, 128
  Install clutch, 129, 157
  Pressure plate, 128, 157
  Release bearing, 129, 157
  Removing clutch assembly, 157
  Retaining springs, 157
Code letters, 32
Coil, 143, 148, 153
Coldstart valve, 8
Compression testing, 15
Connecting rods
  1300/1500, 44
  1600, 44
  40 HP, 44
  Align rods, 93
  Balancing pad, 44
  Balancing, 45, 94
  Big end, 92
  Piston-pin bushing, 92
  Rod damage, 91
  Rod replacement, 93
  Rod side clearance, 93
  Self-aligning, 44

  Shank bump, 44
  Sizing, 92
  Small-end offset, 44
  Swapping rods, 45
  Total weight, 45
  Type 4, 44
  Type 4, interchange, 45
Cooling system, 143
Crankcase assembly
  Barrel shims, 131
  Cam bearings, 121
  Cam plug, 123
  Camshaft backlash, 122
  Case hardware, 125
  Check flywheel, 125
  Crankshaft end play, 125—127
  Crankshaft oil seal, 127
  Crankshaft pulley, 130
  Deck height, 131
  Distributor drive endplay, 119
  End play, 125
  Flywheel shims, 127
  Flywheel/driveplate locking, 126
  Gland nut/pilot bearing, 126
  Install camshaft, 122
  Install crankshaft, 121
  Install cylinder studs, 120
  Install distributor, 120
  Install hardware, 125
  Lube camshaft, 123
  Main bearings & lifters, 121
  Main-bearing dowels, 121
  Oil Screen, 129
  Oil pump, 129
  Oil pump installation, 129—130
  Oil pump pickup, 123
  Oil strainer nut, 129
  Seal case, 124
  Sealing rings & bolts, 123
  Sealing rings, 119
  Torquing techniques, 126
  Windage tray, 123
Crankcase reconditioning, 75—88
  Align-boring, 79
  Bearing saddles, 76
  Cleaning, 76
  Cold tank, 76
  Case cranks, 76
  Crank flex, 78
  Camshaft-bearing bores, 80
  Cam-bearing oil clearance, 80
  Case savers, 81
  Cleaning, 82
  Deep-stud, 81
  End thrust, 79
  Fretting, 76
  Full-flow oil filter, 82

173

Gallery plugs, 77
Hot-tank, 76
Main-bearing bores, 77
Oil valves, 75
Out-of-round, 78
Overheating, 79
Oversize cylinders, 81
Shuffle pin, 81
Sand seal, 82
Welding cases, 77
Cranking vacuum, 13
Crankshaft assembly
  Install rods, 118
  Oil-control pressure, 119
  Plastigage rods, 117
  Rod bearing clearance, 117
  Side-clearance, 118
  Torque rod bolts, 117
Crankshaft oil seal, 149
Crankshaft, 38—43, 82—88
  40 HP, 38
  1300, 39
  1500, 39
  1600, 39
  Balancing, 43
  Counterweights, 42
  Cracks, 85
  Crank flex, 42
  Crank whip, 85
  Crossdrilling, 39
  Dowels, 85
  Edge ride, 86
  End play, 7, 86
  Fillet, 86
  Final cleaning, 87
  Final polishing, 87
  Flanges, 39, 42
  Flywheel-mating flange, 87
  Gland nut, 88
  Hammering, 85
  Inspection, 82
  Install, 121
  Install timing gears, 87
  Journal diameter, 84
  Magnafluxing, 86
  Major dimension, 84
  Minor dimension, 84
  Offset grinding, 42
  Oil clearance, 84
  Oil seal, 7
  Out-of-round, 84
  Polishing, 86
  Press off gears, 82
  Runout, 85
  Seizure marks, 83
  Spotchecking, 86
  Taper, 83—84
  Timing gear fit, 83
  Type 4: 1700, 1800, 2000, 42
  Wobble, 7
Crossmembers, 145, 158
Crossmember bolts, 144
Cylinder heads
  Aftermarket heads, 51
  Bowl area, 49
  CHT sensor, 50
  Combustion chamber ledge, 49
  Dual port, 49
  EGR, 52
  Flycut, 49
  Long stud, 48

Quench area, 49
Rocker-arm studs, 48
Rocker-arm bosses, 48
Rocker-box floor, 48
Short stud, 48
Sparkplug boss, 48
Type 1, 36 HP, 48
Type 1, 40 HP, 48
Type 1, 1300, 49
Type 1, 1500, 49
Type 1, 1600 dual port, 49
Type 2, 40 HP, 50
Type 2, 1500, 50
Type 2, 1600, 50
Type 3, 1500, 50
Type 3, 1600, 50
Type 4, 1700, 51
Type 4, 1800, 52
Type 4, 2000, 52
914, 52
Universal replacement heads, 49
Valve seats, 48
Cylinder head reconditioning, 100—115
  Assembly, 114
  Aluminum threads, 105
  Bead-blasting, 102
  Collets, 101
  Combustion chamber, 104
  Cooling fins, 102
  Cracks, 102
  Disassembly, 101
  Flycutting, 103
  Galling, 104
  Keepers, 101
  Rocker-arm stands, 104
  Rocker-arm stud breakage, 104
  Snowballing, 104
  Sparkplug thread inserts, 104
Cylinder head torque, 134
Cylinder heads, install, 133
Cylinder layout, 14
Cylinders, 132

D
D-Jetronic schematic, 164
Detonation, 10
Diagnostic tests, 13
  Cranking vacuum, 13
  Ignition, Open firing, 13
  Manifold vacuum, 13
  Ported vacuum, 13
  Power-balance test, 13
  Wet test, 15
Dieseling, 10
Distributor cap, 153
Distributor chirping, 12
Distributor driveshaft, 14
Distributor, endplay, 119
Distributor, install, 120
Doghouse fan shroud, 9

E
Electronic control unit (ECU), 8
Electronic ignition, 12
Engine cleaning, 18
Engine compartment seal, 157
Engine descriptions
  36 HP, 34
  40 HP, 34
  1300, 34

  1500, 34
  1600, 34
  1700, 35
  1800, 35
  2000, 35
Engine installation, 157—168
Engine mounts, 149
Engine number, 32
Exhaust heat exchangers, 140
Exhaust leaks, 12
Exhaust manifold, 115
Exhaust valve
  Burned, 8
  Dropped, 9
  Sodium-filled, 9
Exhaust, 152—153

F
Fan, 145, 151
Fan housing, 9, 141, 145, 150
Fan hub, 150
Firing order, 14
First start, 169
  Build oil pressure, 169
  Oil splash, 169
  Raise idle, 169
Flame front, 10
Flywheel
  12 volt, 43
  6 volt, 43
  O-ring, 43
Flywheel/Driveplate locking, 116
Fuel injection, 153
  Airflow sensor, 8
  Coldstart valve, 8
  D-Jetronic schematic, 164
  Electronic control unit (ECU), 8
  Fuel pressure regulator, 154
  L-Jetronic schematic, 165
  Intake-air distributor, 147
  Throttle switch, 8
  Troubleshooting, 8
Fuel line, 115, 143, 148, 158
Fuel pump, 139
Fuel shut-off solenoid, 10
Fuel-pressure regulator, 154

G
Gasket sealer, 117
Generator belt, 143, 146
Generator pedestal, 139
Generator pedestal gasket, 140
Generator, 146
Gland nut, 57—58
Guide sleeve, 156

H
Heat exchanger, 144
Heater cables, 158
Hitler, Adolf, 4
Hydraulic lifters, 9, 135, 137

I
Ignition, disabling, 12
Ignition, open firing, 13
Input shaft seal, 156
Intake manifold, 115, 141
Intake runners, 147
Intake-air distributor, 147
Intake-air hissing, 12
Internal noises, 11

L
L-Jetronic schematic, 165
Leak-down testing, 16
Lifting & lowering tools, 18
Lower bellhousing nuts, 158

M
Main-bearing knock, 12
Manifold, 153
Manifold vacuum, 13
Manual transaxle, 156
Mechanical lifters, 137
Muffler, 147

N
Noise diagnosis, 11
  Abnormal fan noises, 11
  Distributor chirping, 12
  Exhaust leaks, 12
  Intake-air hissing, 12
  Internal noises, 11
  Main-bearing knock, 12
  Piston slap, 11, 12
  Piston-pin noise, 12
  Rod knock, 12
  Spun bearing, 12

O
Oil consumption, 5, 171
Oil cooler grommets, 139
Oil cooler, 7, 55—56, 138, 144
Oil coolers, doghouse, 55
Oil filler, 149
Oil filler tube, 144
Oil filter, 149
Oil leaks, 6
Oil passages, 139
Oil pump, 6
  Auto-stick, 54
  Backlash, 94
  Bearing plate, 95
  Blueprinting, 96
  Disassemble, 95
  Dished gear, 53
  Driven gear, 94
  Driving gear, 94
  Dual-stage, 95
  End clearance, 95
  Final assembly, 96
  Flat gear, 53
  High-volume pumps, 54
  Inspect, 94
  Type 1—3, 53
  Type 4, 55
Oil return thread, 145
Oil-control valves, 40
Oil-cooler placement, 9
Oil-pressure sender, 6, 62, 66, 138 149
Oil-pressure-control valve, 40, 119
Oil-pressure-relief valve, 40, 119

P
Part number, 34
Parts identification & interchange,
PCV valve, 6
Phenolic blocks, 147
Piston slap, 11, 12
Piston-pin noise, 12
Piston-pin offset, 132
Pistons
  Circlip grooves, 91

Collapsed skirt, 89
Inspect domes, 90
Install, 132
Measuring ring grooves, 91
Oil-ring grooves, 91
Piston-pin bore wear, 90
Piston-ring side clearance, 91
Piston-skirt diameter, 90
Ring checking clearance, 91
Ring side clearance, 91
Ring-groove cleaning, 91
Scoring, 89
Scuffing, 89
Taper, 89
Pistons & Cylinders
  1200/1300, 45
  1300, 46
  1500/1600, 46
  Combustion chamber volume, 46
  Dished piston, 45
  Flat-top piston, 45
  Slip-in cylinders, 46
  Type 1—3, 45
  Type 4, 46
Poor performance, 7
Porsche, Ferdinand, 4
Ported vacuum, 13
Power-balance test, 13
Preignition, 10
Pulley, 146
Pushrod ends, 136
Pushrod tube seals, 134, 137
Pushrod tube, 134, 137
Pushrods, 137

R
Release bearing, 156, 157
Rings, 132
Rocker covers, 138
Rocker-arm assembly, install, 135
Rocker-arm studs, 136
Rocker-shaft nuts, 136
Rod knock, 12

S
Sparkplug reading, 14
Sparkplugs, 138, 147, 149
Spun bearing, 12
Starter pilot bushing, 157
Steam cleaning, 18
Stethoscope, 10
Stovepipe, 140

T
Tables and Charts
  Chassis serial numbers, 41
  Connecting rods, 43
  Crankcase specifications, 78
  Crankshaft specifications, 85
  Cylinder head P/Ns, 47
  Engine bore sizes, 45
  Engine displacement, 34
  Engine letter codes, 33
  Flywheels, 43
  Higher altitude & lower compression, 15
  Oil pumps & coolers, 56
  Oil-control valves, 40
  Oil-control valve spring pressures, 42
  Ring side clearance, 91

Ring thickness, 91
Torque specifications, 154
Valve dimensions, 109
Volkswagon P/Ns, 31
Teardown, 57—74
  Clutch disc, 57
  Clutch pressure plate, 57
  Clutch, 57
  Crank-pulley nut, 58
  Crankshaft oil seal, 58
  Driveplate, 58
  Flywheel lock, 57
  Flywheel, 57
  Generator, 58
  Gland nut, 57
Temperature sensor II, 144, 148, 164
Thermo switch, 144
Thermostat, 139, 142, 145, 150
Thermostat cable, 150
Thermostat linkage, 146
Throttle cable, 158
Throttle dashpot, 12
Throttle switch, 8
Torque converter seal, 156
Transaxle seal, 7
Tuneup, 170
Type 1 Engine, install, 160
  Air filter, 161
  Air hoses, 161
  Air sensor, 161
  Auto-stick control valve, 160
  Fuel-injection wiring, 161
  Intake-air distributor, 161
  Rearcover plate, 160
  Throttle cable, 160
Type 1, removing
  Air filter, 19
  Alternator, 20
  Automatic stick shift, 22
  Carburetor, 20
  Clean air hoses, 21
  Coil, 21
  Driveplate bolts, 22
  Fuel injection, 19, 20
  Fuel line, 22
  Generator, 20
  Heat-riser tube, 21
  Heater cables, 21
  Heater ducts, 22
  Heater-control valve, 21
  Injectors, 21
  Intake air sensor, 20
  Lower bellhousing nuts, 22
  Oil-pressure sending unit, 20
  Rear engine cover plate, 21
  Throttle cable, 20
  Throttle positioner, 21
  Upper bellhousing bolts, 22
Type 2 Engine, install, 161
  Air filter, 161, 162
  Crossmember, 161, 162
  Engine compartment seal, 162
  Intake-air distributor, 162
  Rear cover plate, 161
  Torque converter, 162
  Wiring, 161
Type 2 & 4, removing
  Air filter, 24
  Automatic transmission, 24
  Bellhousing bolts, 23
  Coil, 24

Electrical connections, 24
Engine-plate screws, 23
Fuel hoses, 24
Gravel Guard, 24
Heater-control cables, 23
Oil filter, 24
Oxygen sensor, 24
Throttle cable, 24
Transaxle, 23, 24
Vacuum hoses, 24
Type 3 Engine, install, 162
  Air filter, 163
  Cooling-air bellows, 163
  Engine mount, 162
  Throttle cable, 162, 163
Type 3, removing
  Air filter, 25
  Bellhousing bolts, 25
  Crossmembers, 26
  Dipstick bellows, 25
  Electrical connections, 25
  Fuel inlet line, 25
  Fuel line, 25
  Intake air distributor, 25
  Intake bellows, 25
  Oil dipstick, 25
  Raise car, 25
  Support engine, 25
  Throttle cable, 25
Type 4 Engine, install, 163
  Automatic transaxle, 163
  Fuel-injection harness, 163
  Injectors, 163
  Manual transaxle, 163
411/412, removing
  Air filter, 26
  Automatic transmission, 26, 27
  Fuel injection, 26, 27
  Lower engine, 27
  Manual transmission, 26
  Moving driveshaft, 26
  Throttle linkage, 26
  Voltage regulator, 26
914 Engine install, 166—169
  Air filter, 169
  Clutch cables, 167
  Clutch freeplay, 168
  ECU, 168
  Exhaust, 168
  Front crossmember, 167
  Fuel lines, 168
  Ground-strap, 167
  Guide sleeve, 167
  Heat exchangers, 168
  Heating controls, 168
  Raise car, 167
  Raise engine, 167
  Raise transaxle, 167
  Relay board, 168
  Release bearing, 166
  Release-bearing arm, 167
  Shift-rod boot, 168
  Speedometer cable, 167
  Throttle cable, 168
  Transaxle connections, 167
  Transaxle preparation, 166
  Transmission oil seal, 167
  Upper bellhousing bolts, 167
914, removing
  Air filter, 28
  Axles & CV joints, 28

CV joints, 30
CV-joint bolts, 29
Cable pivot, 28
Cables, 28
Carburetors, 27
Clutch cable, 29
Clutch pulley, 29
ECU, 27
Engine cover, 27
Exhaust, 28
Front crossmember, 30
Fuel injection, 27
Heater blower ducting, 27
Heater ducting, 29
Heater valves, 29
Lift car, 28
Lower engine, 30
Rear engine mounts, 30
Reverse-light wiring, 28
Shift linkage, 28
Speedometer cable, 29
Support engine, 29
Throttle cables, 28
Transaxle flanges, 29
Vacuum hoses, 28
Vapor hoses, 28
Voltage regulator, 27

U
Upper bellhousing bolts, 159
Upper cylinder covers, 145

V
Valves
  Adjustment, 136—138
  Bottom cut, 110
  Burned valve, 108
  Dished valve tips, 106
  Grinding, 109
  Guides, 105, 106
  Hardface, 108
  Inspection, 107
  Keeper grooves, 108
  Lapping valves, 110
  Lift, 16
  Margin, 109
  Mushroomed valves, 107
  Pilots, 110
  Pitting, 108
  Reaming, 107
  Reconditioning, 107
  Rocker-arm service, 111
  Rocker-arm shafts, 112
  Seat reconditioning, 110
  Seat sealing, 111
  Sodium-filled valves, 107
  Stems, 105, 106
  Swivel-foot adjusters, 112
  Top cut, 110
Valve Springs
  Coil bound, 113
  Free height, 112
  Inspection, 112
  Load at installed height, 112
  Load at open height, 112
  Load at solid height, 112
  Spring fatigue, 113
  Spring rate, 112, 113
  Spring tests, 113
  Squareness, 112
  Valve float, 113

175

OTHER BOOKS FROM HPBOOKS AUTOMOTIVE

HANDBOOKS
Auto Electrical Handbook: 0-89586-238-7
Auto Upholstery & Interiors: 1-55788-265-7
Brake Handbook: 0-89586-232-8
Car Builder's Handbook: 1-55788-278-9
Street Rodder's Handbook: 0-89586-369-3
Turbo Hydra-matic 350 Handbook: 0-89586-051-1
Welder's Handbook: 1-55788-264-9

BODYWORK & PAINTING
Automotive Detailing: 1-55788-288-6
Automotive Paint Handbook: 1-55788-291-6
Fiberglass & Composite Materials: 1-55788-239-8
Metal Fabricator's Handbook: 0-89586-870-9
Paint & Body Handbook: 1-55788-082-4
Sheet Metal Handbook: 0-89586-757-5

INDUCTION
Holley 4150: 0-89586-047-3
Holley Carburetors, Manifolds & Fuel Injection: 1-55788-052-2
Rochester Carburetors: 0-89586-301-4
Turbochargers: 0-89586-135-6
Weber Carburetors: 0-89586-377-4

PERFORMANCE
Aerodynamics For Racing & Performance Cars: 1-55788-267-3
Baja Bugs & Buggies: 0-89586-186-0
Big-Block Chevy Performance: 1-55788-216-9
Big Block Mopar Performance: 1-55788-302-5
Bracket Racing: 1-55788-266-5
Brake Systems: 1-55788-281-9
Camaro Performance: 1-55788-057-3
Chassis Engineering: 1-55788-055-7
Chevrolet Power: 1-55788-087-5
Ford Windsor Small-Block Performance: 1-55788-323-8
Honda/Acura Performance: 1-55788-324-6
High Performance Hardware: 1-55788-304-1
How to Build Tri-Five Chevy Trucks ('55-'57): 1-55788-285-1
How to Hot Rod Big-Block Chevys:0-912656-04-2
How to Hot Rod Small-Block Chevys:0-912656-06-9
How to Hot Rod Small-Block Mopar Engines: 0-89586-479-7
How to Hot Rod VW Engines:0-912656-03-4
How to Make Your Car Handle:0-912656-46-8
John Lingenfelter: Modifying Small-Block Chevy: 1-55788-238-X
Mustang 5.0 Projects: 1-55788-275-4

Mustang Performance ('79–'93): 1-55788-193-6
Mustang Performance 2 ('79–'93): 1-55788-202-9
1001 High Performance Tech Tips: 1-55788-199-5
Performance Ignition Systems: 1-55788-306-8
Performance Wheels & Tires: 1-55788-286-X
Race Car Engineering & Mechanics: 1-55788-064-6
Small-Block Chevy Performance: 1-55788-253-3

ENGINE REBUILDING
Engine Builder's Handbook: 1-55788-245-2
Rebuild Air-Cooled VW Engines: 0-89586-225-5
Rebuild Big-Block Chevy Engines: 0-89586-175-5
Rebuild Big-Block Ford Engines: 0-89586-070-8
Rebuild Big-Block Mopar Engines: 1-55788-190-1
Rebuild Ford V-8 Engines: 0-89586-036-8
Rebuild Small-Block Chevy Engines: 1-55788-029-8
Rebuild Small-Block Ford Engines:0-912656-89-1
Rebuild Small-Block Mopar Engines: 0-89586-128-3

RESTORATION, MAINTENANCE, REPAIR
Camaro Owner's Handbook ('67–'81): 1-55788-301-7
Camaro Restoration Handbook ('67–'81): 0-89586-375-8
Classic Car Restorer's Handbook: 1-55788-194-4
Corvette Weekend Projects ('68–'82): 1-55788-218-5
Mustang Restoration Handbook('64 1/2–'70): 0-89586-402-9
Mustang Weekend Projects ('64–'67): 1-55788-230-4
Mustang Weekend Projects 2 ('68–'70): 1-55788-256-8
Tri-Five Chevy Owner's ('55–'57): 1-55788-285-1

GENERAL REFERENCE
Auto Math:1-55788-020-4
Fabulous Funny Cars: 1-55788-069-7
Guide to GM Muscle Cars: 1-55788-003-4
Stock Cars!: 1-55788-308-4

MARINE
Big-Block Chevy Marine Performance: 1-55788-297-5

HPBOOKS ARE AVAILABLE AT BOOK AND SPECIALTY RETAILERS OR TO ORDER CALL: 1-800-788-6262, ext. 1

HPBooks
A division of Penguin Putnam Inc.
375 Hudson Street
New York, NY 10014